Studies in Computational Intelligence

Volume 941

The series "Studies in Computational Intelligence" (SCI) publishes new developments and advances in the various areas of computational intelligence—quickly and with a high quality. The intent is to cover the theory, applications, and design methods of computational intelligence, as embedded in the fields of engineering, computer science, physics and life sciences, as well as the methodologies behind them. The series contains monographs, lecture notes and edited volumes in computational intelligence spanning the areas of neural networks, connectionist systems, genetic algorithms, evolutionary computation, artificial intelligence, cellular automata, self-organizing systems, soft computing, fuzzy systems, and hybrid intelligent systems. Of particular value to both the contributors and the readership are the short publication timeframe and the world-wide distribution, which enable both wide and rapid dissemination of research output.

Indexed by SCOPUS, DBLP, WTI Frankfurt eG, zbMATH, SCImago.

All books published in the series are submitted for consideration in Web of Science.

More information about this series at http://www.springer.com/series/7092

Rajiv Pandey · Marcin Paprzycki ·
Nidhi Srivastava · Subhash Bhalla ·
Katarzyna Wasielewska-Michniewska
Editors

Semantic IoT: Theory and Applications

Interoperability, Provenance and Beyond

Springer

Editors
Rajiv Pandey
Amity Institute of Information Technology
Amity University Uttar Pradesh Lucknow Campus
Lucknow, Uttar Pradesh, India

Nidhi Srivastava
Amity Institute of Information Technology
Amity University Uttar Pradesh Lucknow Campus
Lucknow, Uttar Pradesh, India

Katarzyna Wasielewska-Michniewska
Systems Research Institute
Polish Academy of Sciences
Warsaw, Poland

Marcin Paprzycki
Systems Research Institute
Polish Academy of Sciences
Warsaw, Poland

Subhash Bhalla
Database System Laboratory
University of Aizu
Aizu-Wakamatsu, Fukushima, Japan

ISSN 1860-949X ISSN 1860-9503 (electronic)
Studies in Computational Intelligence
ISBN 978-3-030-64618-9 ISBN 978-3-030-64619-6 (eBook)
https://doi.org/10.1007/978-3-030-64619-6

This Springer imprint is published by the registered company Springer Nature Switzerland AG
The registered company address is: Gewerbestrasse 11, 6330 Cham, Switzerland

Preface

With an exponentially increasing number of devices connected to the Internet, the Internet of Things (IoT) is encompassing and connecting multiple domains. As a matter of fact, it becomes difficult to point to domains where IoT-based ideas are not explored. This brings about interesting developments related to, what could be considered as, the "second wave of IoT." Here, local and/or domain-specific IoT deployments have to be joined to deliver more complex services to the users. A typical example could be a seaport IoT ecosystem, being combined with a smart city IoT deployment, and with a truck company IoT platform, to optimize logistics and reduce generation of CO_2 (and other forms of pollution). Observe that we assume, here, that the initial deployments were created independently of each other, and they were dedicated to solve specific problems/deliver unique solutions. Moreover, each of them "belongs to" a different stakeholder/group of stakeholders. In addition, it is possible that the newly instantiated ecosystem connects "competitors." For instance, in the above example, multiple (competing) trucking companies are expected to seamlessly work within the developed logistics ecosystem. It should thus be relatively simple to realize that this process brings about (among others): (a) need for interoperability between joining IoT ecosystems and (b) need for provenance and trust management. The first is relatively obvious; systems that were not meant to work together have to communicate and understand each other to deliver required services. Let us now assume that the provenance can be understood as an ability to record the origin and history of pertinent data (elements). In the case of creating multi-stakeholder IoT ecosystems, being able to automatically manage data sharing/exchange/access is crucial, also as one of the key aspects of "building trust."

Let us keep this in mind and reflect on how IoT deployments are realized. The simplest conceptualization involves hardware devices, such as sensors and actuators, connected to gateways, intermediate computing nodes, and the cloud. Depending on the approach, an IoT ecosystem is realized using concepts anchored, among others, in edge, fog, cloud, mist, dew computing, as well as software-defined networking, network virtualization, and stream processing. Obviously, components realizing machine learning/data analytics are also present. All these elements are combined, to create IoT platforms that facilitate publishing, consuming, and analyzing data, within closed and open ecosystems. Taking into account the way that IoT ecosystems are realized, a number of challenges can be identified. Among them of definite importance are (but this list is, obviously, not exhaustive): (i) How to provide common representation and/or shared understanding of data that will enable analysis across (systematically growing) ecosystems? (ii) How to build ecosystems based on data flows? (iii) How to track data provenance? (iv) How to ensure/manage trust? (v) How to search for things/data within ecosystems? (vi) How to store data and assure their quality?

Semantic technologies are often considered among possible ways of addressing these (and other, related) questions. More precisely, in academic research and industrial practice, semantic technologies materialize in the following contexts (this list is, also, not exhaustive, but indicates the breadth of scope of semantic technology usability): (i) representation of artifacts in IoT ecosystems and IoT networks, (ii) providing interoperability between heterogenous IoT artifacts, (ii) representation of provenance information, enabling provenance tracking, trust establishment, and quality assessment, (iv) semantic search, enabling flexible access to data originating in different places across the ecosystem, (v) flexible storage of heterogenous data. Finally, Semantic Web, Web of Things, and Linked Open Data are architectural paradigms, with which the aforementioned solutions are to be integrated, to provide production-ready deployments.

Nevertheless, even though we can observe a systematic uptake of solutions utilizing semantic technologies, in the IoT domain and across the Internet, the total number of actual deployments is not large enough to provide a real-world grounded guidance and best practices to follow. This provides the context for this book, which aims at presenting the current trends in application of semantic technologies in the IoT domain (and closely related areas, such as Semantic Web or Web of Things) and explores enablers that they provide. Moreover, descriptions of real-life use cases, where semantic technologies have been adopted, are included.

The book is divided into four parts and consists of 18 chapters. The first part has five chapters and provides fundamentals that can help understand the remaining parts of the book. The second part deals with IoT data and interoperability and consists of six chapters. The third and fourth parts are devoted to domain-specific and problem-specific applications, which are addressed, in seven chapters. Let us now, briefly, summarize the content of each part.

The first part starts from the chapter by Angelos Chatzimichail, Evangelos Stathopoulos, Dimos Ntioudis, Athina Tsanousa, Maria Rousi, Athanasios Mavropoulos, George Meditskos, Stefanos Vrochidis, and Ioannis Kompatsiaris. They provide an overview of the current trends in application of semantic technologies in the IoT domain, presenting practical applications in multiple domains, such as health care, disaster management, public events, precision agriculture, intelligent transportation, building and infrastructure management. Additionally, research studies on semantic reasoning, aggregation, fusion, and interpretation that aim to intelligently process and ingest sensor data are provided. Moreover, authors elaborate on how semantic technologies can overcome the limitations of device and data heterogeneity. Finally, issues related to the Web of Things and how it augments the IoT concepts are also discussed.

The second chapter was prepared by Jayashree R. Prasad, Priya M. Shelke, and Rajesh S. Prasad. They introduce the basic ideas, concepts, and technologies of the Semantic Web. Authors illustrate these concepts in the context of implementations of Semantic Web desktop and geospatial Semantic Web, as applied to agriculture, health care, and IoT in general. Finally, issues related to data provenance are also considered, specifically within the scope of the PROV data model.

The third chapter has been authored by Reinaldo Padilha França, Ana Carolina Borges Monteiro, Rangel Arthur, and Yuzo Iano. They have focused their attention on semantic searches. Specifically, they provide an overview of the Semantic Web and technology behind the Semantic Web search engines.

In the fourth chapter of this part, Hemanta Kumar Palo considers the fact that there are three different aspects related to the IoT paradigm: (i) things, (ii) Internet, and (iii) semantics. With this in mind, the author reviews and emphasizes the key emerging trends of the semantic technology impacting the IoT. Particularly, the work focuses on different aspects of information modeling, ontology design, semantic interoperability, machine learning, security policy, and processing of semantic data.

Finally, in the last chapter of the first part, Rajiv Pandey and Mrinal Pande discuss the concepts of provenance and trustability, by outlining available models of trust and tools for trust management. Additionally, the chapter introduces an example of trust implementation in an existing ontology, using provenance assertions based on the PROV-DM, proposed by the World Wide Web consortium.

The fact that the second part of this book is devoted to IoT data and its interoperability should not be surprising. It has been argued many times that interoperability is one of the roadblocks of faster uptake of IoT solutions. As a matter of fact, in 2016, the European Commission has funded six independent research projects devoted to IoT interoperability. Keeping this in mind, let us summarize the next six chapters.

The sixth chapter, authored by Arunima Sharma and Ramesh Babu Battula, focuses on complexity and variance in data materializing in IoT deployments, which make it difficult to apply and query available data, to realize user-centered applications. Here, authors propose to use ontologies to resolve issues in data naming. The most popular IoT and selected application domain ontologies are described, and their use is discussed.

The next chapter has been prepared by Amélie Gyrard, Ghislain Atemezing, and Martin Serrano. They recognize heterogeneity of (1) data format, (2) languages to describe sensor metadata, (3) models for structuring sensor datasets, (4) reasoning mechanisms and rule languages to interpret sensor datasets, and (5) applications. In this context, innovative methodologies, to link and associate the data from different domains to improve knowledge discovery, have been discussed. In this context, the chapter is focused on the ontology quality when building sensor-based applications and describes the PerfectO, a Knowledge Directory Services tool. PerfectO assists ontology designers to improve ontologies to be reused in other projects. It selects and classifies a sub-set of tools providing an online interface or a Web service simple to use, which help to enhance ontologies and synthesize a set of practices.

The eighth chapter was authored by Gianfranco E. Modoni, and Marco Sacco and focuses on RDF stores that can be used, among others, in IoT deployments. While multiple RDF stores exist, and may be appropriate and usable for some tasks, they will not fit others. Moreover, a one-size-fits-all solution is not available and, likely, will not be delivered. In this context, a methodological approach to evaluate and rank the relevant functional and non-functional features of the RDF stores is presented. The proposed approach is to help software architects to select which RDF stores best fit their application scenario(s).

The following chapter, written by Vitalina Babenko, Igor Shostak, Mariia Danova, and Olena Feoktystova, deals with creation of ontological knowledge bases in the Semantic Web. Specifically, various tabular structures are considered as sources of knowledge, with the main problem being a contradiction between the wide variety of tabular structures used in knowledge sources and the insufficient efficiency of classical methods for analyzing sources of this type. The implementation of the theoretical results of the study, in the form of algorithmic, mathematical support, as well as experimental studies conducted to determine the upper bound and the nature of the growth of complexity of the method of forming the ontological knowledge bases based on targeted enumeration, has been presented and confirms the validity of the proposed approach.

The tenth chapter, contributed by Beniamino Di Martino and Antonio Esposito, is focused on seamless interoperability of sensor-generated data, which is needed to achieve specific goals. The context of the work is provided by the lack of a universally accepted standard for sensor communications. In the chapter, a prototype tool for the analysis of sensors' API tries to overcome the interoperability issues in a sensor network and provides an instrument to support sensors' orchestration and management. The tool aims not only at automatically analyzing the APIs, but also to derive a semantic representation of them, which can be then used to support the manual annotation with external domain ontologies.

Finally, the last chapter in this part was written by Pratiyush Guleria and Manu Sood. Here, authors focus their attention on interoperability in the healthcare sector. Specifically, they propose to use the foundations of the Semantic Web, in a three-layered framework for IoT interoperability, and have framed a Web ontology structure for semantic interoperability in IoT for the healthcare sector. This is combined with a text analytics model, which performs semantic data classification on a synthetic healthcare dataset, to predict the patient diagnosis using machine learning techniques.

The third part of the book deals with semantic IoT in the context of domain-specific applications. The first chapter was contributed by Gaurav Kant Shankhdhar, Richa Sharma, and Manuj Darbari. Their contribution is focused on the agriculture/farming industry. The considered solution provides a lightweight IoT framework, focused on farming in developing countries like India. In this context, authors have developed a semantically enriched agent-based model

(ABSMSA), which uses SAGRO-Lite lightweight ontology, proposed by the authors, along with the IoT-Lite and the Complex Event Service Ontology (CESO).

The thirteenth chapter was authored by Mahda Noura, Amélie Gyrard, Benjamin Klotz, Raphael Troncy, Soumya Kanti Datta, and Martin Gaedke. It is focused on the automotive industry and attempts at answering the question: How to better understand the knowledge provided by Google results to build future "smart vehicle-centric" applications? Authors present an exhaustive systematic literature survey which is a basis for building a "common sense knowledge" dataset for the automotive sector. The proposed methodology (KEAS: Knowledge Extraction for the Automotive Sector) aims to analyze the most popular "knowledge elements," required to build smart vehicle applications and delivers: (1) keyphrase synonyms, (2) synonyms used to build a corpus of scientific publications, (3) smart car ontologies, and (4) the extraction of the most common terms for the automotive sector.

The next chapter was authored by Regel Gonzalez-Usach, Matilde Julian, Manuel Esteve, and Carlos E. Palau and deals with both interoperability and Ambient Assisted Living (AAL) and Active and Healthy Aging (AHA). Here, results from the European project ACTIVAGE, concerning benefits obtained through the use of IoT in elderly smart homes, by enabling semantic interoperability and cooperation across smart home clusters located in 12 different cities across Europe are presented. The challenges that were solved in the project using real-time semantic translation include: providing interoperability and semantic integration, the management of massive real-time streams of IoT data.

In the fifteenth chapter, authored by Gennady Chuiko, Yaroslav Krainyk, Olga Dvornik, and Yevhen Darnapuk, semantic provenance management for medical data is considered. Specifically, authors consider the presence of semantic data on different levels of a complex health monitoring system. The model of the semantics-based system, for medical data provenance, has been proposed along with the revision of the whole set of available technologies to employ in the semantic engine and analysis of its behavior under different circumstances. Here, inclusion of semantic information into the general dataflow should not only allow evaluating data quality but also discover behavioral patterns to identify problems inside a specific part of the system. Authors consider several scenarios that involve different device types that measure a patient's state.

The last part of this volume is devoted to problem-specific applications. In the sixteenth chapter, Matthew Weber and Edward A. Lee consider semantic localization for the IoT. They base their work on an interesting observation that Euclidean geometry and Newtonian time, with floating-point numbers, may not be the best choice for IoT ecosystems. Hence, they survey location models from robotics, the Internet, cyber-physical systems, and philosophy. As a result, a logical framework, wherein a spatial ontology is defined as a model-theoretic structure, is proposed. It is then argued that space-aware IoT services gain advantages for privacy and interoperability when they are designed for the most abstract spatial-ontologies possible.

Next, in the seventeenth chapter, Arun Kumar and Sharad Sharma consider simplifying trigger-action programming, for a control within an IoT scenario. Authors propose an IoT-based semantic interoperability model, with EUPont Semantic Web ontology, to deliver semantic interoperability among heterogeneous IoT devices for control of end-user applications. They believe that the EUPont could serve as the core information layer for the future IoT end-user programming solutions.

Finally, in the last chapter, written by Jayashree R. Prasad, Shailesh P. Bendale, and Rajesh S. Prasad, authors consider interrelation between IoT ecosystems and software-defined network (SDN), network function virtualization (NFV), cloud technologies, and semantic technologies. Here, the key underlying problem is, again, interoperability between devices. The context is provided by the observation that deployment of 5G (and later, 6G) networks will result in an almost unbound growth in the number of connected (heterogeneous) devices.

We would like to express our gratitude to all reviewers who have evaluated all submissions and helped improve the ones that have been selected for the inclusion in this volume. Thank you for your hard work.

Lucknow, India	Rajiv Pandey
Warsaw, Poland	Marcin Paprzycki
Lucknow, India	Nidhi Srivastava
Aizuwakamatsu, Japan	Subhash Bhalla
Warsaw, Poland	Katarzyna Wasielewska-Michniewska

Contents

IoT Data and Interoperability

Domain-Specific Applications

Problem-Specific Applications

Fundamentals

Semantic Web and IoT

Angelos Chatzimichail, Evangelos Stathopoulos, Dimos Ntioudis, Athina Tsanousa, Maria Rousi, Athanasios Mavropoulos, George Meditskos, Stefanos Vrochidis, and Ioannis Kompatsiaris

Abstract In this chapter, we provide an overview of the current trends in using semantic technologies in the IoT domain, presenting practical applications and use cases in different domains, such as in the healthcare domain (home care and occupational health), disaster management, public events, precision agriculture, intelligent transportation, building and infrastructure management. More specifically, we elaborate on semantic web-enabled middleware, frameworks and architectures (e.g. semantic descriptors for M2M) proposed to overcome the limitations of device and data heterogeneity. We present recent advances in structuring, modelling (e.g. RDFa, JSON-LD) and semantically enriching data and information derived from sensor environments, focusing on the advanced conceptual modelling capabilities offered by semantic web ontology languages (e.g. RDF/OWL2). Querying and validation solutions on top of RDF graphs and Linked Data (e.g. SPARQL, SPIN and SHACL) are also presented. Furthermore, insights are provided on reasoning, aggregation, fusion and interpretation solutions that aim to intelligently process and ingest sensor information, infusing also human awareness for advanced situational awareness.

A. Chatzimichail (✉) · E. Stathopoulos · D. Ntioudis · A. Tsanousa · M. Rousi · A. Mavropoulos · G. Meditskos · S. Vrochidis · I. Kompatsiaris
A. Chatzimichail (✉)
e-mail: angechat@iti.gr

E. Stathopoulos

D. Ntioudis

A. Tsanousa

M. Rousi

A. Mavropoulos

G. Meditskos

S. Vrochidis

I. Kompatsiaris
Centre For Research and Technology Hellas (CERTH), Information Technologies Institute, Thessaloniki, Greece
e-mail: angechat@iti.gr

R. Pandey et al. (eds.), *Semantic IoT: Theory and Applications*, Studies in Computational Intelligence 941, https://doi.org/10.1007/978-3-030-64619-6_1

1 Introduction

The era of the Internet of Things (IoT) is upon us, with a huge number of IoT devices already in everybody lives. A large number of applications in Smart Cities, E-health, Security and other domains are exploiting the IoT technologies, like sensors, smartphones and actuators. One of the most significant aspects of IoT is the things interconnection providing an interconnected system of different services and applications. Everyday huge amounts of data are being generated and these data provide extremely valuable knowledge databases. The IoT tries to estimate a situation based on the data knowledge in order to enable services to make smart decisions.

However, several challenges arose with the existing IoT technologies regarding the interoperability of those different IoT technologies since their data is based on predetermined formats without following a catholic vocabulary to describe the interoperable data. The basic structure of the IoT is the Machine-to-Machine (M2M) communication [1]. For example, the measurements of sensors are required to be distributed and analyzed by other devices or sensors and not being human readable without any kind of processing. Therefore, those measurements should be understandable from one machine to another. Although, towards the rendering capable of global machine communication by autonomous information discovery and analysis, it is mandatory to struct and group data respectively.

Semantic Web (SW) technologies have been widely used to interpret and integrate data deriving from a plethoric variety of resources on the Web. The main objective of the Semantic Web is the provision of a new form of content that is understandable and can be edited by both humans and computers. IoT domain has recently adopted several Semantic Web technologies in order to enhance the content of data and interoperability [2]. This is achieved by virtualization IoT data based on reusable vocabularies that can be interpreted by distinct software modules. This semantic annotation employs various semantic web standards, such as RDF, RDFs and OWL to construct intellectual models, like ontologies to describe different domain concepts and the connections that exist among them. Through this way, the semantic annotation transforms the human oriented Web to a machine-interpretable Web. Except for that, semantic web provides several protocols and query languages, which can be deployed to query and reason over RDF datasets to infer new knowledge from them.

The ubiquity of Web technologies renders them a viable solution for managing data coming from things. The Internet of Things is transforming into Web of Things to leverage the advantages of the Web. The Web Thing Model imposes software modules that allow the things to easily integrate the Web (with JSON messages). Consequently, the IoT systems can benefit through the representation and description of things and their environments and through the semantic annotation of the data coming from things, leading to better understand them.

The book chapter is organized as: On Sect. 2 Applications, Related Work and Research Challenges in Semantic Web and IoT are presented. On Sect. 3 IoT knowledge Representation with Semantic Web technologies is analysed with more emphasis in Modelling sensors (Sect. 3.1), multimodal events and observations (Sect. 3.2) and Ontology-based reasoning (Sect. 3.3). In Sect. 4 The Semantic Web of Things is presented and how it augments the IoT.

2 Applications, Related Work and Research Challenges in Semantic Web and IoT

In this section we describe relevant surveys on the research area. Also, we describe the research challenges and the applications that Semantic Web can be combined with IoT. How Semantic Web technologies help IoT systems to address different issues.

2.1 Applications on Different Domains

2.1.1 Smart Home

A useful guide on IoT applications for smart home can be found in [3]. It summarizes the related work and systems of smart home applications using Internet of Things. The articles examined were collected and created a taxonomy of four classes, formed based on the content of the articles. Ramparany and Cao [4] deals with the issue of data homogeneity in IoT applications. The authors propose a framework that uses semantic technology to combine IoT data obtained from a smart home. The goal is to utilize the data in order to recognize presence and abnormal activities. Huang et al. [5] focuses on safety services of smart home applications and proposes a semantic approach for recognizing risk events. Zolfaghari et al. [6] proposes an Ontology-based framework for activity recognition. The framework, named SOnAr, consists of two approaches, that use ontologies for context-based activity recognition. In the first approach, activities are recognized with the use of Ontology Web Languages for reasoning. The second approach combines ontologies, statistics and Fuzzy Logic in order to recognize more complex activities and overcome issues of scalability and sensors' noise.

2.1.2 Smart Cities

One of the most common and fast-growing concepts in information technology is the smart city concept. The definitions and ideas used to characterize a smart city, however, are just as numerous as the cities themselves. The word "smart city" is used with various definitions because there is no specific description to match all of them. In fact, a smart city is known as a network of heterogeneous systems that should be connected with each other, be orchestrated and also be intelligent. In order to provide interoperability between such large and diverse systems several common knowledge representation models are used [7].

A number of projects have been developed like SeeClickFix [8], FixMyStreet [9] and ImproveMyCity [10] with the aim of providing the community and public authorities with an online service to link every issue within their own region. An ontology is developed for the definition of smart city environments in [11], known as Smart City Services Ontology (SCSO). It was created on the basis of four intelligent urban applications including Smart Parking, Smart Garbage Control, Smart Streetlight and Smart Complaint. Smart Parking was designed to provide information regarding free parking areas in real time. The Smart Streetlight system provides actual-time auditing of streetlights so that they can operate only when needed. Smart garbage control system is primarily designed to identify and inform the accountable authorities via intelligent sensors, waste level in the garbage containers. Finally, the complaint management system enables citizens to submit reports regarding municipal problems and allows officials to arrange their settlement.

The Open Traffic Lights (OTL) [12] project proposes a methodology for publishing traffic lights using an ontology as the knowledge representation model. The ontology [13] is available under an open license describing the topology of an intersection as well the signal timing of traffic signals. Live traffic lights from crossing points in Antwerp, Belgium are shown in a web application [14]. In addition, in [15], they tested how to estimate a dynamically changing phase of the traffic lights with the live OTL dataset of Antwerp.

Smart Cities also consistently support the transition to viable and efficient energy systems through the promotion of energy efficient policies regarding renewable and smart energy management and production. Big data analytic services to predict energy consumption and manage usage patterns were made available through the Internet of Things (IoT) smart grids. A context-aware framework is proposed by the authors of [16], which uses sensor information to create a context-aware model. They created an ontology and a reasoning service for the management of power equipment.

Efficient urban water management is also a growing challenge due to the increasing population density and the competing pressures towards sustainability that is related not only with water conservation but also with energy expenditure. The energy consumption of clean water systems, even in locations with water abundance, mandates efficient management and waste minimization. The authors of [17] present a novel cloud platform, that brings together clean and waste systems to deliver demanding responsive water management. That is achieved by integrating several sensing and analytical components via a semantic knowledge base. This knowledge base reuses GIS data alongside dynamic sensor data, social concepts and inference rules.

2.1.3 Transportation

As cities' population increments, efficient transportation grows critical for quality of life, economic productivity and the environment. Traffic monitoring, which includes tracking of a moving or immobilized car and also involves traffic operations to infer the traffic within the city streets, is an important concern of smart traffic systems. In the IoT domain, heterogeneous sensors and devices, such as cameras etc., provide media content that facilitate global interoperability between the physical and the virtual world.

In [18], the Multimedia Web Ontology Language (MOWL) is used to store data coming from heterogeneous sensors, such as cameras, WSN, GPS and other components thus encoding the smart traffic domain knowledge. Incoming data are first processed and then analyzed to predict real-time traffic conditions and trigger the appropriate actions.

Similarly, connected cars can easily turn a simple car from a transportation object to a digital platform for integrating humans with a city. To this end, the cars need to be informed of their environmental connected services, how they are linked with each other and finally how to connect with them so that they can benefit from their use. The authors of [19] developed a semantic repository in which they preserve data consisting of the car's view on the real world. The semantic repository regularly acquires new location-specific information from sensors and devices to satisfy user requests for service discovery, such as parking spaces and restaurants.

In [20], an ontology was developed for collaborative route planning based on data coming from navigation sensors (e.g GPS). The model describes interactions between various types of travelers such as drivers, autonomous cars and pedestrians, given that they all have some kind of navigation sensors and share navigational information. Those different kind of travelers share their navigational needs and desires for meeting various objectives. An algorithm for self driving tourists was proposed in another study [21]. The model describes the experiences of drivers and services requested by tourists. It examines traffic information, schedules of touristic places (i.e. museums), the vehicle's capacity in fuel as well as the type of fuel and the time of departure.

ST4RT project [22] aims at providing transformation between various standards and protocols based on an ontology, resulting in improved interoperability among different legacy systems of transport organisations. Whenever two systems which adopt different standards are required to exchange information, the semantic transformation takes place. The main idea is to create associations from the data models of the two systems to a global reference model.

2.1.4 Health and Well-Being

INTER-Health is a healthcare platform, developed as part of the INTER-IoT platform, a (Horizon 2020) European project. The platform is actually an integration of other existing heterogeneous IoT platforms, namely UniversAAL and BodyCloud [23]. This integration intended to provide services that the individual platforms could not support. The Inter-Health platform focuses on people with food and physical disorders and observes their lifestyle in order to prevent health problems. Goossen [24] provides an overview of Detailed Clinical Models (DCM). A DCM is an information model that expresses clinical concepts and requirements for clinical information. The paper discusses the advantages of DCMs compared to other approaches, like the two level modeling. A framework for distributed e-health records is presented in [25].

The system is a unified semantic interoperability framework that uses fuzzy ontology and consists of three layers. Each layer performs the following operations respectively: (a) storing of heterogeneous health records, (b) mapping local ontologies to global ones, by using relevant algorithms or human expertise and (c) user interface through which medical experts send queries.

2.1.5 Security—Safety

Ontologies have become a trend recently for making decisions easier during a climate crisis (such as floods, earthquakes, forest fires etc.). The enormous flow of information from humans and sensors is one of the most difficult challenges that the authorities face during such crisis events. A lightweight ontology was proposed in [26] to manage climate crisis and combine all relevant aspects of crisis management: crisis representation, sensor analysis, crisis incidents and related impacts as well as unit allocation of first responders.

The authors of [27] have suggested a system that lies in the intelligent combination of devices and human information against human and situational awareness, in order to encourage security and a secure ecosystem for people. In order to make the use of deductive reasoning over the gathered information feasible to tackle a range of urgent situations, such as health-related problems and missing children in crowded places, the DESMOS ontology has been developed that covers most of the principles that are related to the identification of critical incidents and the implementation of risk management processes.

Significant developments have recently been made in technology for autonomous vehicles, without direct human control. The safety evaluation of the automated driving functions is an important subject in the automotive industry. Methodically defined scenarios by experts can help to strengthen engineering and safety research. Numerous studies have shown that ontologies provide an effective context for various autonomous vehicles' applications. Several scenarios for the development of automated driving services are proposed in [28]. The authors of [29] concentrated on designing a model for supporting vehicle communication. They described an ontology that encompasses all possible on-road scenarios. They derived situation-aware routing protocols based on this model, and created simulated traffic and unique scenarios that are likely to cause accidents.

A research was performed on the detection of health hazards in metro construction sites in [30]. Security risk detection in metro construction is a knowledge-intensive process and is one of the most important tasks while managing risk. The information is collected mainly in non-structured formats from various sources. Additionally, each project typically develops its own information system to facilitate decision making. In the study, an ontology offers a way to standardize and codify knowledge related to safety risk that can also be distributed between different actors involved in the project as well as between difference computer systems. The ontology can also be used in the production of smart applications that can support the detection of safety risk.

2.1.6 Earth Observation

Combining semantics with deep learning techniques into Earth Observation data has been a hot topic during the past few years. The primary reason behind this tendency is that while Earth Observation data are increasing rapidly, semantics offer an intelligent fusion between data deriving from heterogeneous sources and high-level solutions in decision-making issues, as they enhance knowledge discovery. In most cases, the problems are associated with physical resources management, environmental protection and monitoring. For this purpose, several ontologies and systems have been developed in literature. Some of them are presented below.

Most systems combine semantics in order to achieve knowledge discovery by fusing data from heterogeneous sources [31–35] while others are responsible for semantic querying for image retrieval [36]. The main challenge encompasses, have to do with combining data from heterogeneous sources (e.g. OpenStreetMaps [32, 34], satellite and aerial images [32], Google Earth images [34]) which is achieved using different fusing techniques (like FuseNet architecture [32]). The scope of using semantics in earth observation data has to do with semantic labeling in most cases [32, 33].

The ontologies developed to map earth observation data are associated with hydrological data [37] and environmental monitoring data [35]. More specifically, the ontology presented in [37] represents sensors, observation and hydrological events classes, while Modular Environmental Monitoring ontology (MEMOn) [35] builds a more extensive structure which contains a large amount of aspects like disaster, temporal, environmental material, sensor, environmental process, geospatial, observation and measurement and infrastructure modules.

Intelligent Interactive Image Knowledge Retrieval (I^3KR) [31] is a system which conducts semantic-based Knowledge discovery by using EO data archives. The system uses a hybrid ontology approach to interconnect data from different sources and DL reasoning services to apply semantic restrictions. PREDICAT [35] is a system which aims in interconnecting data from heterogeneous monitoring systems using ontologies, data integration and reasoning techniques to produce knowledge from existing natural disasters and predict possible future natural catastrophes.

2.1.7 Creative Industries

According to the Department of Culture, Media and Sport, there is a plethora of creative sectors that are identified as belonging to the creative industries [38]. This subchapter encompasses state-of-the-art related work and applications from all sectors except for those related to Cultural Heritage which are presented thoroughly in the following subsection.

There are great opportunities in creative industries for administering digital content. In that aspect, an ontology-based framework named as V4Ann was developed as part of the V4Design platform [39]. Its main purpose was the knowledge representation, the semantic aggregation from multiple sources and the combination of annotations deriving from multimodal analysis results of digital content. Regarding architectural designs in urban spaces, MindSpaces aims to provide solutions for creating functionally and emotionally attractive environments [40]. Furthermore, the i-Treasures platform, based on multimodal fusion and semantic media interpretation, has created an open and extendable platform which contains a wide range of captured intangible cultural expressions available in digital form [41]. Finally, the IoT field includes also wearable devices in which in that scope the WEAR Sustain network assessed the issues of sustainability and ethics among all creative actors of the industry into a unified knowledge base [42].

2.1.8 Cultural Heritage

In the Cultural Heritage (CH) domain, the concern and actions about cultural artefacts protection are of great significance, thus semantics are deployed, spanning from semantic analysis of sounds deriving from sensors [43], preventive conservation with automatic detection of potential hazards and execution of suggested possible solutions [44], weather, environmental and structural monitoring [45], advanced predictive analytics [46], wireless low-cost non-invasive systems for monitoring parameters of space [47], merge of different ontologies with advanced reasoning capabilities on top [48], to a system for automated regulation of microclimate parameters [49].

Intertwined with the CH domain is also the field of tourism. Efforts towards smart discovery of cultural routes incorporating data from heterogeneous sources and geospatial semantics have been conducted [50–52]. Another aspect within the correlation of CH with tourism, is the provided user experience, where attempts to improve the UX and enhance the visitor's interaction with cultural objects have been committed [53, 54], in some cases even by deploying Augmented Reality features [55]. In that direction, personalization techniques that mine user's interests into CH [56] or extraction and identification of personas were investigated [57, 58].

The modeling to predict and describe the dynamics of interaction processes was within research scope [59, 60]. The necessity of adequate ontological modeling led to further semantically enriching museum collections [61] and hidden knowledge extraction and representation [62], even to development of semantically enriched 3D models [63].

Finally, big data infrastructures to administer cultural items digitally were constructed and tested, tendering a variety of services unto the point of incorporation of several aforementioned modules [64].

2.1.9 Manufacturing

Industry 4.0, known as the Fourth Industrial Revolution, combines automation and digitalization in manufacturing technologies. Many of the previous old manual devices are now supposedly supplanting by new devices and autonomous systems. The devices (physical or virtual) are continuously getting smarter and Artificial Intelligence (AI) capabilities are being placed in objects, robots and spaces, enabling them to comprehend their environment and reason, interpret and learn. As the number and intelligence of "things" increases, there is a need to shift from statically interconnected IoT nodes to autonomous and collaborative entities in Industry to harness intelligence and support dynamic connectivity, interactivity and decision making augmenting the operational value of the industry.

The combination of AI and Semantic Web technologies will lead to a solution for a number of complicated problems related to interoperability, automated and self-configurable systems such as those from Industry 4.0. Through this combination it can be achieved an holistic view of a Factory of the Future (FoF) enabling better decision making across different management layers to reduce overall complexity. This holistic approach includes the interconnection of heterogeneous data sources, the production chain, business processes and so on. Another example for the integration of Semantic technologies is on the Autonomous systems. Machines can communicate and exchange information with other machines under the same vocabulary, in order to succeed a common target. New machines can participate easily in the production chain without the necessity delegate a heavy work force on this task, by simply creating interoperable services. Faulty devices are easily being substituted by discovering new devices with similar functionality to prevent downtime during the production process.

The semantic technologies in an Industry 4.0 platform can be integrated at the edge or at the cloud layer, depending on the use case application. The semantic knowledge layer receives data after a middleware uses standard protocols like MQTT, OPC UA or HTTP(S) and formatting the data using open standards, like OPC UA, PPMP, PackML. Then, it employs defining and sharing of semantic information to allow for easier analysis across different systems [65].

There are many studies which [66, 67] model industrial products and services. In [68] an approach is presented for integrating IoT to a MAS (Multi Agent Systems) based manufacturing environment, semantically enriched the relevant ontology and its validation through a Hardware-in-a-Loop simulation utilizing a gamification system. In [69] the SAREF ontology was extended with the creation of the SAREF4INMA ontology for describing the Smart Industry and Manufacturing domain. SAREF4INMA is based on several standards and IoT initiatives, as well as on real use cases, and includes classes, properties and instances specifically created to cover the industry and manufacturing domain. Alvanou et al. [70] proposes the implementation of MTConnect as machine-interpretable ontology (OWL) to achieve two things: Firstly, to preserve the semantics of the reference within the model and secondly, to enable its interlinking with other datasets to form the basis of the Industry 4.0 vision. MTConnect defines specific data patterns to facilitate healthcare monitoring of machine tools. Thus, it provides the foundations for predictive maintenance to reduce the possibly premature exchange of expensive machine parts or to prevent entire machine outages due to ruptured parts based on sensor data. Authors utilized the existing ontologies such SSN, SAREF and SEM for their ontology. Some queries in SPARQL as a proof of concept for ontology completion were also mentioned. Industrial robots used in manufacturing kitting stations are modelled in ontologies presented in [71] in a project from the National Institute of Standards and Technology (NIST).

The main challenges for the semantic technologies in IoT in Industry are in setting ontologies for integration and interoperability between the already existing old industry standards, use of existing ontologies (e.g. SSN, SAREF etc) to make applications for industry 4.0 using sensor data provided by autonomous systems and the execution of exemplary use case scenarios in real industrial conditions.

2.1.10 Agriculture—Farming

In agriculture, there is very fast-growing trend with smart devices entering the fields and helping farmers to comprehend better the crops and their production. This trend is called precision agriculture and currently is generating huge volumes of raw data from IoT sources such as: chemical sensors, electro-chemical sensors, drones, weather stations and so on. Those thousand of lines raw data are meaningless and isolated, and therefore they do not add extra knowledge to the farmer. Agriculture activities are based on a multiparameter knowledge with many interconnections between the parameters. The efficacy of data derives from context and meaning, as well as its combination with other data from different agriculture sources. Semantic technologies can provide practicality to the agriculture/farming data by providing common interchange data formats. Also, through SW new knowledge can be provided through the use of reasoners.

Semantic resources are typically divided in two different categories for general agriculture or specialized domains of agriculture. Several significant agriculture ontologies are OntoAgroHidro [72], Crop ontology [73], GCP ontology [74], Agroportal [75], Agricultural Technology Ontology [76], Citrus Ontology [77], Agriculture Activity Ontology (AAO) [78], AgOnt [79], Agronomy Ontology [80]. Projects like Agrovoc [81] consists of +36,000 concepts and +750,000 terms in up to 35 languages, have provided with structured vocabularies for the agriculture domain. The Global Agricultural Concept Scheme (GACS) [82] contains in its f iles of interoperable concepts the schemes related to agriculture from AGROVOC multilingual agricultural thesaurus (35,000 concepts), the CAB Thesaurus [83] (140,000 concepts) and the NAL Thesaurus [84] (53,000 concepts).

Agriculture requires common data schemes for semantic web technologies to render plausible the transfer of semantically described data and the development of common ontologies. One such a standard is known as: The Agricultural Metadata Element set (AgMes) [85]. AgroRDF [86] is one of the major standards for data exchange, which is designed specifically for agricultural data. The applications with agriculture semantic technologies are divided mainly into those different categories: Knowledge based systems, Remote Sensing, Decision Support and Expert Systems [87].

2.2 Main Research Challenges

The idea of integrating physical objects and communicating with each other is not a new one. Various technologies and standards have been proposed until today. Many of those technologies have been used and established the vision of Web of Things. Although, one of the most crucial challenges until now is to address the interoperability of those things, technologies and standards under a common framework.

The spread of IoT and consequently the WoT is expected to fetch a huge amount of real-time sensing and not only, data to the Web services. This leads to a vast number of information and services that needs to be interpreted. In contrast to the traditional web pages and documents that are accessible to the present web, WoT will bring dynamic content that will rapidly change due to the nature of IoT data. The search engines of WoT should efficiently provide real-time data and discover dynamic services. In order to enable, a catholic web of things framework, research community will need to be based on open standards independent of particular vendors, in which every developer can extend and enrich the different developed technologies.

One other important research challenge is the research on smart object integration on creating context-aware ambient environments. New solutions need to be suggested in order to address this challenge. The research here is basically focused on health domain, security and safety, retail market and smart homes.

Security, privacy and trust among different smart objects and users are a vital issue on WoT. The widely utilization of REST interfaces has enabled the use of similar web security approaches in WoT also (based on HTTP protocol). As the significance of IoT data is growing rapidly, the research in trustworthiness is a very important issue, as everyday human applications are based on those data. Trust issues are integrating the interaction issues among smart objects. Advancements in the social WoT have suggested new solutions to those research challenges [88].

Nowadays web applications and services are based on software for specific tasks. However, they lack flexibility when they take into consideration human in the loop. One of the most crucial future research challenges is when the smart object will have to interact with humans. Over the next years, there will be virtual smart objects that will understand human emotions, experience and reactions, providing common sense. This will become true by extensive research combination of cognitive psychology, social IoT and advanced artificial intelligence techniques. Semantic web can augment this research field by capturing human knowledge, feelings and experiences in different domains (health, social life, relationships and so on.)

3 IoT Knowledge Representation with Semantic Web Technologies

In this section we describe the semantic web technologies and standards, that are most popular for IoT representation. How can the use of powerful formats add structure and meaning to the content of data coming from IoT devices and interlink related data.

3.1 Modelling Sensors

Sensors embedded in devices that are attached on the human body or sensors directly placed on the human body, are called wearable sensors [89]. The existence of such sensors in mobile and wearable devices has led to their extensive use in activity recognition and fall detection tasks. Accelerometers and gyroscopes are the most popular ones, with accelerometers being the most effective in recognising activities when used individually. Gyroscopes are also quite popular, however, they are mostly used in combination with the accelerometers. Accelerometers are known to perform well in recognizing activities in general but they are more successful in activities with repetitive movement [89], since they measure a moving object's magnitude and direction. They usually fail to recognize similar activities when used individually [90], thus it is more effective to use them along with other inertial sensors to improve the performance of a human activity recognition system. Gyroscopes perform well in recognizing an object's orientation because they measure the rotation speed [91]. Gyroscopes are widely used in activity recognition studies, but most of the times not as the only sensor. Fusion of such sensors, whether performed before or after the classification algorithm, is found to improve the recognition rates of a human activity recognition system, since one sensor may capture movements not well detected by the other [92]. Magnetometers are a third kind of wearable sensor that is also explored in activity recognition studies; their individual performance though is poor and they are mostly used in combination with the other sensors. The aforementioned sensors are found in all smartphones and smartwatches, which is the main reason they are widely utilized for activity recognition studies since it is easy to extract their measurements. The sensors are most often triaxial and they produce three vectors of raw signals, one for each axis of the Cartesian reference system [93].

The raw data consist of three vectors of values, each one relevant to one axis of the Cartesian system. After the extraction of the raw data, is the preprocessing stage, which may include filtering and/or the normalization of the data to eliminate signal noise. Features are afterwards extracted by a time window, which is employed because it makes two signals comparable. Feature extraction retains valuable information from the signals [94]. The two basic categories of extracted features are time domain and frequency domain and a list of the ones computed in most studies can be found in [94]. After the extraction of features, a feature selection method may be applied to identify which features will potentially assist in the improvement of the recognition of activities and to eliminate large feature sets [94]. The HAR framework is concluded with the classification process, where a classification algorithm is applied to recognize the activities.

Activity recognition tasks are actually multiclass classification problems. The choice of the classification algorithm is driven by various parameters like the types of the recorded activities, the type of data and the extracted features. Some classification algorithms are found to perform very well in the majority of such studies, like Support Vector Machines (SVM), Naive Bayes (NB) and Decision Trees [94]. If a system consists of different sensors and there is a need to utilize the information provided by all of them, fusion methods are applied. Fusion is the combination of information and that can be performed after the classification process to combine the classification results of each sensor or at earlier stages, before the data enter a classifier, where it combines the extracted features of different sensors [95].

3.2 Modelling Multi-modal Events and Observations

3.2.1 Location

Data semantics are extensively used in location-based services (LBS), in order to find and integrate the information related with the users. Several LBS were analyzed and recorded in [96]. First they have classified the LBS data based on relevant definitions and use. A distinction was made between Domain Data, Content data and Application data where the Domain Data include spatial and temporal concepts (e.g. location, position, time etc.), Content data mainly describe specific content, and finally, Application data comprising of the actual services and user profiles.

3.2.2 Activities—Events

Event-Model-F [97] describes a process for identifying and describing real word events. It is based on DUL and follows the "descriptions and situations" (DnS) ontology design framework [98] for modelling different concepts of events, such as object attendance, relationships, and different meanings of the same event by introducing six ontology design patterns. In addition to the DnS model, Event-Model-F implements a number of internal representations to describe relationships among events, such as causality and correlation. Figure 1 [99] describes the pattern of Event-Model-F correlation of Events.

The Simple Event Model Ontology (SEM) [100] is an attempt to establish an ontology model for events with no extreme semantic restrictions. The open nature of the Web itself and the necessity to design various perspectives of the same event, support this decision. The proposed ontology has core classes such as Event, Actor, Place and Time and corresponding properties that allow us to model fundamental facts. This also involves means to express some restrictions related to different points of view, namely: (1) Event bounded roles, (2) time bounded validity of facts (e.g. type dependent type or roles) and (3) attribution of the authoritative source of a statement.

3.2.3 Video

A minimum collection of properties capable of defining media resources such as videos, images and audio files has been developed by the W3C Media Annotation Working Group (MAWG). The implemented annotation model [101] consists of several descriptive and technical properties that can be used to describe any kind of multimedia resources along with their technical characteristics. Among the descriptive properties are the title, the language, the creator, the publisher etc. The technical properties refer to more technical aspects of the a media resource like the compression rate, the format etc. The descriptive properties have been defined in such a way that the ontology can be considered as media-agnostic since they can describe any multimedia object. On the other hand the technical properties are specific to particular types of multimedia objects. Figure 2 provides a summary of the key concepts of the given ontology [102].

3.2.4 Text—Social Media

Named entity linking (NEL) or Entity linking (EL) is the task of mapping an entity appearing in a document to a respective Knowledge Base (KB) identifier and is considered as a fundamental step to semantic language understanding. Considering the abundance of text circulating online and the quantity of pieces of information to be represented in text analysis systems, being able to correctly discern among homographic entities determines the KB's quality and subsequently, the entire system's efficacy. Word polysemy, abbreviations and acronyms, spelling variations and synonyms are some of the factors that pose challenges towards entity disambiguation and correct linking. The application possibilities are numerous and range from

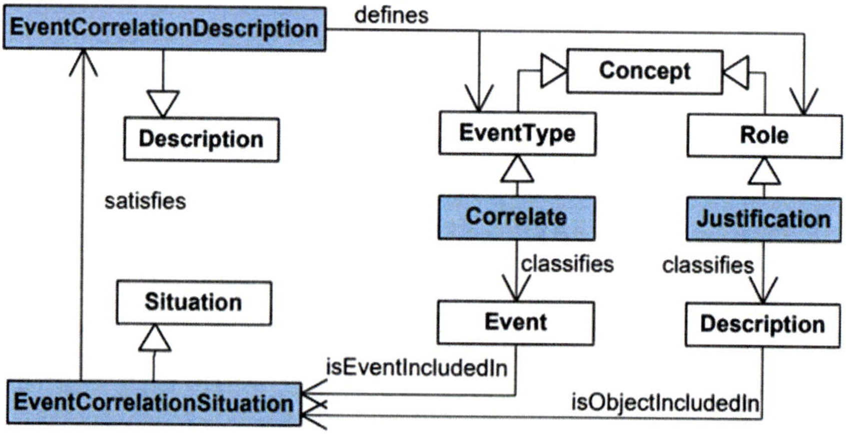

Fig. 1 Event-model F [99]

knowledge graph construction [103] to question-answering systems [104] and information retrieval [105]. Traditional NEL approaches relied on text-based models which leveraged linguistic hand-engineered features [106] and machine learning classifiers (SVM) [107]. Modern systems exploit large knowledge bases (DBpedia, Wikipedia, WordNet) to create knowledge graphs [108] and deep learning techniques [109] which leverage both global and local features to achieve document level disambiguation; character and word embeddings, an attention mechanism and a CRF layer. Local and global features are combined to tackle disambiguation in [110], which proposes a Personalised PageRank-based approach, a popular Random Walk (RW) algorithm. Lastly, in [111] a RW variant (random walk with restart—RWR) and a high-coherence densest subgraph algorithm are combined to create Babelfy, an integrated approach to EL and Word Sense Disambiguation (WSD).

3.2.5 Modelling Domain—Context

The primacy in cultural heritage domain is owned by CIDOC Conceptual Reference Model (CRM) [112] which is responsible as both a theoretical and a practical tool to integrate information in the field of cultural heritage. It provides definitions and structures depicting concepts in the domain enabling querying and investigation of such data. It is the nurture of over 20 years of development and maintenance by the CIDOC Documentation Standards Working Group and more recently by the CIDOC GRM SG. Since 2006 it has been recognized as an official ISO standard (ISO 21127:2014).

Ontologies in health domain can capture information like patient profile, including physiological information, personal information, activities and information specific to the health status of the patient. An abundance of ontologies have been developed for this scope, to support smart home capabilities, provide smart assisting living

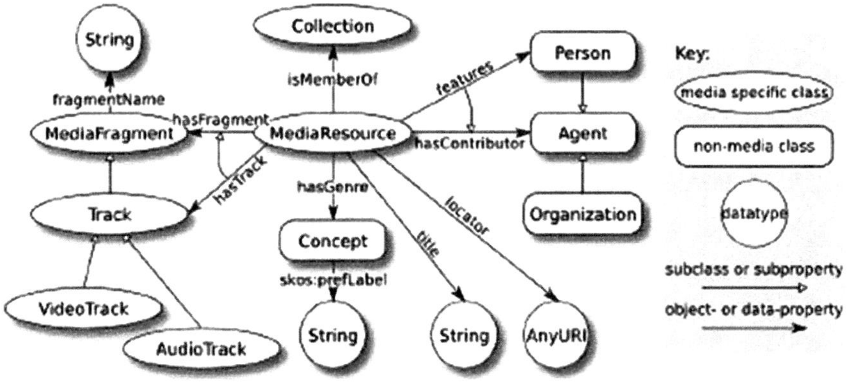

Fig. 2 Ontology for media resources

technologies, improve patients everyday activities and enhance patients residence in healthy environments. Some of them are presented below.

A plain vocabulary to describe people, activities and information about their relations is provided by Friend Of A Friend (FOAF) [113, 114] ontology. This ontology is often used to describe personal information, social connections and networks between people, for instance their membership in groups. People, described as instances of foaf: Person class, can contain many properties like name, email address, image and age. On the other hand, General User Model Ontology (GUMO) [115] offers a more uniform representation of user models. The ontology describes a user from many different perspectives: contact information, demographics, ability, personality, characteristics, motion, role, nutrition, facial expression and emotional, physiological and mental state.

A more extensive ontology for capturing patient information is AHA [116]. AHA ontology uses information coming from wearables to support Ambient Assisting Living environments. The main scope is monitoring activities and extending smart home abilities to assist in lifestyle profiling and healthy ageing issues. The ontology captures body measurements information (such as weight, height), activity-specific information (such as activity levels, energy expenditure, body position that each activity affects) and health state information (such as general health, heart rate, temperature). AHA ontology schema is presented below.

3.3 Ontology-Based Reasoning

This section is related to semantic complex event processing, queries, reasoning rules and algorithms as well as different reasoning frameworks.

3.3.1 Reasoning Frameworks

Description Logics (DLs) [117] are a group of knowledge representation frameworks characterized by logically driven semantics and well-defined inference structures. The key components are classes representing sets of objects (e.g. *Person*), properties representing entity associations (e.g. *livesIn*) and individuals representing individual objects (e.g. *Tom*). Description Logics provide a powerful set of reasoning frameworks. Pellet [118] Racer [119], Fact++ [120] and Hermit [121] are examples of state of the art implementations of such frameworks. Starting from basic definitions, such as Person, it is possible to describe more complicated concepts. For example, the concept $\exists hasFather.Person$ describes those objects that are related through the $hasFather$ property with an object from the class Person. A DL knowledge base K typically consists of a *TBox T* (terminological knowledge) and an *ABox A* (assertional knowledge). The TBox includes axioms describing possible ways of associating domain objects. For example, the TBox axiom $Cat \sqsubseteq Animal$ asserts that all objects that belong to the class Cat, are members of the class Animal too. The ABox

Table 1 TBox and ABox axioms

Name	Syntax	Semantics
Concept inclusion	$C \sqsubseteq D$	$C^I \sqsubseteq D^I$
Concept equality	$C \equiv D$	$C^I = D^I$
Role equality	$R \equiv S$	$R^I = S^I$
Role inclusion	$R \sqsubseteq D$	$R^I \subseteq S^I$
Concept assertion	$C(a)$	$a^I \in C^I$
Role assertion	$R(a, b)$	$(a^I, b^I) \in R^I$

contains axioms which define entities of the real world, for example *Cat(Daisy)* and *isLocated(Daisy,garden)* express that Daisy is a cat and she is located in the garden. Table 1 summarizes the set of TBox and Abox axioms.

The OWL language is commonly used within the community for ontology development and data representation. DLs have greatly influenced the design of OWL and especially the formalization of the semantics and the choice of language constructors. OWL comes in three highly articulate dialects: OWL Lite, OWL DL and OWL Full. The most concise of the three is OWL Full: it does not place any limitations on the use of OWL constructors, nor does it raise the distinction between individuals, properties and class. However, this kind of expressiveness comes at a cost, namely the loss of decisiveness which makes it difficult for the language to be implemented.

An updated version of OWL (that is known as OWL 1) is the OWL 2 language [122]. It extends OWL 1 with eligible constraints on cardinality; thus one may argue, for example, that a social event is an event with more than one actor: $SocialEvent \equiv Event \sqcap \geq 2\ hasActor.Person$. Another notable characteristic of OWL 2 is the expanded relational expressiveness provided by the implementation of axioms (property chains) of complex-property-inclusion. To preserve decisiveness, these axioms are subject to a regularity constraint, which cyclically disallows the concept of properties.

A lot of effort has been dedicated to incorporating OWL with rules. A suggestion for this aim is the Semantic Web Rule Language (SWRL) [123], in which rules are represented under the standard first order logic semantics. Allowing class and property predicates to exist without any constraints in the head and body of a rule, SWRL maximizes the connections between the OWL and rule elements, while at the same time making the synthesis undecidable. Many ideas have addressed syntactic constraints on rules [124, 125] as well as their descriptive intersection of Description Logic Programs (DLP) [126]. The DL-safe rules implemented in [124] for example, specify that rules apply only over known individuals. It should be mentioned that DL reasoners offering support for SWRL actually implement a subset of SWRL based on this notion of DL-safety in practice.

From a separate point of view, a variety of methods have studied the fusion of annotation models and rules based on mappings on rules engines of a sub-set of ontology semantics. For example, [127] describes the grammar of pD^* as a weaker version of OWL Full. In this grammar classes can also be considered as instances and they are generalized to refer to a broader subset of OWL vocabulary. Driven by the entailments of the pD^* grammar and DLP, the OWL 2 RL profile of semantics is realised as a partial axiomatisation of the OWL 2 semantics in the form of first-order, known as OWL 2 RL/RDF rules. Rules defined by users over the ontology allow richer semantic relationships to be articulated outside the descriptive capabilities of OWL, combined with ontological awareness and rules.

SPARQL [128] is a declarative language which the W3C recommends to extract and update information within RDF graphs. It is an expressive language, which describes complex relations between entities. The syntax and difficulty of the SPARQL query language have been studied relatively technically, showing that both SPARQL algebra and relational algebra share the same expressive power [129]. SPARQL is mainly known as a query language for RDF, however it can define SPARQL rules by using the CONSTRUCT graph format, which can generate new RDF statements by merging existing RDF graphs. These rules are described in terms of a CONSTRUCT and a WHERE clause: CONSTRUCT specifies the graph patterns, that is the set of RDF triple patterns that should be ingested to the underlying RDF graph when the graphs in the WHERE clause fit successfully.

Finally, the SPARQL Inferencing Notation (SPIN) [130] is an attempt to simplify the interpretation and execution of SPARQL rules on top of RDF graphs. Using SPIN, SPARQL queries can be stored as RDF triples along with any RDF ontology, allowing RDF instances to be connected to the related SPARQL queries, as well as sharing and reuse of SPARQL queries. SPIN follows the interpretation of SPARQL inference rules that can be used by iterative rules implementations to extract new RDF statements from existing ones.

3.3.2 Fusion

One of the most important issues in the IoT sensor networks is the data management. Sensor networks are facing resource constraints problems due to low battery power, limited data processing capabilities, limited communication resources and a small amount of memory. Furthermore, in many applications there are data coming from many heterogeneous data sources that needs to be compared, combined and correlated between each other. In this way, depending on the applications appropriate data aggregation systems must be implemented for the processing of the data at the edge or in the cloud.

Semantic Web helps enabling interoperability among data from different sources through the content annotation. To become retrievable, data coming from sensor network systems should be annotated.

The fusion through semantic technologies is realised through different structures. For example, in text fusion, it is important to fuse statements and assertions from different sentences, tables, or paragraphs to define definitions, objects, and their semantic relationships. Another important role for semantic fusion is when there is a need to fuse ontologies. The majority of researchers are using available ontologies and in most cases there is a need to combine existing ontologies under the same framework. When multiple ontologies are used under the same development, a mapping of ontologies is important to define the concept representations of the various ontologies relating to the same domain [118]. This can be done by using specific properties like owl:equivalentClass and rdfs:subClassOf.

There are many studies in semantic fusion. In [131] summarizes the implementation of a functional semantic fusion system for live content from the Web. In [132] a service-oriented platform dedicated to fusion processes has been presented. The underlying common language for services is focused on a collection of ontologies that allow for the representation and reasoning of various objects, circumstances and possible threats, and so on. In [133], a use case is proposed that represents the development of a current and future consumer knowledge base, leveraging of social and connected open data on the basis of which any company could infer useful information as a decision-making support. Semantic technologies perform semantic aggregation, persistence, reasoning and retrieval of information, as well as the triggering of alerts over the semantized information.

3.3.3 Validation

The following section presents two approaches for validating RDF data, the Shape Expressions Language [134] and the Shapes Constraint Language [135]. Both share the same goal, that is to provide a framework for validating RDF data.

ShEx is a language for describing RDF graph structures. The basic model of this language, also known as a ShEx schema, contains all the requirements that the RDF data graphs under investigation must fulfill in order to be considered as valid. For example, a requirement could be the datatype of the involved subjects or the combination of subjects, predicates and objects. Based on a list of predefined requirements, the RDF data is tested against it and a validation report is being produced consisting of the parts of the RDF data that do not align.

Another method for validating RDF graphs is called Shapes Constraint Language. Similarly to ShEx, in SHACL a list of pre-defined properties define the requirements that an RDF graph should fulfill in order to be considered as valid. Those requirements are called shapes graphs in SHACL and the data that is validated against a shape graph are called data graphs. Given a shapes graph and a data graph the result of the validation process is also an RDF graph that reports the conformance of the data graph to the shapes graph.

3.3.4 Temporal Reasoning—Stream Reasoning—CEP (Complex Event Processing)

Incorporating the time dimension aspects in both modeling and reasoning, implicitly is granting supplementary temporal assets in objects and knowledge representation in general, thus enhancing the feasibility in exploiting information in order to realize complex event processes to a greater extent, encompassed within the domain of time.

The basic principle that needs to be abided by so as to indite the essence of time is the time instant, an infinitesimal moment in time, based on which more compound temporal concepts are able to be defined, such as time intervals, duration, commencement and conclusion of events, periodicality, schedules and so on [136]. Based on those structural entities and with the adjustment and application of advanced reasoning techniques, one can monitor the temporal flow of occurrences of events, time-irrelevant instances alterations over time and evolution [137].

What differentiates the temporal reasoning from stream reasoning is the rapid frequency in which novel data are acquired and/or metamorphosed and the urgent need for live reasoning. In more detail, it requires fully-automated fault-tolerant pipelines along with heuristic optimization techniques to be able to address to nearly instantaneous rearrangement demands based on live triggers [138].

3.3.5 Querying—Linked Data

The last decades a great amount of data has been available through web technologies. Most of these data are associated with geolocation information. Since geolocated data are rapidly increasing, there comes the need to use and combine these information to extract hidden knowledge. Semantic web technologies are responsible for such tasks as they use reasoning techniques to combine data from heterogeneous sources, supporting in that way more complex semantic queries. Some of the most popular sources are DBpedia, Open Street Maps and Wikidata and their usage in semantic reasoning systems is shown below.

In [139] the authors propose a system which utilizes Open Street Map data to support more complex reasoning rules. The system uses an information broker to apply rule-based reasoning and extract topological relations among entities. More specifically, OWL is used to represent semantically the information and SQWRL rules and vertical plane sweeping technique are used as spatial reasoners. The vertical plane sweeping technique calculates the overlapping polygon, given two polygons. SQWRL rules build on top of the ontology to specify some standards that extract hidden knowledge. In this work the standards are associated with travel planning and footways retrieval.

In [140] OSM is used to gather georeferenced information about points of interest (POIs). The collected information contain points, polylines or polygons and a combination of both indicating the relation between them. The purpose is to detect human activities happening in nearby locations. The methodology creates a connection between human activities and POIs or periods of time. A DL reasoning service defines rules for grouping a number of human operations per category of point of interest and a number of human operations in a specific period of time. The system also predicts human activities according to the popularity of POIs using High Level Representation of Behavioral Model (HRBModel).

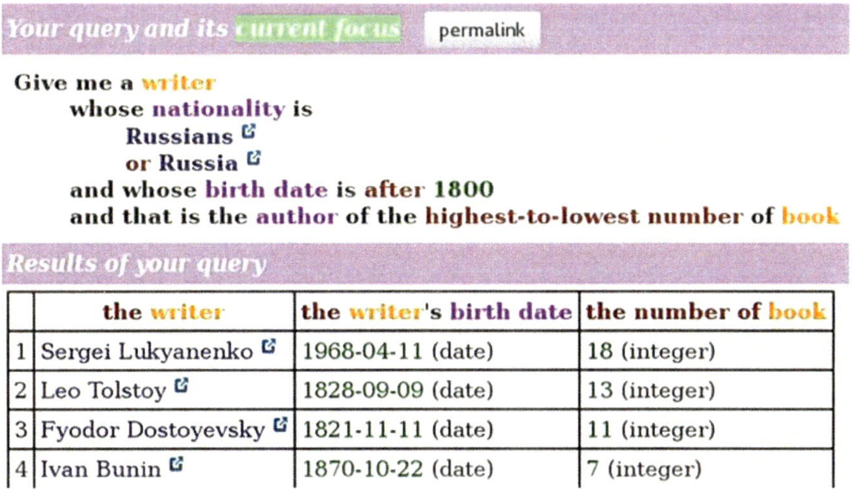

Fig. 3 SPARKLIS natural language queries and results

```
PREFIX n1: <http://dbpedia.org/ontology/>
PREFIX n2: <http://dbpedia.org/resource/>
SELECT DISTINCT ?Writer_1 ?birthDate_3
    (COUNT(DISTINCT ?Book_4) AS ?number_of_Book_5)
WHERE {
    ?Writer_1 a n1:Writer .
    { ?Writer_1 n1:nationality n2:Russians . }
    UNION { ?Writer_1 n1:nationality n2:Russia . }
    ?Writer_1 n1:birthDate ?birthDate_3 .
    FILTER ( str(?birthDate_3) >= "1800" )
    ?Book_4 a n1:Book .
    ?Book_4 n1:author ?Writer_1 . }
GROUP BY ?Writer_1 ?birthDate_3
ORDER BY DESC(?number_of_Book_5)
LIMIT 1000
```

Fig. 4 SPARQL query in DBpedia expressing SPARKLIS natural language query

SPARKLIS [141] is an online tool which combines Natural Language Processing (NLP) with linked data (such as DBpedia, Wikidata, etc.) to apply semantic search. Natural language expressions are used to represent semantic queries and an auto-complete function is provided to fill the queries with possible information. More specifically, the tool consists of three parts: the query as expressed in natural language, the query related terms and the results of the executed query. The subsequent example (Fig. 3) displays the natural language query and results, while Fig. 4 shows the same query as executed in SPARQL.

3.3.6 Handling Noise, Uncertainty and Imperfect Information

As masses of data have grown tremendously during the past years due to, but not exclusively, the eruption of the IoT field, it only seems logical to have arisen issues regarding the quality of such data. Towards this direction several state-of-the-art frameworks have been proposed to address malfunctions deriving from data noise, data uncertainty and imperfect information [142].

According to [143], a logical reasoning model has been used to predict missing data. Grounded on ontological domain knowledge along with a satisfactory dataset of statements, logical reasoning can track inconsistencies and infer new statements as predictions of missing data.

Fuzzy reasoning, encompassing all the properties defined in fuzzy logic theory described in [144], and more specifically non-monotonic reasoning is based on the concept that an assertion can be generated from premises not entirely specified, but in the occasion of an exception emerging the conclusion can be withdrawn [145]. Unfortunately, experiential research regarding adding non-monotonic layers upon reasoning to deal with uncertainty and conflicting data is sparse and not systematic as in the case of [146], where a rule base compression approach is suggested for the decrease of non-monotonic rules, or in the case of [147] where a framework, called FUSE, integrating fuzzy reasoning and semantic reasoning was developed towards a unified reasoning process for the provision of personalized learning recommendations adaptively and semantically. In addition, a proposal presenting fuzzy analogical reasoning has been conducted where the case study of MiMo incorporating soft computing showcases the evaluation [148].

Finally, at the exertion of tackling the nuisance of imperfect information upon reasoning, several investigations were completed, such as in alternating-time temporal logic (ATL) about responsibility in multiagent systems [149] or agents with perfect recall where the past is not forgotten in nested games [150]. Furthermore, investigations towards Graded Computation Tree Logic with finite path semantics (GCTL*f) under imperfect information settings were performed [151].

4 The Semantic Web of Things and How It Augments the IoT

With the advent of IoT hundreds of sensors, smart devices and smartphones have been deployed in our everyday lives. The result of this is tremendous amounts of data with great differences in formats and domains. This has posed great challenges for machines to understand information and extract knowledge from those data. For better representation of IoT different data research studies have proposed different techniques to enable machines to intelligently understand heterogeneous data. Semantic Web of Things (SWoT) is a continuation of World Wide Web that tries to solve the problems arised from the heterogeneous systems and provides a better understanding of the different IoT domains. Web of Things' main purpose is to enable interoperability across IoT platforms and application domains. Overall, the WoT's purpose is to uphold and complement existing IoT standards and solutions [152].

With the semantic technology in Web of Things the domain knowledge and background information are combined with sensor data, making machines easier to understand and process. Moreover, semantics provide a coherent description architecture that enhance information and the exchange of knowledge between variable sensor nodes. Before WOT, sensors and Web world have been completely disconnected. With WOT IoT related data on the Web would help users in different domains by accessing directly sensor data and monitoring the real world parameters integrated with similar context information from the Web.

In order to meet WOT the IoT world, large-scale open interfaces and data formats need to be optimized and incorporated with their relevant IoT counterparts [153]. Generally, IoT users are interested in real-world situations and knowledge, rather than in sensing systems and their raw data. Through SWoT, there are the appropriate abstractions to map sensors and their raw output to real-world entities with real semantics. To realise the SWoT researchers extend the IoT with all the remarkable features of Semantic Web: (a) widely use URIs and HTTP, (b) connecting of domain models through interoperable references, (c) use of common standard languages and (d) domain expressiveness through extrapolation of logical sequences. Some other challenges that SWoT tries to solve are: (i) gradually growing IoT ecosystems with many individual devices; (ii) ability to interconnect devices from different vendors; (iii) ability of open source developers to develop software applications for IoT environments; (iv) develop applications for generic domains exploiting data from various sensors.

5 Conclusion

IoT gains increasingly popularity and its implementations are facing large advancements leading to a new digital era. IoT platforms are essentially the main component of a comprehensive IoT solution as they allow the collection and analysis of data produced at endpoints, resulting the growth of big data analytics artificial intelligence at the edge and other applications. The great progress in the number of network-enabled devices deployed in real world, enhanced by advanced processing techniques, has created vast quantities of databases. As IoT largely relies on a wide variety of heterogeneous systems and technologies, there is no standardized language for data representation and processing. This has contributed to a large number of IoT systems that are incompatible. Thus, it is very difficult for data scientists to extract information from the huge number of data provided by the IoT applications every second.

Semantic web technologies try to overcome such challenges. Semantic web leverages web standards and semantic technologies to interconnect all types of devices by transforming raw sensor data into high-level knowledge that is understandable by humans and machines. Interoperability is one of the most important challenges in an IoT environment, where different devices, services and entities try to connect each other. Semantic modelling produces a definite scheme of the data meaning in a structured way by combining application knowledge and context-relevant information with sensor data. The ontology—based development, which is a domain of semantic modelling, of IoT frameworks can lead to universal IoT solutions multiplying the benefits of IoT.

In this book chapter, we provided an overview of the current trends in application of semantic technologies in the IoT domain. We provided research studies on reasoning, aggregation, fusion and interpretation solutions that aim to intelligently process and ingest sensor information, infusing also human awareness for advanced situational awareness. Finally, issues around Web of Things and how it augments the IoT are discussed.

References

1. Swetina, J., Lu, G., Jacobs, P., Ennesser, F., Song, J.: Toward a standardized common M2M service layer platform: introduction to oneM2M. IEEE Wirel. Commun. **21**(3), 20–26 (2014)
2. Noura, M., Atiquzzaman, M., Gaedke, M.: Interoperability in internet of things: taxonomies and open challenges. Mobile Netw. Appl. **24**(3), 796–809 (2019)
3. Alaa, M., Zaidan, A.A., Zaidan, B.B., Talal, M., Kiah, M.L.M.: A review of smart home applications based on Internet of Things. J. Netw. Comput. Appl. **97**, 48–65 (2017)
4. Ramparany, F., Cao, Q.: A semantic approach to IoT data aggregation and interpretation applied to home automation. In: 2016 International Conference on Internet of Things and Applications (IOTA), pp. 23–28. IEEE (2016)
5. Huang, X., Yi, J., Zhu, X., Chen, S.: A semantic approach with decision support for safety service in smart home management. Sensors **16**(8), 1224 (2016)
6. Zolfaghari, S., Zall, R., Keyvanpour, M. R.: SOnAr: smart ontology activity recognition framework to fulfill semantic web in smart homes. In: 2016 Second International Conference on Web Research (ICWR), pp. 139–144. IEEE (2016)
7. Eine, B., Jurisch, M., Quint, W.: Ontology-based big data management. Systems **5**(3), 45 (2017)
8. SeeClickFix | 311 Request and Work Management Software. https://en.seeclickfix.com/. Cited 25 2019
9. FixMyStreet. https://www.fixmystreet.com/. Cited 25 2019
10. Tsampoulatidis, I., Ververidis, D., Tsarchopoulos, P., Nikolopoulos, S., Kompatsiaris, I., Komninos, N.: ImproveMyCity: an open source platform for direct citizen-government communication. In Proceedings of the 21st ACM International Conference on Multimedia, p. 839–842. ACM (2013)
11. Qamar, T., Bawany, N. Z., Javed, S., Amber, S.: Smart city services ontology (SCSO): semantic modeling of smart city applications. In: 2019 Seventh International Conference on Digital Information Processing and Communications (ICDIPC), pp. 52–56. IEEE (2019)
12. Van de Vyvere, B., Colpaert, P., Mannens, E., Verborgh, R.: Open traffic lights: a STRATEGY for publishing and preserving traffic lights data. In: Companion Proceedings of the 2019 World Wide Web Conference, pp. 966–971. ACM (2019)
13. Open traffic lights ontology. https://w3id.org/opentrafficlights. Cited 25 2019
14. Wachttijd tot groenlicht https://codepen.io/brechtvdv/full/BMQPNX Cited 25 2019

15. Van de Vyvere, B., D'haene, K., D'haene, K., Colpaert, P., Verborgh, R.: Predicting phase durations of traffic lights using live open traffic lights data. In: Joint Proceedings of the 1st International Workshop On Semantics For Transport and the 1st International Workshop on Approaches for Making Data Interoperable co-located with 15th Semantics Conference (SEMANTiCS 2019), pp. 1–7 (2019)
16. Choi, C., Esposito, C., Wang, H., Liu, Z., Choi, J.: Intelligent Power Equipment Management Based on Distributed Context-Aware Inference in Smart Cities. IEEE (2018)
17. Howell, S.K., Rezgui, Y., Beach, T., Zhao, W., Terlet, J., Li, H.: Smart water system interoperability: integrating data and analytics for demand optimized management through semantics. In: ICCCBE, pp. 1–9 (2016)
18. Goel, D., Pahal, N., Jain, P., Chaudhury, S.: An ontology-driven context aware framework for smart traffic monitoring. In: 2017 IEEE Region 10 Symposium (TENSYMP), pp. 1–5. IEEE (2017)
19. Weber, M., Akella, R., Lee, E. A.: Service discovery for the connected car with semantic accessors. In: 2019 IEEE Intelligent Vehicles Symposium, vol. IV, pp. 2417–2422. IEEE (2019)
20. Syzdykbayev, M., Hajari, H., Karimi, H. A.: An ontology for collaborative navigation among autonomous cars, drivers, and pedestrians in smart cities. In: 2019 4th International Conference on Smart and Sustainable Technologies (SpliTech), pp. 1–6. IEEE (2019)
21. Barzegar, M., Sadeghi-Niaraki, A., Shakeri, M., Choi, S.M.: A context-aware route finding algorithm for self-driving tourists using ontology. Electronics **8**(7), 808 (2019)
22. Carenini, A., Ugo, D. A., Stefanos, G., Kallehbasti, P., Mehdi, M., Rossi, M. G., Riccardo, S.: ST4RT–semantic transformations for rail transportation. In: Transport Research Arena TRA 2018, pp. 1–10 (2018)
23. Pace, P., Aloi, G., Caliciuri, G., Gravina, R., Savaglio, C., Fortino, G., Corona, M.: INTER-Health: an interoperable IoT solution for active and assisted living healthcare services. In: 2019 IEEE 5th World Forum on Internet of Things (WF-IoT), pp. 81–86. IEEE (2019)
24. Goossen, W.T.: Detailed clinical models: representing knowledge, data and semantics in healthcare information technology. Healthc. Inf. Res. **20**(3), 163–172 (2014)
25. Adel, E., El-Sappagh, S., Barakat, S., Elmogy, M.: A unified fuzzy ontology for distributed electronic health record semantic interoperability. In: U-Healthcare Monitoring Systems, pp. 353–395. Academic Press (2019)
26. Kontopoulos, E., Mitzias, P., Moßgraber, J., Hertweck, P., van der Schaaf, H., Hilbring, D., Lombardo, F., Norbiato, D., Ferri, M., Karakostas, A., Vrochidis, S.: Ontology-based representation of crisis management procedures for climate events. In: ISCRAM (2018)
27. Chatzimichail, A., Chatzigeorgiou, C., Tsanousa, A., Ntioudis, D., Meditskos, G., Andritsopoulos, F., Karaberi, C., Kasnesis, P., Kogias, D.G., Gorgogetas, G., Vrochidis, S.: Internet of things infrastructure for security and safety in public places. Information **10**(11), 333 (2019)
28. Bagschik, G., Menzel, T., Maurer, M.: Ontology based scene creation for the development of automated vehicles. In: 2018 IEEE Intelligent Vehicles Symposium (IV), pp. 1813–1820. IEEE (2018)
29. Bista, H., Yen, I. L., Bastani, F., Mueller, M., Moore, D.: Semantic-based information sharing in vehicular networks. In: 2018 IEEE International Conference on Web Services (ICWS), pp. 82–289. IEEE (2018)
30. Xing, X., Zhong, B., Luo, H., Li, H., Wu, H.: Ontology for safety risk identification in metro construction. Comput. Ind. **109**, 14–30 (2019)
31. Durbha, S.S., King, R.L.: Semantics-enabled framework for knowledge discovery from Earth observation data archives. IEEE Trans. Geosci. Remote Sens. **43**(11), 2563–2572 (2005)
32. Audebert, N., Le Saux, B., Lefèvre, S.: Joint learning from earth observation and openstreetmap data to get faster better semantic maps. In: Proceedings of the IEEE Conference on Computer Vision and Pattern Recognition Workshops, pp. 67–75 (2017)
33. Audebert, N., Le Saux, B., Lefèvrey, S. Fusion of heterogeneous data in convolutional networks for urban semantic labeling. In: 2017 Joint Urban Remote Sensing Event (JURSE), pp. 1–4. IEEE (2017)

34. Yao, W., Marmanis, D., Datcu, M.: Semantic segmentation using deep neural networks for SAR and optical image pairs. In: Proceedings of Big Data from Space, pp. 1–4 (2017)
35. Masmoudi, M., Taktak, H., Lamine, S.B.A.B., Karray, M.H., Zghal, H.B., Archimede, B., Mrissa, M., Guegan, C.G.: PREDICAT: a semantic service-oriented platform for data interoperability and linking in earth observation and disaster prediction. In: 2018 IEEE 11th Conference on Service-Oriented Computing and Applications (SOCA), pp. 194–201. IEEE (2018)
36. Tiede, D., Baraldi, A., Sudmanns, M., Belgiu, M., Lang, S.: Architecture and prototypical implementation of a semantic querying system for big Earth observation image bases. Eur. J. Remote Sens. **50**(1), 452–463 (2017)
37. Wang, C., Wang, W., Chen, N.: Building an ontology for hydrologic monitoring. In: 2017 IEEE International Geoscience and Remote Sensing Symposium (IGARSS), pp. 6232–6234. IEEE (2017)
38. DEPARTMENTFORcULTURE, M.S.: Creative Industries Economic Estimates. https://www.gov.uk/government/statistics. Cited 2015
39. Meditskos, G., Vrochidis, S., Kompatsiaris, I.: V4Ann: representation and interlinking of atom-based annotations of digital content. In: International Conference on Semantic Systems, pp. 124–139. Springer, Cham (2019)
40. Alvanitopoulos, P., Diplaris, S., de Gelder, B., Shvets, A., Benayoun, M., Koulali, P., Moghnieh, A., Shekhawat, Y., Stentoumis, C., Hosmer, T., Anadol, R., Borreguero, M., Martin, A., Sciama, P., Avgerinakis, K., Petrantonakis, P., Briassouli, A., Mille, S., Tellios, A., Fraguada, L., Sprengel, H., Kalisperakis, I., Cabanas, N., Nikolopoulos, S., Skouras, S., Vogler, V., Zavraka, D., Piesk, J., Grammatikopoulos, L., Wanner, L., Klein, T., Vrochidis, S., Kompatsiaris, I.: MindSpaces: art-driven adaptive outdoors and indoors design. In: Ninth International Conference on Digital Presentation and Preservation of Cultural and Scientific Heritage—DiPP2019, vol. 8, pp. 391–400 (2019)
41. Dimitropoulos, K., Tsalakanidou, F., Nikolopoulos, S., Kompatsiaris, I., Grammalidis, N., Manitsaris, S., Denby, B., Crevier-Buchman, L., Dupont, S., Charisis, V., Hadjileontiadis, L.: A multimodal approach for the safeguarding and transmission of intangible cultural heritage: The case of i-Treasures. IEEE Intell. Syst. **33**(6), 3–16 (2018)
42. Sametinger, F., Baker, C., Ranaivoson, H., Bryan-Kinns, N.: WEAR sustain. Wearable technologists Engage with Artists for Responsible innovation), Sustainability Strategy Toolkit (2019)
43. Kasnesis, P., Tatlas, N.A., Mitilineos, S.A., Patrikakis, C.Z., Potirakis, S.M.: Acoustic sensor data flow for cultural heritage monitoring and safeguarding. Sensors **19**(7), 1629 (2019)
44. Jouan, P.A., Hallot, P.: Digital twin: a Hbim-based methodology to support preventive conservation of historic assets through heritage significance awareness. Int. Arch. Photogrammetry Remote Sens. Spat. Inf. Sci. **42**(2019), 609–615 (2019)
45. Kasnesis, P., Kogias, D.G., Toumanidis, L., Xevgenis, M.G., Patrikakis, C.Z., Giunta, G., Calsi, G.L.: An IoE architecture for the preservation of the cultural heritage: the STORM use case. In: Harnessing the Internet of Everything (IoE) for Accelerated Innovation Opportunities, pp. 193–214. IGI Global (2019)
46. Mousheimish, R., Taher, Y., Zeitouni, K., Dubus, M.: Smart preserving of cultural heritage with PACT-ART. Multimedia Tools Appl. **76**(24), 26077–26101 (2017)
47. Maksimović, M., Ćosović, M.: Preservation of cultural heritage sites using IoT. In: 2019 18th International Symposium INFOTEH-JAHORINA (INFOTEH), pp. 1–4. IEEE (2019)
48. Moraitou, E., Aliprantis, J., Caridakis, G.: Semantic Preventive conservation of cultural heritage collections. In: SW4CH@ ESWC (2018)
49. Anatoly, K., Rezeda, K., Maxim, L., Feng, L., Hu, L., Chen, M., Igor, B.: CHPC: A complex semantic-based secured approach to heritage preservation and secure IoT-based museum processes. Comput. Commun. **148**, 240–249 (2019)
50. Stathopoulos, E.A., Paliokas, I., Meditskos, G., Diplaris, S., Tsafaras, S., Valkouma, E., Pehlivanides, G., Riggas, C., Vrochidis, S., Votis, K., Tzovaras, D.: Smart discovery of cultural and natural tourist routes. In: IEEE/WIC/ACM International Conference on Web Intelligence-Companion, vol. 208–214. ACM (2019)

51. Moraitou, E., Konstantakis, M., Kontaki, C., Aliprantis, I., Kalatha, E., Kalavrytinos, P., Tsigris, A., Tsougkrianis, P., Anagnostopoulos, C., Caridakis, G.: Travelogue with augmented cultural and contemporary experience. In: CIRA@ EuroMed, pp. 66–75 (2018)
52. Nishanbaev, I., Champion, E., McMeekin, D.A.: A survey of geospatial semantic web for cultural heritage. Heritage **2**(2), 1471–1498 (2019)
53. Chianese, A., Piccialli, F.: A smart system to manage the context evolution in the Cultural Heritage domain. Comput. Electr. Eng. **55**, 27–38 (2016)
54. Piccialli, F., Chianese, A.: A location-based IoT platform supporting the cultural heritage domain. Concurrency Comput. Pract. Experience **29**(11), e4091 (2017)
55. Chianese, A., Piccialli, F., Jung, J.E.: The internet of cultural things: towards a smart cultural heritage. In: 2016 12th International Conference on Signal-Image Technology and Internet-Based Systems (SITIS), pp. 493–496. IEEE (2016)
56. Marulli, F., Benedusi, P., Racioppi, A., Ungaro, L. F.: What's the matter with cultural heritage tweets? An ontology–based approach for CH sensitivity estimation in social network activities. In: 2015 11th International Conference on Signal-Image Technology and Internet-Based Systems (SITIS), pp. 789–795. IEEE (2015)
57. Konstantakis, M., Michalakis, K., Aliprantis, J., Kalatha, E., Moraitou, E., Caridakis, G.: A methodology for optimised cultural user personas experience-CURE architecture. In: Proceedings of the 32nd International BCS Human Computer Interaction Conference, vol. 32, pp. 1–8 (2018)
58. Chianese, A., Piccialli, F.: A perspective on applications of in-memory and associative approaches supporting cultural big data analytics. Int. J. Comput. Sci. Eng. **16**(3), 219–233 (2018)
59. Cuomo, S., De Michele, P., Piccialli, F., Sangaiah, A.K.: Reproducing dynamics related to an Internet of Things framework: a numerical and statistical approach. J. Parallel Distrib. Comput. **118**, 359–368 (2018)
60. Konstantakis, M., Aliprantis, J., Teneketzis, A., Caridakis, G.: Understanding user experience aspects in cultural heritage interaction. In: Proceedings of the 22nd Pan-Hellenic Conference on Informatics, pp. 267–271. ACM (2018)
61. Korzun, D., Varfolomeyev, A., Yalovitsyna, S., Volokhova, V.: Semantic infrastructure of a smart museum: toward making cultural heritage knowledge usable and creatable by visitors and professionals. Pers. Ubiquitous Comput. **21**(2), 345–354 (2017)
62. Petrina, O.B., Korzun, D.G., Volokhova, V.V., Yalovitsyna, S.E., Varfolomeyev, A.G.: Semantic approach to opening museum collections of everyday life history for services in internet of things environments. Int. J. Embedded Real-Time Commun. Syst. (IJERTCS) **8**(1), 31–44 (2017)
63. Maietti, F., Piaia, E., Mincolelli, G., Di Giulio, R., Imbesi, S., Marchi, M., Giacobone, G.A. and Brunoro, S.: Accessing and understanding cultural heritage through users experience within the INCEPTION project. In: Euro-Mediterranean Conference, pp. 356–365. Springer, Cham (2018)
64. Castiglione, A., Colace, F., Moscato, V., Palmieri, F.: CHIS: a big data infrastructure to manage digital cultural items. Future Gener. Comput. Syst. **86**, 1134–1145 (2018)
65. Eclipse IoT | IoT development made simple. https://iot.eclipse.org. Cited by 3 2020
66. Hepp, M.: eclassowl: A fully-edged products and services ontology in owl. In: Poster Proceedings of ISWC, Galway (2005)
67. Kharlamov, E., Grau, B.C., Jimenez-Ruiz, E., Lamparter, S., Mehdi, G., Ringsquandl, M., Nenov, Y., Grimm, S., Roshchin, M., Horrocks, I.: Capturing industrial information models with ontologies and constraints. In: The Semantic Web—ISWC 2016—15th International Semantic Web Conference, Kobe, Japan, pp. 17–21, 2016, Proceedings, Part II (2016)
68. Alexakos, C., Anagnostopoulos, C., Kalogeras, A. P.: Integrating IoT to manufacturing processes utilizing semantics. In: 2016 IEEE 14th International Conference on Industrial Informatics (INDIN), pp. 154–159. IEEE (2016)
69. De Roode, M., Fernández-Izquierdo, A., Daniele, L., Poveda-Villalón, M., García-Castro, R. SAREF4INMA: A SAREF Extension for the Industry and Manufacturing Domain

70. Alvanou, G., Lytra, I., Petersen, N.: An MTConnect Ontology for Semantic Industrial Machine Sensor Analytics
71. Kootbally, Z., Kramer, T.R., Schleno , C., Gupta, S.K.: Overview of an ontology- based approach for kit building applications. In: 2017 IEEE 11th International Conference Semantic Computing (ICSC), pp. 520–525. IEEE (2017)
72. Bonacin, R., Nabuco, O.F., Junior, I.P.: Ontology models of the impacts of agriculture and climatechanges on water resources: scenarios on interoperability and information recovery. Future Gener. Comput. Syst. **54**, 423–434 (2016)
73. Shrestha, R., Davenport, G.F., Bruskiewich, R., Arnaud, E.: Development of crop ontology for sharing crop phenotypic information. In: Drought Phenotyping in Crops: From Theory to Practice, pp. 167–176 (2011)
74. Shrestha, R., Senger, M., Ramil, M., Davenport, G., Arnaud, E.: Development of gcp ontology for sharing crop information. Nat. Prec. (2010)
75. Jonquet, C.: Agroportal: an ontology repository for agronomy. In: European Conference Dedicated to the Future Use of ICT in the Agri-Food Sector, Bioresource and Biomass Sector, EFITA'17, Demonstration Session (2017)
76. International Food Policy Research Institute: Linked Open Data—Agricultural Technology Ontology (2017). http://data.ifpri.org/lod/at. Cited by 3 2020
77. Wang, Y., Wang, Y., Wang, J., Yuan, Y., Zhang, Z.: An ontology-based approach to integration of hilly citrus production knowledge. Comput. Electron. Agric. **113**, 24–43 (2015)
78. Joo, S., Koide, S., Takeda, H., Horyu, D., Takezaki, A., Yoshida, T.: Agriculture activity ontology: an ontology for core vocabulary of agriculture activity. In International Semantic Web Conference (Posters & Demos), vol. 33 (2016)
79. Hu, S., Wang, H., She, C., Wang, J.: AgOnt: ontology for agriculture internet of things. In: International Conference on Computer and Computing Technologies in Agriculture, pp. 131–137. Springer, Berlin, Heidelberg (2010)
80. Aubert C., Buttigieg P.L., Laporte M.A., Devare M., Arnaud E.: CGIAR Agronomy Ontology (2017). http://purl.obolibrary.org/obo/agro.owl. Cited by 3 2020
81. CTA: Agrovoc Multilingual Agricultural Thesaurus. http://aims.fao.org/standards/agrovoc/concept-scheme. Cited by 3 2020
82. Agriculture Semantics. https://agrisemantics.org/. Cited by 3 2020
83. Cab thesaurus. https://www.cabi.org/cabthesaurus/. Cited by 3 2020
84. Agriculture Class. https://agclass.nal.usda.gov/. Cited by 4 2020
85. FAO: Agricultural Metadata Element set (agmes) (2018). http://aims.fao.org/standards/agmes. Cited by 3 2020
86. Martini, D., Schmitz, M., Mietzsch, E.: agrordf as a semantic overlay to agroxml: a general model for enhancing interoperability in agrifood data standards. In: CIGR conference on Sustainable Agriculture Through ICT Innovation (2013)
87. Drury, B., Fernandes, R., Moura, M.F.: A survey of semantic web technology for agriculture. In: Information Processing in Agriculture (2019)
88. Atzori, L., Iera, A., Morabito, G., Nitti, M.: The social internet of things (siot)-when social networks meet the internet of things: concept, architecture and network characterization. Comput. Netw. **56**(16), 3594–3608 (2012)
89. Chen, L., Hoey, J., Nugent, C.D., Cook, D.J., Yu, Z.: Sensor-based activity recognition. IEEE Trans. Syst. Man Cybern. Part C (Appl. Rev.) **42**(6), 790–808 (2012)
90. Wang, A., Chen, G., Wu, X., Liu, L., An, N., Chang, C.Y.: Towards human activity recognition: a hierarchical feature selection framework. Sensors **18**(11), 3629 (2018)
91. Luštrek, M., Kaluža, B.: Fall detection and activity recognition with machine learning. Informatica **33**(2) (2009)
92. Ustev, Y. E., Durmaz Incel, O., Ersoy, C.: User, device and orientation independent human activity recognition on mobile phones: Challenges and a proposal. In; Proceedings of the 2013 ACM Conference on Pervasive and Ubiquitous Computing Adjunct Publication, pp. 1427–1436. ACM (2013)

93. Brezmes, T., Gorricho, J.L., Cotrina, J.: Activity recognition from accelerometer data on a mobile phone. In: International Work-Conference on Artificial Neural Networks, pp. 796–799. Springer, Berlin, Heidelberg (2009)
94. Lara, O.D., Labrador, M.A.: A survey on human activity recognition using wearable sensors. IEEE Commun. Surv. Tutor. **15**(3), 1192–1209 (2012)
95. Mangai, U.G., Samanta, S., Das, S., Chowdhury, P.R.: A survey of decision fusion and feature fusion strategies for pattern classification. IETE Tech. Rev. **27**(4), 293–307 (2010)
96. Tryfona, N., Pfoser, D.: Data semantics in location-based services. In: Journal on Data Semantics vol. III, pp. 168–195. Springer, Berlin, Heidelberg (2005)
97. Scherp, A., Franz, T., Saathoff, C., Staab, S.: F–a model of events based on the foundational ontology dolce+ DnS ultralight. In: Proceedings of the Fifth International Conference on Knowledge Capture, pp. 137–144. ACM (200)
98. Gangemi, A., Mika, P.: Understanding the semantic web through descriptions and situations. In: OTM Confederated International Conferences on the Move to Meaningful Internet Systems, pp. 689–706. Springer, Berlin, Heidelberg (2003)
99. Scherp, A., Franz, T., Saathoff, C., Staab, S.: A core ontology on events for representing occurrences in the real world. Multimedia Tools Appl. **58**(2), 293–331 (2012)
100. Van Hage, W.R., Malaisé, V., Segers, R., Hollink, L., Schreiber, G.: Design and use of the simple event model (SEM). Web Semant. Sci. Serv. Agents World Wide Web **9**(2), 128–136 (2011)
101. Media Ontology. https://www.w3.org/TR/mediaont-10/. Cited by 4 2020
102. Stegmaier, F., Bailer, W., Bürger, T., Suárez-Figueroa, M.C., Mannens, E., Evain, J.P., Hóffernig, M., Champin, P.A., Dóller, M., Kosch, H.: Unified access to media metadata on the web. IEEE Multimedia **20**(2), 22–29 (2012)
103. Paulheim, H.: Knowledge graph refinement: a survey of approaches and evaluation methods. Semantic Web **8**(3), 489–508 (2017)
104. Dubey, M., Banerjee, D., Chaudhuri, D., Lehmann, J.: Earl: joint entity and relation linking for question answering over knowledge graphs. In: International Semantic Web Conference, pp. 108–126. Springer, Cham (2018)
105. Cifariello, P., Ferragina, P., Ponza, M.: Wiser: A semantic approach for expert finding in academia based on entity linking. Inf. Syst. **82**, 1–16 (2019)
106. Zhang, W., Su, J., Tan, C.L., Wang, W.T.: Entity linking leveraging: automatically generated annotation. In: Proceedings of the 23rd International Conference on Computational Linguistics, pp. 1290–1298. Association for Computational Linguistics (2010)
107. Bunescu, R., Paşca, M.: Using encyclopedic knowledge for named entity disambiguation. In: 11th Conference of the European Chapter of the Association for Computational Linguistics (2006)
108. Lehmann, J., Isele, R., Jakob, M., Jentzsch, A., Kontokostas, D., Mendes, P.N., Hellmann, S., Morsey, M., Van Kleef, P., Auer, S., Bizer, C.: DBpedia-a large-scale, multilingual knowledge base extracted from Wikipedia. Semantic Web **6**(2), 167–195 (2015)
109. Ganea, O. E., Hofmann, T.: Deep joint entity disambiguation with local neural attention. arXiv preprint arXiv:1704.04920 (2017)
110. Pershina, M., He, Y., Grishman, R.: Personalized page rank for named entity disambiguation. In: Proceedings of the 2015 Conference of the North American Chapter of the Association for Computational Linguistics: Human Language Technologies, pp. 238–243 (2015)
111. Moro, A., Raganato, A., Navigli, R.: Entity linking meets word sense disambiguation: a unified approach. Trans. Assoc. Comput. Linguist. **2**, 231–244 (2014)
112. CIDOC Conceptual Reference Model. http://www.cidoc-crm.org/. Cited by 4 2020
113. Golbeck, J., Rothstein, M.: Linking social networks on the web with FOAF: a semantic web case study. AAAI **8**, 1138–1143 (2008)
114. Brickley, D., Miller, L.: FOAF Vocabulary Specification **91**, (2007)
115. Heckmann, D., Schwartz, T., Brandherm, B., Schmitz, M., von Wilamowitz-Moellendorff, M.: Gumo–the general user model ontology. In: International Conference on User Modeling, pp. 428–432. Springer, Berlin, Heidelberg (2005)

116. Díaz-Rodríguez, N., Grönroos, S., Wickström, F., Lilius, J., Eertink, H., Braun, A., Dillen, P., Crowley, J., Alexandersson, J.: An ontology for wearables data interoperability and ambient assisted living application development. In: Recent Developments and the New Direction in Soft-Computing Foundations and Applications, pp. 559–568. Springer, Cham (2018)
117. Baader, F., Calvanese, D., McGuinness, D., Patel-Schneider, P., Nardi, D. (eds.): The Description Logic Handbook: Theory, Implementation and Applications. Cambridge University Press, Cambridge (2003)
118. Sirin, E., Parsia, B., Grau, B.C., Kalyanpur, A., Katz, Y.: Pellet: A practical owl-dl reasoner. Web Semant. Sci. Serv. Agents World Wide Web **5**(2), 51–53 (2007)
119. Haarslev, V., Möller, R. (2003). Racer: a core inference engine for the semantic web. In: EON, vol. 87
120. Tsarkov, D., Horrocks, I.: FaCT++ description logic reasoner: System description. In: International Joint Conference on Automated Reasoning, pp. 292–297. Springer, Berlin, Heidelberg (2006)
121. Glimm, B., Horrocks, I., Motik, B., Stoilos, G., Wang, Z.: HermiT: an OWL 2 reasoner. J. Autom. Reasoning **53**(3), 245–269 (2014)
122. Grau, B.C., Horrocks, I., Motik, B., Parsia, B., Patel-Schneider, P., Sattler, U.: OWL 2: The next step for OWL. Web Semant. Sci. Serv. Agents World Wide Web **6**(4), 309–322 (2008)
123. Horrocks, I., Patel-Schneider, P. F., Boley, H., Tabet, S., Grosof, B., Dean, M. (2004). SWRL: A semantic web rule language combining OWL and RuleML. W3C Member Submission **21**(79), 1–31
124. Motik, B., Sattler, U., Studer, R.: Query answering for OWL-DL with rules. Web Semantics Sci. Serv. Agents World Wide Web **3**(1), 41–60 (2005)
125. Rosati, R.: DL+ log: tight integration of description logics and disjunctive datalog. KR **6**, 68–78 (2006)
126. Grosof, B.N., Horrocks, I., Volz, R., Decker, S.: Description logic programs: combining logic programs with description logic. In: Proceedings of the 12th International Conference on World Wide Web, pp. 48–57. ACM (2003)
127. Ter Horst, H.J.: Extending the RDFS entailment lemma. In: International Semantic Web Conference, pp. 77–91. Springer, Berlin, Heidelberg (2004)
128. Harris, S., Seaborne, A.S., Prud'hommeaux, E.S.: 1.1 Query language. W3C Recommendation **21**, 10 (2013)
129. Pérez, J., Arenas, M., Gutierrez, C.: Semantics and complexity of SPARQL. In: International Semantic Web Conference, pp. 30–43. Springer, Berlin, Heidelberg (2006)
130. Knublauch, H., Hendler, J.A., Idehen, K.: SPIN-overview and motivation. W3C Member Submission **22**, W3C (2011)
131. Lenders, V.: Semantic fusion of live web content: system design and implementation experiences. In: 2013 Workshop on Sensor Data Fusion: Trends, Solutions, Applications (SDF), pp. 1–6. IEEE (2013)
132. Bellenger, A., Lerouvreur, X., Gatepaille, S., Abdulrab, H., Kotowicz, J.P.: An information fusion semantic and service enablement platform: the FusionLab approach. In: 14th International Conference on Information Fusion, pp. 1–8. IEEE (2011)
133. Torre-Bastida, A.I., Villar-Rodriguez, E., Del Ser, J., Gil-Lopez, S.: Semantic information fusion of linked open data and social big data for the creation of an extended corporate CRM database. In: Intelligent Distributed Computing, vol. VIII, pp. 211–221. Springer, Cham (2015)
134. Shape Expressions. http://shex.io/shex-primer/. Cited by 5 2020
135. Shapes Constraint Language. https://www.w3.org/TR/shacl/ Cited by 5 2020
136. Hobbs, J.R., Pan, F.: Time ontology in OWL. W3C Working Draft **27**, 133 (2006)
137. Li, S., Chen, S., Liu, Y.: A method of emergent event evolution reasoning based on ontology cluster and Bayesian network. IEEE Access **7**, 15230–15238 (2019)
138. Dell'Aglio, D., Eiter, T., Heintz, F., Le Phuoc, D.: Special issue on stream reasoning. Semantic Web **10**(3), 453–455 (2019)

139. Mobasheri, A.: A rule-based spatial reasoning approach for OpenStreetMap data quality enrichment; case study of routing and navigation. Sensors **17**(11), 2498 (2017)
140. Dashdorj, Z., Sobolevsky, S., Lee, S., Ratti, C.: Deriving human activity from geo-located data by ontological and statistical reasoning. Knowled. Based Syst. **143**, 225–235 (2018)
141. Ferré, S.: Sparklis: an expressive query builder for SPARQL endpoints with guidance in natural language. Semantic Web **8**(3), 405–418 (2017)
142. Bamgboye, O., Liu, X., Cruickshank, P.: Towards modelling and reasoning about uncertain data of sensor measurements for decision support in smart spaces. In: 2018 IEEE 42nd Annual Computer Software and Applications Conference (COMPSAC), vol. 2, pp. 744–749. IEEE (2018)
143. Hadhiatma, A.: Improving data quality in the linked open data: a survey. J. Phys. Conf. Ser. **978**(1), 012026. IOP Publishing (2018)
144. Zadeh, L.A.: Information and control. Fuzzy Sets **8**, 338–353 (1965)
145. Longo, L.: Argumentation for knowledge representation, conflict resolution, defeasible inference and its integration with machine learning. In: Machine Learning for Health Informatics, pp. 183–208. Springer, Cham (2016)
146. Gegov, A., Gobalakrishnan, N., Sanders, D.: Rule base compression in fuzzy systems by filtration of non-monotonic rules. J. Intell. Fuzzy Syst. **27**(4), 2029–2043 (2014)
147. Cuong, N.D.H., Arch-Int, N., Arch-Int, S.: FUSE: a fuzzy-semantic framework for personalizing learning recommendations. Int. J. Inf. Technol. Decis. Making **17**(04), 1173–1201 (2018)
148. D'Onofrio, S., Müller, S.M., Papageorgiou, E.I., Portmann, E.: Fuzzy reasoning in cognitive cities: an exploratory work on fuzzy analogical reasoning using fuzzy cognitive maps. In: 2018 IEEE International Conference on Fuzzy Systems (FUZZ-IEEE), pp. 1–8. IEEE (2018)
149. Yazdanpanah, V., Dastani, M., Jamroga, W., Alechina, N., Logan, B.: Strategic responsibility under imperfect information. In: Proceedings of the 18th International Conference on Autonomous Agents and MultiAgent Systems, pp. 592–600. International Foundation for Autonomous Agents and Multiagent Systems (2019)
150. Bulling, N., Jamroga, W., Popovici, M.: Reasoning about strategic abilities: agents with truly perfect recall. ACM Trans. Comput. Logic (TOCL) **20**(2), 10 (2019)
151. Murano, A., Parente, M., Rubin, S., Sorrentino, L.: Model-checking graded computation-tree logic with finite path semantics. Theoret. Comput. Sci. **806**, 577–586 (2019)
152. Web of Things (WoT) Architecture. https://www.w3.org/TR/wot-architecture/. Cited by 7 Jan 2020
153. Pfisterer, D., Romer, K., Bimschas, D., Kleinc, O., Mietz, R., Truong, C., Hasemann, H., Krø'ller, A., Pagel, M., Hauswirth, M., Karnstedt, M.: SPITFIRE: toward a semantic web of things. IEEE Commun. Mag. **49(11)**, 40–48 (2011)

Semantic Web Technologies

Jayashree R. Prasad, Priya M. Shelke, and Rajesh S. Prasad

Abstract The overarching aim of the Semantic Web is to allow computers to perform more valuable research and to construct structures that can facilitate reliable connections around the Network. Semantic Web applications help people to establish data stores on the Internet, build vocabulary, and compose data handling guidelines. Using the details of metadata, the semantic web recovers efficiently the web page user is searching for. Authors have introduced Semantic web, its development and technologies in brief, in this work. Provenance is the most crucial feature in for trustworthiness of semantic web. This feature is focused with the help of provenance data model. Authors discussed semantic web implementations such as semantic web desktop, geospatial semantic web etc. and their applications in different fields such as agriculture, healthcare, and IoT.

Keywords Linked open data · Ontologies · Resource description framework(RDF) · Web ontology language (OWL) · Provenance

1 Introduction

Semantic web is the key evolutionary step inside the world of web. Linked Data Web, Web 3.0, the Web of Data are commonly used names to represent semantic web. In order to facilitate our browsing experience, semantic web connects data sources in such a way that it will be process-able and understandable by machines [1]. The Semantic Web (SW) has been considered as a way to develop semantic

J. R. Prasad
Sinhgad College of Engineering, Pune, India
e-mail: jrprasad.scoe@sinhgad.edu

P. M. Shelke
Vishwakarma Institute of Information Technology, Pune, India
e-mail: priya.shelke@viit.ac.in

R. S. Prasad (✉)
Amity School of Engineering and Technology, Amity University Rajasthan, Jaipur, India
e-mail: rsprasad@jpr.amity.edu

R. Pandey et al. (eds.), *Semantic IoT: Theory and Applications*, Studies in Computational Intelligence 941, https://doi.org/10.1007/978-3-030-64619-6_2

spaces over Web published contents in order to sophisticate the effective retrieval and processing of Web information by both humans and machines for varied purposes [2]. There is not much difference between the semantic web and the World Wide Web rather it enhances the current web for greater utility. Semantic Web extends various mechanisms provided by the current web. It supports appropriate visual display along with precise semantics of web contents so that computer and its users can work in better collaboration. Semantic web comes into action when individuals working in a specific field or occupation consent to common patterns for representation of their crucial information. When these taxonomies are build up by more number of individuals, Semantic Web enables to connect to common schemes and interpret their details, bit by bit extending the quantity of individuals and networks whose Web software can comprehend each other naturally [3]. Abstracting away the physical storage and networking layers concerned in data exchange between two machines is the greatest advantage of web 1.0. Direct connectivity in between the documents can be experienced by this development. When we click a link, within a fraction of second we are on that content, which might reside on different machine, different network or even different continent! Abstraction of the network and physical layers is carried out by Web 1.0. In the network, data and application layers play important role while exchanging the information. Abstraction on these layers is carried out by the Semantic Web. Facts are connected using it. So, if you need to connect to specific piece of information in a document or application, you can directly link to it instead of getting linked to whole container document or application. All updates in the information are automatically reflected to the user.

2 Semantic Web-Future of Internet

In 1989, invention of internet and web was carried out by Sir Tim Berners-Lee, working with CERN. Most basic, fundamental innovation was the hyperlink. With just a click on hyperlink, a very quick access can be obtained by the client over that linked document. The basic building block of this World Wide Web is Web page. It is the simplest way of sharing/exchanging information. Typically, HTML is used to code the web pages and hyperlinks are used to connect them with each other. Web is growing with an astonishing speed and billions of web pages are already present. However, most of the web pages are still designed only from the perspective of human users and processing them by machines is challenging task. In such cases, job of machines is restricted only to decode the information on the web pages such as color schema, headers, links etc. Furthermore, web search engines can offer humans web pages according to the search strings but cannot offer support to interpret results. User has to select most relevant of informative link as per his/her knowledge and interest. With billions of web pages, hundreds of relevant links for the desired topic, situation of user to retrieve exact information is becoming pathetic. Although, some search engines are indexing the search results, the enormous size of search results are making it tedious to get relevant information and it has no more the easy task now as desired. Today's web is a kind of syntactic web in which information on the web page is presented by computers and human users are responsible for interpretation and identification of significant information out of it [4]. However, due to huge volume of digital data, identification and interpretation of relevant data has become more demanding from the perspective of human users. So, in such scenario, assistance from the computers is expected to support this demanding process. However, in order to make the computers supportive for identification and interpretation of relevant data, we need to develop the web content which are understandable by computers. Therefore, the credit of roots of Semantic web-the revolutionary concept was given to Sir Tim Berners-Lee in 2001. According to Berner-Lee semantic web should be as decentralized as possible like syntactic web. With the help of software programs provided by Semantic Web, extraction of metadata from published information has become possible. This metadata is interpretable by machines. In other words, we're applying new data descriptors to web content and data already existing. This allows computers to create concrete interpretations close to the way in which information is interpreted by humans to attain their goals.

As per the Tim Berners Lee, the ultimate goal of semantic web is to allow computers to exploit knowledge more effectively on behalf of humans. "Semantic Web" phrase has two expressive words in it. Here, Semantic" refers to computer-processed data or what the system may do with data. Whereas "Web" conveys the concept of an interconnected object navigable space with URI mappings to resources.

Information Retrieval (IR), Internet of Things (IOT) and Personal Assistants (PA) are the three things under the umbrella of Semantic web used to realize the original vision of it. However, to actualize the todays' vision Linked Open Data and Semantic Metadata is evolved as two primary things.

Semantic Web can function effectively if data stored on computers is well structured as well as well-organized. Automatic reasoning is possible with the help of complete set of inference rules.

2.1 Linked Open Data (LOD)

A strong hybrid of Linked Data and open data is Linked Open Data This LOD makes use of open sources. According to Tim Berners-Lee's clearest description, LOD is Linked Data published under an open license which does not obstruct its free reuse. Semantic web tries to establish links between the datasets which can be processed by not only humans but also computers and linked data provides ways to establish these links. Linked Data is a set of design standards for the exchange of interlinked machine-readable data over the Internet. When use and distribution of data can be made free (subject to condition), it is called Open Data. However, open data and linked data are not equal. It is possible that open data is not linked and linked data is not open for reuse and distribution. Hence, this new blend of Linked Open Data was introduced by W3C community [5].

LOD represents structured organization of data in the form of graph which can be interlinked across the servers. Following lines states, the four principles of Linked Data outlined by Tim Berners-Lee in 2006 [6]:

Use URIs as a name to identify things.

Use HTTP URIs in order to look up the things.

Use open standards (RDF, SPARQL etc.) to offer useful information when a thing is referenced with URI.

Use HTTP URI based names to refer other things on web.

LOD empowers the human users and machines to get to information crosswise over various servers and translate its semantics all the more effectively. Therefore, the Semantic Web rises above from a space containing linked documents to a space including linked information. This, thus, creates the powerful network of machine-processable interconnected information.

Linked Open Data includes:

Realistic data about specific entities and concepts

Ontologies—semantic schemata defining:

Classes of objects (e.g., Vehicle, Department, Document);

Relationship types (e.g., a child of, contained in);

Attributes (e.g., the DoB of an employee, location of department).

One striking example of a LOD set is DBpedia—a publicly supported network exertion to concentrate organized data from Wikipedia and make it accessible on the Web.

Today, large number of datasets are distributed as LOD crosswise over various sectors, for example, encyclopedia, reference book, geographic information, government information, logical database and articles, entertainment, voyaging, and so on. In Life Sciences alone, there are in excess of 100 logical databases distributed as LOD.

2.2 Semantic Metadata

Semantic metadata expresses the "meaning" of data. By using Semantic metadata search engines can facilitate their work, as they get more information to search the related contents. Semantic metadata makes everything simpler to organize and associate. Also, when everything is interlinked, components are all the more effectively remixed, set up together, repurposed and eventually understands. Semantic Metadata adds up to semantic tags to normal Web pages so that their meaning is described with more significance.

For instance, the home page of the National Board of Accreditation (NBA) can be semantically marked with references to a number of suitable conceptions and things, e.g., technical programmes, Washington Accord.

Such metadata makes it a lot simpler to discover Web pages' dependent on semantic criteria. It settles any potential equivocalness and guarantees that when we look for National Board of Accreditation (NBA), we won't get pages about National Basketball Association.

3 Semantic Web Technologies

Functionalities like integration, standardization, development of tools, and adoption by users are required for a more advanced Semantic Web. Following technologies helps to achieve it.

3.1 *Explicit Metadata*

Current formatting of web pages is suitable for human readers and not for programs and machines. Most common language used for creating web pages is HTML (Hyper Text Markup Language). HML defines Web page layout. It is composed of a set of elements instructing the browser about presentation of the content on the screen. XML stands for eXtensible Markup Language and its development is driven by HTML deficiencies. XML was designed for the data storage and transmission. XML was designed to be readable by both humans and computers. Let's have a look at following example in HTML [6].

```
<h1>Flora Institute of Engineering</h1>
An autonomous engineering college located in Pune,
Maharashtra, India. The college has NAAC and NBA
accreditation. It offers various engineering courses at
undergraduate and post graduate level. Dr. Sinha, Dr.
Rajwade and Dr. Mehta (Admission In charge) can guide
you for the admissions.
<h2>Courses offered</h2>
Mechanical Engineering<br>
Computer Engineering<br>
Electronics Engineering<br>
Civil Engineering<br>
```

For human readers, above example might provide satisfactory output, however machines have problems processing it. Keyword based search will produce results based on keywords like Engineering, courses etc., but distinguishing between staff is ambiguous. In order to solve such problems, Semantic web approach attacks the problem from web pages. Instead of HTML, usage of other appropriate language which could carry content of web page along with formatting information is expected. XML is adopted by semantic web for this purpose.

The same example is written in XML as follows-

```
<institute>Flora Institute of Engineering
<Staff>
```

```
    <Director>Dr. Sinha</Director>
    <Dean>Dr. Rajwade</Dean>
    <Admission Incharge>Dr. Mehta</Admission Incharge>
  </Staff>
  <Course>
    <Undergraduate>
      <course1>Mechanical Engineering</course1>
      <course2>Computer Engineering</course2>
      <course3>Electronics Engineering</course3>
      <course4>Civil Engineering</course4>
    </Undergraduate>
    <Postgraduate>
      <course5>Mechanical Engineering</course5>
      <course6>Civil Engineering</course6>
    </Postgraduate>
  </Course>
    <location>Pune, Maharashtra, India</location>
  </institute>
```

Machines can easily access and process this XML representation. It is nothing but metadata. It represents data about data i.e. it expresses meaning of data which is what exactly meant by Semantic word.

Syntactic interoperability is supported by XML by means of structuring the web documents. DTDs and XML schema are used in order to make these web documents machine accessible. The structure of XML web document is defined by writing DTD (Document Type Definition), a bit restricted way used in old approaches or writing XML schema, more extensive way, in latest approaches.

3.1.1 DTDs

There are two types of DTDs-external and DTD. If a separate file is used to define the elements of DTD, it's an external DTD and if they are included in the XML document itself it's an internal DTD. Use of external DTDs are preferred over internal DTDs as they can be refereed across multiple documents. In case of internal DTDs, duplication can't be avoided and maintenance becomes an issue [6].

Consider the Element

```
<Employee>
<name>Prasad Rai</name>
<email>Prasad.rai80@gmail.com</email>
</Employee>
```

A DTD for this element type is written as-

```
<! ELEMENT Employee (name, email)>
<! ELEMENT name (#PCDATA)>
<! ELEMENT email (#PCDATA)>
```

Below is the explanation of this DTD-

- The document can use the element types Employee, name, and email.
- The order of a name element and an email element contained in Employee element is followed strictly.
- Contents of name element and an email element can be varied. In DTDs, elements are expressed using #PCDATA atomic type only.

However, there are certain practical limitations of writing such DTDs.

3.1.2 XML Schema

Structure of XML documents can be represented in a broader and richer way using XML schema. Syntax of XML schema is based on XML itself. The key features of XML schema are as follows-

Definition of new types either by extending or restricting existing ones.

Well defined set of data types.

Other schemas can be built upon existing schemas.

Reusing and refining existing schemas.

Now, let's see simple example of schema element which contains the definition of element and attribute type which are defined using data type.

```
<complexType name="employeeType">
<sequence>
<element name="fname" type="string"/>
<element name="emailaddress" type="string"/>
</sequence>
</complexType>
```

New elements or attributes can be extended from existing data types. The original type is related to extended type by a hierarchical (Super to sub, parent to child) relationship. Objects instantiated from the extended type are quiet obvious to contain all the properties of the original type. In addition to that, they many hold extra information, but they do not contain less or wrong type information.

The meaning is that employeeType element in an XML document include exactly one **fname** elements of type string and must include exactly one emailaddress element of type string.

Tree representation of XML documents is appreciated; A formal data model for XML is represented by trees. This representation is often instructive.

3.2 Ontologies

Ontologies are basic components of the semantic web. These are files which define the terms relationships. Ontologies help in the classification of data and knowledge as classes or taxonomies. As per definition of ontology suggested by T. R.

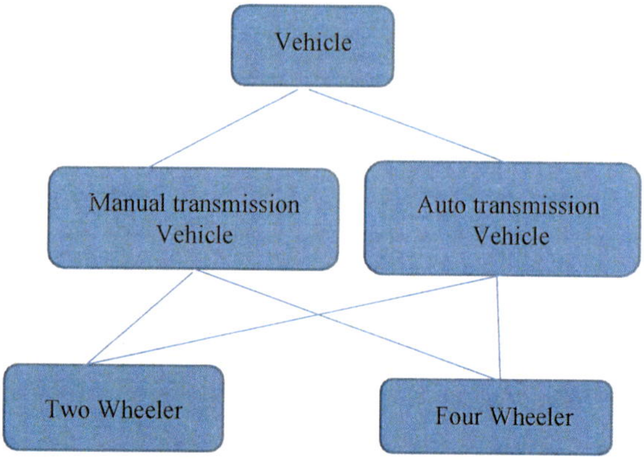

Fig. 1 Hierarchies in vehicle domain

Gruber's and refined by R. Studer: *An ontology is an explicit and formal specification of a conceptualization.* Throughout ontology there is a finite number of terms and the relationships between certain terms. Essential domain concepts (classes of objects) are demonstrated by terms. for example, in a vehicle domain vehicle, manual transmission vehicle, auto transmission vehicle, two-wheeler, four-wheeler, engine, wheels, driver, passenger, license etc. are concepts. These classes are related to each other by hierarchical relationship. In vehicle domain two-wheeler, four wheelers are subclasses of vehicle as all the properties of vehicle are present in two-wheeler and four wheelers. Subclass has all the properties of super class plus it may have its own specific properties. Other than hierarchical relationship, ontologies may have extra information about properties, disjoint or overlapped concepts, any restrictions or constraints and other logical relationships (Fig. 1).

Ontologies seems to be very useful in supporting the semantic interoperability. Ontologies are helpful for the organization as well as navigation of the web sites. Generally, web pages provide menu, in terms of hierarchies. Use clicks on it go navigate through it. Moreover, accuracy of web searches can be improved with the help of ontology. The search engine can filter out and display the pages based on precise ontology term within less time. Hierarchical relationships between terms of ontology can be exploited by the web engine in order to accomplish the web search. E.g. in case specific term related searches are not available, engine can display results of more general term in hierarchy.

3.3 RDF

Resource Description Framework is a system to describe resources. RDF's basic concepts are resources, properties, and statements. RDF has very simple yet elegant data model. Data model consists of resources connected to each other via properties. According to RDF, anything is resource which can be distinguished by Uniform

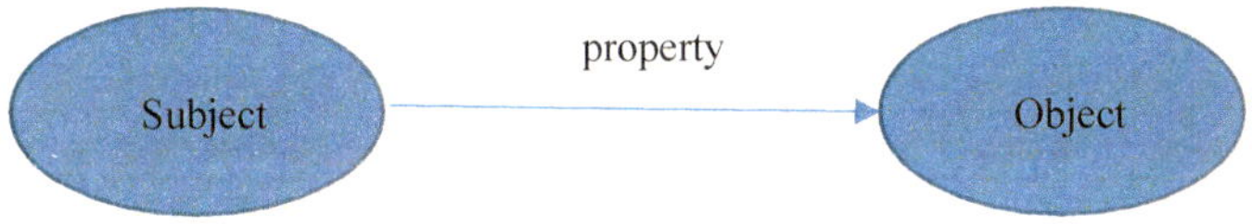

Fig. 2 A graph of RDF statement

Resource Identifier (URI) reference. A property may also be treated as a resource but it is used to define the relationship between two resources. RDF statement is a fundamental unit of RDF. It is represented in the form of (subject, property, object). A simple graph connecting two nodes (subject and object) via directed arc (property) models the RDF statement (Fig. 2). Directed Labelled Graph (DLG), which comprises set of such graphs is used to extract the domain knowledge. RDF model as a graph is insensitive both to syntax and semantics. Yet in either XML or N3 syntax, or even a specialized graphical notation language such as direct labelled graph (DLG2), a RDF model can be serialized. At the other hand, the semantics of an RDF model is obtained by relation to RDF schema language (RDFS) and ontology web language (OWL). Two other semantic web technologies are RDFS and OWL. Both languages are layered on top of RDF to provide inference and axiom support for two features that render semantic web technologies a step from data representation to information representation [7].

3.4 RDF Schema

The RDF Schema specifies the terminology used in RDF data models. RDF Schema includes simple modeling for expressing the information of RDF like classes, properties, hierarchies etc. A number of modeling primitives are used to organize RDF vocabularies in typed hierarchies. RDFS extends RDF to include "schema vocabulary", e.g. Class, Property, type, subClassOf, subPropertyOf, range, domain etc.

3.5 OWL

Web Ontology Language (OWL) offers more modeling primitives compared to RDF Schema. It is a Semantic Web language with clean, formal semantics. It reflects rich and diverse knowledge of things, groups of things and interrelationships. OWL is a language which is based on computational logic. Computer programs can leverage information expressed in OWL. Three species of OWL (1) OWL full is union of OWL syntax and RDF (2) OWL DL restricted to FOL (First Order Logic) fragment (3) OWL Lite is 'easier to execute' OWL DL subset.

3.6 Logic

It is the discipline having study of reasoning conducted or assessed according to strict principles of validity. Logic provides formal languages with well understood formal semantics. Most important about logic is conclusions can be inferred from given knowledge with the help of logic. It is more general than ontologies. Intelligent agents can make use of it to make decisions and select further courses of action. Logic must be machine-process able and must be usable in conjunction with other data for the purpose of Web.

3.7 Agents

Agents represent pieces of software which work independently and proactively to make the decision making process simpler for a web user. Agents are not the replacement for the human users in semantic web. They are not going to make decisions as well. However, they can collect and organize data, provide multiple and preferable solutions to the user and thus facilitates the human user to take appropriate decisions while surfing the web. Semantic web agents use metadata to find and retrieve information from Web documents. Boosting of Web searches, easy understanding of recovered information, and effective communication with other agents is possible by using ontologies. Information is processed and conclusions can be drawn using logic by agents.

4 The Semantic Web Stack

A semantic web stack is shown in Fig. 3. In a conference held at Washington DC, in Dec 2000, Tim Berners-Lee presented this layered structure of semantic web stack [8]. The layers of this system are structured in such a way that higher levels are developed by exploiting the syntax and semantics of lower levels. This primary discussion in this paper is about ontology and software agent based computations. Higher levels further empower the functionality.

W3C standards have specified following technologies for semantic web:

- Resource Description Framework (RDF),
- RDF Schema (RDFS),
- Simple Knowledge Organization System (SKOS),
- SPARQL, a query language for RDF,
- Notation3 (N3), with sophisticated human-readability,
- N-Triples, a data storage and transmission format,
- Turtle (Terse RDF Triple Language),
- Web Ontology Language (OWL), a family of languages representing information,

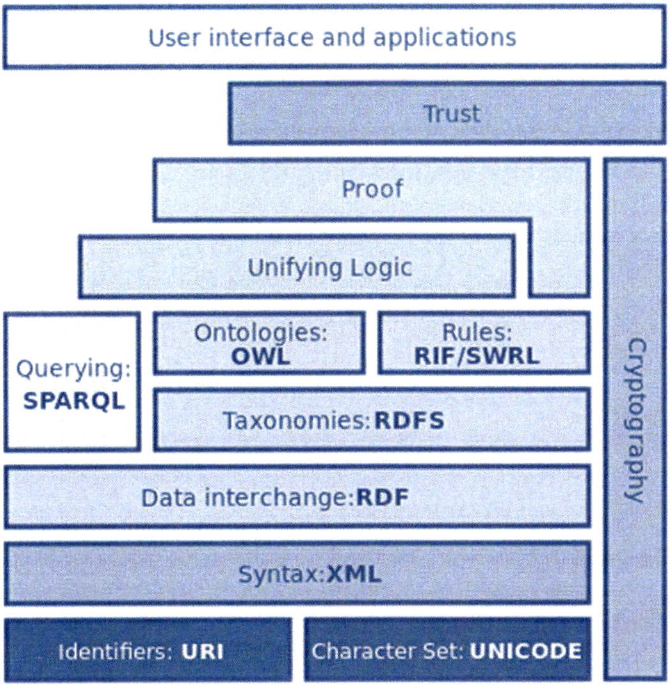

Fig. 3 A semantic web stack

- Rule Interchange Format (RIF), a system of web rule language dialects enabling interchange of rules on the Web.

The semantic web stack layers are as follows:

- Syntax Layer
- RDF(s) Layer
- Ontology Layer
- Logical Layer
- Proof and trust Layer

To the base, we get XML in the syntax layer.

- XML—It provides syntactical structure of content elements. No semantics is associated with it.
- XML Schema—Structure and contents of XML documents are provided and restricted by XML Schema.

RDF and RDF Schema are part of *RDF layer*.

- RDF-Using RDF web resources and their relationships are represented as data models. RDF/XML, N3, Turtle, and RDFa etc. are for representation. Most basic standard of the Semantic Web is RDF.

- RDF Schema is an extension of RDF. It provides the vocabulary that defines the RDF-based re-sources properties and classes. It also provides semantics of both objects and their relationships.

The *Ontology layer* defines various Web Ontology Languages.

- OWL-More vocabulary elements like disjoint, equality, symmetry, enumeration, cardinality etc. are provided by OWL.
- SPARQL is a Network data source protocol and query language.
- RIF is the format of the W3C interchange statute. It is nothing but expression of web rules using XML language. Specific variants of RIF are the Basic Logic Dialect (RIF-BLD) and RIF Production Rules Dialect (RIF PRD).

The *Proof layer* comprises the real inferential process and the representation of proofs in Web languages as well as evidence confirmation.

Eventually, usage of digital signatures and other forms of Trust layer information is created. This form of expertise comes from approvals from reliable agents, certification agencies and government bodies.

5 Provenance in Semantic Web

Although by means of Sematic web ontologies, pedagogical agents manage to retrieve knowledge from the huge amount of web data its trustworthiness is questionable. If bits of information are to be reused or merged with other resources, it should be checked for authenticity and due consideration should be provided to its originators. Provenance helps users trust the information and its sources in web environment. It mandates the amalgamation of trust or provenance information in the semantic web. Even if the information is retrieved easily from the data, to believe that information answers to who, when, how and where that particular piece of data item was developed are crucial. In order to develop the reliability of the semantic web, provenance seems to be vital feature. In order to assess the trustworthiness of semantic web several trust models and metrics have been developed by many researchers till date, however it still lacks standard prescribed form to implement it. PROV-DM data model is the first conceptual model introduced by the worldwide web consortium and acts as a framework for simple provenance assertions in Ontologies, generating and embedding them. PROV data model at the abstract stage comprises of three main types (Entity, Activity, and Agents) and their relationships [9]. PROV data model contents are enlisted in Table 1.

An **entity** is something physical, digital, conceptual, or other with certain fixed aspects; entities may be real or imaginary. For example: Car, book, paper, table, etc.

An **activity** is one that takes place over a span of time that operates in or through entities; activities and entities are related to each other in two distinct ways: activities that utilize entities and create entities. For example: cut, paste, draw, drag, drive from source to destination, paint an object, move a file etc.

Table 1 Elements of PROV data model

Type/relation	Representation in the PROV-N notation	Details
Entity	entity(id, [attr1=val1, ...])	id: an identifier for an entity attributes: an optional set of attribute-value pairs ((attr1, val1), ...)
Activity	activity(id, st, et, [attr1=val1, ...])	id: an identifier for an activity startTime: an optional time (st) for the start of the activity endTime: an optional time (et) for the end of the activity attributes: an optional set of attribute-value pairs ((attr1, val1), ...)
Agent	agent(id, [attr1=val1,...])	id: an identifier for an agent attributes: a set of attribute-value pairs ((attr1, val1),...)
Generation	wasGeneratedBy (id;e,a,t,attrs)	id: an optional identifier for a generation entity: an identifier (e) for a created entity activity: an optional identifier (a) for the activity that creates the entity time: an optional "generation time" (t), the time at which the entity was completely created attributes: an optional set (attrs) of attribute-value pairs
Usage	used(id;a,e,t,attrs)	id: an optional identifier for a usage activity: an identifier (a) for the activity that used an entity entity: an optional identifier (e) for the entity being used time: an optional "usage time" (t), the time at which the entity started to be used attributes: an optional set (attrs) of attribute-value pairs
Communication	wasInformedBy (id;a2,a1,attrs)	id: an optional identifier identifying the relation informed: the identifier (a2) of the informed activity informant: the identifier (a1) of the informant activity attributes: an optional set (attrs) of attribute-value pairs

(continued)

Table 1 (continued)

Type/relation	Representation in the PROV-N notation	Details
Derivation	wasDerivedFrom (id; e2, e1, a, g2, u1, attrs)	id: an optional identifier for a derivation generatedEntity: the identifier (e2) of the entity generated by the derivation usedEntity: the identifier (e1) of the entity used by the derivation activity: an optional identifier (a) for the activity using and generating the above entities generation: an optional identifier (g2) for the generation involving the generated entity (e2) and activity (a) usage: an optional identifier (u1) for the usage involving the used entity (e1) and activity (a) attributes: an optional set (attrs) of attribute-value pairs
Attribution	wasAttributedTo (id;e,ag,attr)	id: an optional identifier for the relation entity: an entity identifier (e) agent: the identifier (ag) of the agent whom the entity is ascribed to, and therefore bears some responsibility for its existence attributes: an optional set (attrs) of attribute-value pairs
Association	wasAssociatedWith (id;a,ag,pl,attrs)	id: an optional identifier for the association between an activity and an agent activity: an identifier (a) for the activity agent: an optional identifier (ag) for the agent associated with the activity plan: an optional identifier (pl) for the plan the agent relied on in the context of this activity attributes: an optional set (attrs) of attribute-value pairs

(continued)

Table 1 (continued)

Type/relation	Representation in the PROV-N notation	Details
Delegation	actedOnBehalfOf (id;ag2,ag1,a,attrs)	id: an optional identifier for the delegation link between delegate and responsible delegate: an identifier (ag2) for the agent associated with an activity, acting on behalf of the responsible agent responsible: an identifier (ag1) for the agent, on behalf of which the delegate agent acted activity: an optional identifier (a) of an activity for which the delegation link holds attributes: an optional set (attrs) of attribute-value pairs

Generation is an activity which completes the creation of a new entity. This entity did not exist before generation, and after this generation becomes available for use. For example: completed creation of a file, completed editing of a particular document etc.

Usage is the initiation of an activity utilizing an entity. The activity has not started to utilize this entity until use, and may not have been influenced by the entity. For example: a program starting to copy values to database, a process starting to print a text, etc.

Communication is called the creation of an entity by an activity, and its eventual usage by another activity. For example: The ability of holding a webinar was told of the activity of sending invitation mails (a communication instance).

A **derivation** is a transformation of an entity into another, an alteration of an entity resulting in a new one, or the development of a new entity dependent on a pre-existing entity. For example: transformation of water into vapor, transformation of the relational table into linked data set.

An **agent** is anything which assumes some sort of responsibility for an activity taking place, for an entity's life, or for the activity of another agent. An agent can be an entity or activity of a specified form. This implies that the model may be used to articulate the agents' provenance themselves.

Attribution is the assigning of an entity to an agent.

An **association** of activity is an assignment of responsibility to an agent for an activity, suggesting that the agent has a role in the activity.

Delegation is the transition of authority and responsibility to an agent (by itself or another agent) to conduct a certain activity as a delegate or representative, while the agent working on behalf of the agent holds certain accountability for the results of the delegated function [9].

Figure shows provenance of a document entity with some agents and activities in the form of graph. Entities, activities and agents are represented as nodes, with oval, rectangular, and pentagonal shapes, respectively. Directed edges represent the generation, association relation (Fig. 4).

This document "WD-prov-dm-20200722", is the fourth version of original. There is an editing activity. The document is generated as an outcome of the editing activity (instance of Generation). Multiple agents (Priya, Jaya) can be seen. Agents were allocated different roles in the editing activity: contributor and editor [9].

Following partial class diagram (Fig. 5) provides the brief data model for provenance. An Entity class is connected to itself via self-association "WasDerived-From" relationship. Association class "WasDerivedFrom" acts as a superclass having subclasses as WasRevisionOf (Revision), WasQuotedFrom (Quotation), and HasPrimarySource (Primary Source).

A revision is a derivation for which the resulting entity is a revised version of some original.

A quotation is the repetition of (some or all of) an item, such as text or picture, by another who may or may not be the original author.

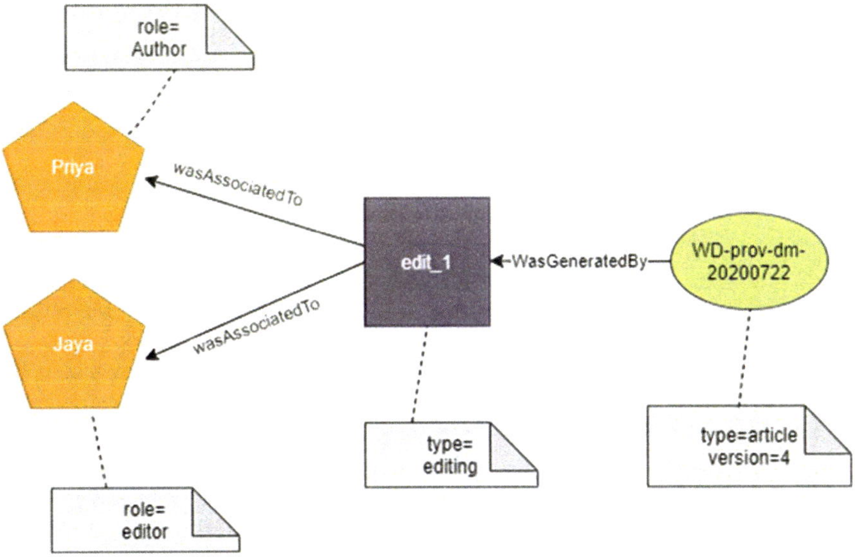

Fig. 4 Graphical illustration of provenance of a document

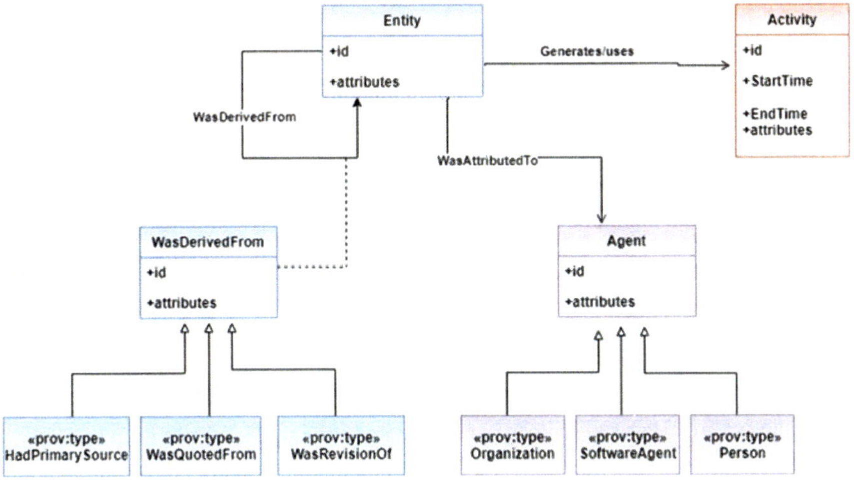

Fig. 5 UML class diagram for provenance data model

A primary source for a topic relates to something created by any person with direct expertise and information regarding the topic, at the time of the subject's analysis, without gain from hindsight [9].

Entity class is connected to Agent class via "WasAttributedTo" relation. Agent is superclass having Organization, Software-Agent, Person subclasses.

Entity class is associated with Activity class which either generates or uses entity.

In paper [10] authors have created reliable semantic web by exploiting provenance assertions and have also provided the mechanism to verify this trust ability using provenance of provenance descriptions. PROV-DM data model is used for deployment. In this paper provenance is created using the concept of entity, agent and activity and these provenance descriptions are connected to each other via Bundle and Mentionof concepts using Prov-store tool for University people program ontology application. Authors have demonstrated that to render the trustworthiness of document provenance of provenance must be ensured and this can be effectively accomplished with the help of Bundles data structure. Also Mentionof relation offers the opportunities to stitch provenance definitions provided by one group to be utilized by another group.

Pandey et al. describes action research methodology using OWL functional syntax [11] and using XML/OWL syntax in Protégé [12] to add provenance to Amity university program ontology. This provenance is confirmed using Hermit Reasoner. Provenance allows processes to be reproduced, and also provides new reasoning information. There, authors explored how interactions between activities, entities, and agents are interested in constructing new entities from the initial by the use of their ontology at the university. They showed the procedures for forming, embedding and reasoning provenance in relation to their Ontology, and also claimed that it can also be used effectively in other applications.

6 Semantic Web Implementations and Applications

6.1 Software Agents

More effective work communication is possible between human users and machines by using semantic web and agents. In order to facilitate automatic web discovery based on customization of user requests, users' constraints and preferences are used by semantic web. Software agents have been defined as prospective customers of semantic web services in order to communicate with semantic SWS specifications in order to independently find, search, write, activate and execute services based on user requirements. There is, though, a communication gap between the two. AgentWeb gateway is an initiative for dynamic and seamless interoperation of multi agent systems and Web services [13]. By taking into consideration engineering student's learning preferences and their specific needs one framework is proposed by authors in [14]. This system is useful for personalized learning and can be thought of as a concept of a tailored intelligent multi-agent learning system. Authors used Semantic Web, Ontologies, recommender system and Intelligent Software Agents for its development. In [15] test environment that is intended to support e-learning in software engineering education is discussed in which automatic questions generation is done by software agents using ontologies.

6.2 Semantic Desktop

Semantic Desktop is a common term for concepts relating to improving the user interface and data processing abilities of a machine so that data can be more easily exchanged between various programs or functions, including data that could not be handled automatically by a computer once. It also incorporates some ideas about being able to instantly exchange knowledge between different individuals. Existing Semantic Desktops are still accused of being too complicated to use or not scaling well. Authors presented new prototype concerned with semantic desktop inspired by NEPOMUK (social semantic desktop),which exploits the contextual information and can dynamically reorganize itself [16].

6.3 Geospatial Semantic Web

A concept of placing geospatial information at the heart of the Semantic Web to enable extraction of knowledge and incorporation of knowledge is termed as Geospatial Semantic Web. To define geographical incidences, it uses popular science terms, semantic gazetteers and geospatial ontologies. Geospatial data semantics, next generation cyber networks, standardized geo domain vocabulary and geographic information extraction are primary fields of interest for the Geospatial Semantic Web. In addition, recent methodologies that promote semantic interoperability without reducing heterogeneity must be taken into account [17]. Geospatial Semantic Web provides a number of uses. One prominent application is from the domain of cultural heritage. Nowadays, modern and evolving digital cultural heritage repositories are growing. Such repositories, utilizing geospatial semantic web technologies, aim to resolve the difficulties of maintaining and retaining scattered and poorly linked cultural heritage documents that do not have structured search interfaces [18].

6.4 Semantic Web in Agriculture

While Agriculture and its related industry have a number of significant semantic resources and standards of data interchange, the application based on semantic web technologies in agriculture is underused. The adoption of semantic web technologies in the agricultural domain is dependent on available semantic resources in agricultural field. For example, AGROVOC is the major and most detailed semantic resource comprising 35,000 concepts and 40,000 terms about agriculture, and also about food, nutrition, fisheries, forestry and the environment. The key areas of agriculture where semantic web technologies were applied are classified mainly into: knowledge-based systems, remote sensing, decision support, and expert systems. Knowledge-based systems are applications which cause information to be stored in a knowledge base to solve complicated issues. Inside the knowledge-based systems the most common application fields are: Question-Answering and Semantic Knowledge Retrieval. In Question-Answering Schemes in the Agricultural Sector, users are informed about agricultural problems such as crops, plant diseases and insect invasion etc. In agriculture, information retrieval systems are usually configured to retrieve specific information about crops, pests, etc. Expansion of the keyword and knowledge management areas in information retrievals was supported by ontology. The agricultural realm is rapidly using remote sensing to gather data from farms such as temperature and soil pH. Users can use that information to infer future crop health. The semantic sensor web is the application of semantic Web technologies to remote sensing domain. By exploiting the large sensor web, sensor web applications can query and draw inferences. It can help farmers by providing real-time input into decision-making systems. crop planning and production and food production are the most popular area of decision support system. Usually the crop production programs provide farmers with actionable knowledge that they may use to minimize crop damage. An ontology is created using aggregated information about crops, pests, diseases, land preparation, growing and harvesting methods that converts this information into actionable information. Decisions on food production support systems used to: control, manage, or assist direct food production. Semantic web technologies had the role of acting as a knowledge base or assisting in the integration of data sources. Expert systems typically infer a disease based on crop sample observations. In these systems an ontology works as a base of knowledge from which to make inferences [19].

6.5 Semantic Web in Healthcare

Huge amounts of data are generated on a regular basis in hospitals, clinics and other medical institutions. Patients' medical reports, their data monitoring and monitoring needs to be properly managed in order to cater for better medical facilities, enhanced healthcare services and biomedical products. While there is an overwhelming volume of data available, it is fragmented and distributed. IOT and sematic web technologies play important role in addressing the key challenge of handling the interoperability in health related data. Internet of Things (IoT) applications residing on the Web are the next logical development. Various ontologies are developed in healthcare domain which can be accessed from bioportal. E.g. Medical Dictionary for Regulatory Activities Terminology (MedDRA) (MEDDRA), Current Procedural Terminology (CPT), SNOMED CT etc. [20]. Multi-agent software systems such as AOIS, MASE, MET4, Karthika, and Cancer Search Engine promote the exchange of knowledge across diverse user groups linked via the Web [21]. Examples of IOT-Semantic web-based healthcare applications are Smart Appliances REFerence for Health (SAREF4Health), Health and Alarm Ontology, and Integrated Health Management Technology (TIHM) etc.

6.6 Semantic Web and IoT

IoT applications have the potential to contribute in many areas such as healthcare, agriculture, automobile etc. which can facilitate to improve human life. It can be predicated that IoT platforms are responsible for are transforming the landscape of information and communication technology. This is clear from the growing usage of internet-enabled gadgets in our everyday lives. These devices, via sensors, produce large quantities of data that are analyzed by cloud platforms in order to develop IOT applications. However, heterogeneity is the basic feature of the components of IOT systems, which is also reflected in their generated data. Due to this heterogeneity of data, interpretation, the proper exploitation and integration of IOT systems becomes very difficult, leading to interoperability between different IOT systems. It is important to infer information from raw data collected to create interoperable, successful IoT applications. Not only can the application of semantic techniques to IoT enable interoperability, but it will also support effective data access, knowledge extraction and integration [22].

OpenIot is a kind of first open source IoT project that enables IOT services in the cloud platform to be semantically interoperable. Based on W3C Semantic Sensor Networks (SSN), OpenIoT can collect data from any type of sensor (including mobile sensors) with its proper semantic annotation. Using OpenIoT's visual tools, users can develop and deploy IoT applications with near zero programming [23].

FIESTA-IoT addresses seven-level semantic interoperability issues. It integrates previously developed semantic-based projects such as OpenIoT, IoT-est, IoT-A, IoT6 etc. as well as non-semantic-based projects such as SmartSantander EU. Data level ensures interoperability by annotating it semantically, Model level aligns existing IoT ontologies. Query level queries unified bases of knowledge, reasoning level unifies meaningful information. Service/Application level operates on "Experimentation-as-a-Service (Eaas)" focused on "Linked Open Services" inspired by Linked Data. Applicative domain level develops cross-domain/vertical applications [24].

In [25], Gergely Marcell Honti and Janos Abonyi took a brief analysis of the Internet of Robotic Things (IoRT) related ontologies such as RoboEarth, Smart and Networking Underwater Robots in Cooperation Meshes (SWARM) and Robotics and Automation Core Ontology (CORA). They also focused on IoT applications such as SCRIBED, Knowledge Model for city (Km4City),READY4SmartCities based on semantic web technologies.

7 Conclusion

This chapter presents basic ideas, concepts and technologies used in Semantic Web initiative. Semantic web is interpreted as web of data. Its main intension is to enable structured and unstructured data source's integration. To integrate them, these data sources are represented in RDF format and their semantics is expressed using RDF schema. Another aim of semantic web is to improve and enrich the current world wide web. Better search engine efficiency, dynamic website personalization and material enrichment of online documents are the main scenarios in this case that are performed utilizing ontology-based technologies. Detailed discussion on provenance in semantic web is carried out with Prov data model.

Authors have also validated the usage of Semantic Web technologies in various application domains such as E-learning, geospatial, agriculture and healthcare and IoT.

References

1. https://www.cambridgesemantics.com/blog/semantic-university/intro-semantic-web/
2. Abelló, A., Romero, O., Pedersen, T.B., Berlanga, R., Nebot, V., Aramburu, M.J., Simitsis, A.: Using semantic web technologies for exploratory OLAP: a survey. IEEE Trans. Knowl. Data Eng. **27**(2), 571–588 (2014)
3. Feigenbaum, L., Herman, I., Hongsermeier, T., Neumann, E., Stephens, S.: The semantic web in action. Sci. Am. **297**(6), 90–97 (2007)
4. Breitman, K., Casanova, M. A., Truszkowski, W.: Semantic Web: Concepts, Technologies and Applications. Springer Science & Business Media (2007)
5. https://www.ontotext.com/knowledgehub/fundamentals/linked-data-linked-open-data/
6. Antoniou, G., van Harmelen, F.: A Semantic Web Primer, MIT Press. Cambridge, MA (2004)
7. Wang, X., Gorlitsky, R., Almeida, J.S.: From XML to RDF: how semantic web technologies will change the design of 'omic' standards. Nat. Biotechnol. **23**(9), 1099 (2005)
8. Berners-Lee, T.: Linked data-design issues. http://www.w3.org/DesignIssues/LinkedData.html (2006)
9. https://www.w3.org/TR/2013/REC-prov-dm-20130430/#term-attribute-type
10. Pandey, M., Pandey, R.: Provenance linking using bundles in OWL ontology. Int. J. Comput. Appl. **975**, 8887 (2017)
11. Pandey, M., Pandey, R., Darbari, M.: Provenance descriptions using the OWL functional syntax in Protégé. Int. J. Innov. Technol. Exploring Eng. (IJITEE), **8**(8), 2421–2428 (2019)
12. Pandey, M., Pandey, R., Darbari, M.: Provenance use and its application in education domain using owl/XML syntax in Protégé. Int. J. Innov. Technol. Exploring Eng. (IJITEE) **8**(9), 1037–1045 (2019)
13. Shafiq, M.O., Ding, Y., Fensel, D.: Bridging multi agent systems and web services: towards interoperability between software agents and semantic web services. In: 2006 10th IEEE International Enterprise Distributed Object Computing Conference (EDOC'06), pp. 85–96. IEEE (2006, October)
14. Melesko, J., Kurilovas, E.: Personalised intelligent multi-agent learning system for engineering courses. In: 2016 IEEE 4th Workshop on Advances in Information, Electronic and Electrical Engineering (AIEEE), pp. 1–6. IEEE (2016, November)
15. Stancheva, N.S., Popchev, I., Stoyanova-Doycheva, A., Stoyanov, S.: Automatic generation of test questions by software agents using ontologies. In: 2016 IEEE 8th International Conference on Intelligent Systems (IS), pp. 741–746. IEEE (2016, September)
16. Jilek, C., Schröder, M., Schwarz, S., Maus, H., Dengel, A.: Context spaces as the cornerstone of a near-transparent and self-reorganizing semantic desktop. In: European Semantic Web Conference, pp. 89–94. Springer, Cham (2018, June)
17. Janowicz, K., Hitzler, P.: Geospatial Semantic Web. Int. Encycl. Geogr. People Earth Environ. Technol. People Earth Environ. Technol. 1–6 (2016)
18. Nishanbaev, I., Champion, E., McMeekin, D.A.: A survey of geospatial semantic web for cultural heritage. Heritage **2**(2), 1471–1498 (2019)

19. Drury, B., Fernandes, R., Moura, M.F., de Andrade Lopes, A.: A survey of semantic web technology for agriculture. Inf. Process. Agric. **6**(4), 487–501 (2019)
20. Mokgetse, T.L.: Need of ontology-based systems in healthcare system. Ontology-Based Inf. Retrieval Healthc. Syst. 257 (2020)
21. Shah, P., Thakkar, A.: Comparative analysis of semantic frameworks in healthcare. In Healthcare Data Analytics and Management, pp. 133–154. Academic Press (2019)
22. Al-Osta, M., Ahmed, B., Abdelouahed, G.: A lightweight semantic web-based approach for data annotation on IoT gateways. Procedia Comput. Sci. **113**, 186–193 (2017)
23. Soldatos, J., Kefalakis, N., Hauswirth, M., Serrano, M., Calbimonte, J. P., Riahi, M., Aberer, K., Jayaraman, P.P., Zaslavsky, A., Žarko, I.P., Skorin-Kapov, L.: Openiot: Open source internet-of-things in the cloud. In: Interoperability and Open-Source Solutions for the Internet of Things, pp. 13–25. Springer, Cham (2015)
24. Gyrard, A., Serrano, M.: Fiesta-iot: federated interoperable semantic internet of things (iot) testbeds and applications. In ICT (2015)
25. Honti, G.M., Abonyi, J.: A review of semantic sensor technologies in internet of things architectures. Complexity **2019** (2019)

A Look at Semantic Web Technology and the Potential Semantic Web Search in the Modern Era

Reinaldo Padilha França, Ana Carolina Borges Monteiro, Rangel Arthur, and Yuzo Iano

Abstract The Semantic Web supports a set of technologies that exploit the standardization of the semantic representation of informational resources available on the web, representing the evolution of the current web. It provides a mechanism for formatting data in a machine-readable manner. Helping people in certain activities are done manually and end up consuming a lot of time in human daily life, linking individual properties of these data with globally accessible schemes. Since with so much information evolution in digital searches is inevitable, which with this technology provides ease and provides inferences about sates in scalable activities and modes. Therefore, this chapter aims to provide an overview of the semantic web and technology behind the semantic web search Engines, showing and approaching its success relation, with a concise bibliographic background, categorizing and synthesizing the potential of both technologies.

Keywords Semantic · Semantic web search · Ontology · Web · Data

1 Introduction

Semantic Web is a new concept of the internet that comes to change everything we know through the network and our interactivity with it within the next few years. The growth of the internet was huge and in less than two decades, what was a

R. P. França (✉) · A. C. B. Monteiro · Y. Iano
School of Electrical and Computer Engineering (FEEC), University of Campinas – UNICAMP, Av. Albert Einstein – 400, Barão Geraldo, Campinas, São Paulo, Brazil
e-mail: padilha@decom.fee.unicamp.br

A. C. B. Monteiro
e-mail: monteiro@decom.fee.unicamp.br

Y. Iano
e-mail: yuzo@decom.fee.unicamp.br

R. Arthur
Faculty of Technology (FT), University of Campinas (UNICAMP), Limeira, São Paulo, Brazil
e-mail: rangel@ft.unicamp.br

R. Pandey et al. (eds.), *Semantic IoT: Theory and Applications*, Studies in Computational Intelligence 941, https://doi.org/10.1007/978-3-030-64619-6_3

project to share information from all over the world eventually created unimaginable proportions. This is certainly very good because of the breadth and quantity of sources with diverse content, but it ended up causing real chaos of information. Because anyone publishes their own information, the volume is so absurd that it pollutes the results and makes browsing increasingly confusing and scattered [1].

Semantic Web is not a new information network, but a project to apply intelligent concepts to the current Internet. In it each information comes with a well-defined meaning and is no longer loose in the sea of content, allowing a better interaction with the user. New search engines, innovative interfaces, creation of thesaurus, and intelligent content organization are some examples of enhancements. This way, it will no longer be necessary to have to mine the internet for in search of the desired information, it will behave as a whole, no longer as a pile of information piled up [2].

It enriches research and exploitation of results on the web, semantic web technologies continue to rise as a progressive integration of search engines and social networks, the development of search engines tailored to the specific needs of companies, and even the growing adoption for e-commerce sites, since it's about optimizing the target of relevant information [1–3].

The Semantic Web introduces several standards RDF, which model for describing all data, and even considering RDF Schema, which creates of vocabularies and descriptive term sets. Even evaluating OWL as a language for creating ontologies supporting logical processing (inference, clustering). Still pondering SPARQL that allows obtaining information from RDF graphs [4].

OWL (Web Ontology Language), explicitly relates relationships as constraints and hierarchy between concepts in a (semi) formal way. Allowing both people and machines to understand the concepts that represent the data available, brings up structured knowledge allowing the use of agents for search and inference. OWL is one of the most robust forms of representation currently occurs through the use of ontologies [5].

The interaction between man and computer has become increasingly common, through the development of methodologies that improve both telecommunication systems [6–8], operational system [9] and human health [10–14]. In this context, semantic search describes the possibility for computers and humans to work in cooperation, linking meanings of words, and giving meaning to content published on the Internet. In short, the news made search results more accurate for users, with search engines going beyond the keyword and considering not only the search terms, but synonyms, context, and more complex factors such as user location and search previous performed [1, 15].

Offering less spammy search results, more knowledge of user intent, and more conversation enables users to have a better experience. That way, today when trying to produce good content, the user becomes authoritative in your market and are much more likely to be referenced by search engines [16, 17].

The Semantic Web is the employment concept of web data defined and linked. So, it can be used by machines not just for display purposes, but for automation, integration, and reuse between different applications. What reflects a current Web extension, where digital information is given well-defined meaning, allowing computers (machines) and people (users) to work better together. However, the characteristic of the Semantic Web that best defines it is that it is machine-readable content [18].

Therefore, this chapter aims to provide an overview of the semantic web and semantic web search Engines, showing and approaching its success relation, with a concise bibliographic background, categorizing and synthesizing the potential of both technologies.

For this purpose, this study was based on the research of **58** scientific papers and books that address the theme of the present research. Exploring mainly a historical review and applicability of techniques related to Semantic Web Technology. These papers were analyzed based on the publication date of fewer than **5** years, with emphasis on publications with dates greater than 2014 as well as publications and indexing in renowned databases, such as IEEE, SCOPUS, and Scholar Google

2 Semantic Web

The semantic web is capable of interconnecting word meanings, and it is an extension of the current world web (according to Fig. 1), which allows computers (machines) and humans (users) to work in cooperation. This process is able to give meaning,

Fig. 1 Current world web

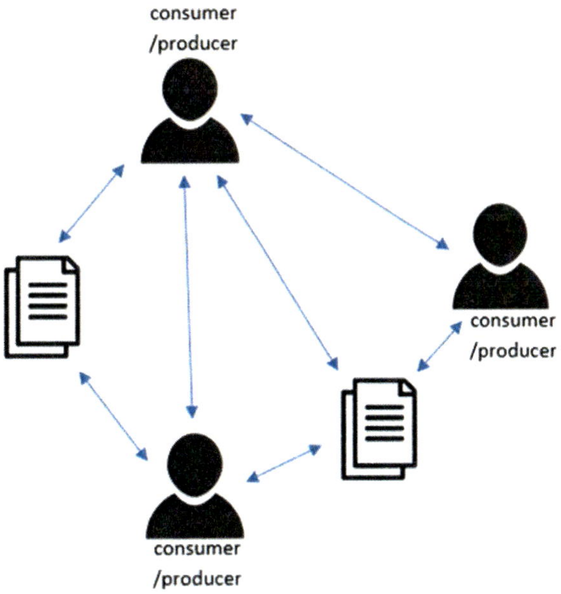

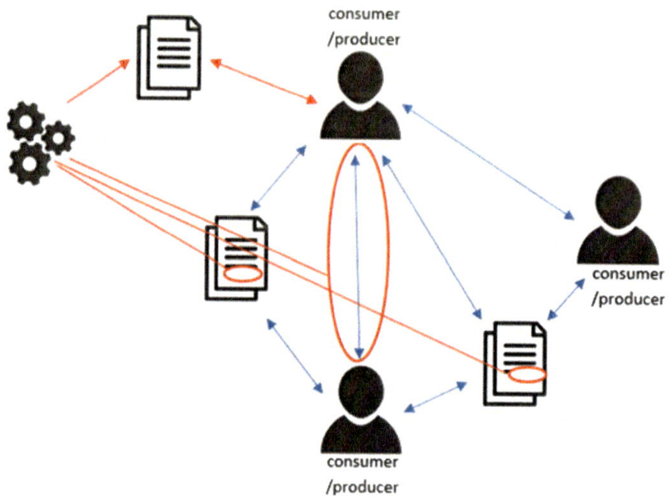

Fig. 2 Semantic web

make sense to content published on the Internet so that it is noticeable by both computer and human. This technology is not a new information network, but a way of applying intelligent concepts to today's Internet, where each information comes with a well-defined meaning. So, is not only found looser in the sea of content that exists today but allows thus a better interaction with the user. From the user's point of view, it will no longer be necessary to mine the internet for something desired, which will now behave as a whole, and not as a pile of information piled up as a result of new search engines with innovative interfaces, and due to intelligent content organization and the creation of thesaurus [19, 20].

The semantic web (according to Fig. 2) uses web metadata, a concept that emerged in 1994, concerning a set of technologies designed to make web resource content (videos, photos, audio, among other media) more accessible and usable by users. The purpose of this technology is to extend the principles of web documents to data, which can be accessed using the Uniform Resource Identifier (URI) web architecture, and relate to each other in the same way as documents are. Since understand and link data placed on the network with semantics, that is, making computers through ontologies and inferential rules find more accurate answers to queries, discarding what is irrelevant to the user, i.e., making information "understandable" to the computer [21–23].

World Wide Web Consortium (W3C) is considered a standardization body of web languages, consisting of a system that allows the linking of information that was previously treated separately. So, the functioning of the semantic web gives the current web the ability to aggregate multiple related data, by the attributes that determine them or even "by the semantics". Making a simple web search activates a metadata engine capable of producing a list of items that exactly match the requested search criteria, not just a raw, unstructured data set. Semantic Web technology introduces several standard standards such as Resource Description Framework (RDF), which is a model for describing all data. Even RDF Schema (RDFS), which creates vocabularies and descriptive term sets. Or yet OWL (Web Ontology Language), which is a language for creating ontologies that support logical processing, clustering, and data inference, and Protocol. Or the same RDF Query Language (SPARQL), which allows obtaining information from RDF graphs [24–26].

However, there is not yet a consensual definition of Semantic Web, it is seen as a vision that through the idealization of having web data defined and linked can be used by machines for intentions of integration, automation, and reuse between distinct applications, and not for display purposes only. The semantic web extends the principles of documents found on the web to data. In short, Semantic Web is a new step in the development of the Internet marked by intelligent user interaction with the web as well as the material made available on it mainly by the organization of all this content. It is the technology of a new step in the internet transforming the virtual network. of information in an increasingly human environment [27].

The characteristic of the Semantic Web that best defines it is that it is machine-readable content in which information is given well-defined meaning. Allowing computers and people to work better together to enrich research and the exploitation of results on the web. The development of search engines and their progressive integration with social networks tailored to the specific needs of companies, along with the growing adoption by e-commerce sites optimizing the target of relevant information. Still improving referral of a site through Search Engine Optimization (SEO), helping to optimize marketing targeting, are aspects of semantic web technologies that continue to rise, encompassing a huge and beneficial environment of interest to technology giants [28, 29].

The reality seen and known is that most of the Internet content is meant to be consumed by possibly not fully understandable by programs. What through computers can analyze web pages in structural terms, such as checking the web of links that connect the pages to each other but not yet reliably processing their meaning. As far as the limitations of today's web are concerned, the fact that through the Semantic Web being evolved and widespread will result in structuring the content of most "machine-readable" web pages. This context creates an environment where software can perform sophisticated tasks for users when scrolling through the pages.

Just as for this technology to be fully operational, computers must have sets of inference rules and access to structured collections of information. So, they can have "automated reasoning", but technologies for this purpose have evolved over the years. Reflecting on years of studies and research of Artificial Intelligence, as well as its Machine and Deep Learning aspects, together with Big Data, making this reality of digital reasoning ever more real [1, 30].

Unlike the initial bias due to its idealization, related to research focused on centralized systems, which required all users to share the same definitions of concepts and rules. Which was redirected to this focus giving the breadth of knowledge that is desired to represent on the Web, and the Semantic Web focused on integrating distinct representations and decentralizing knowledge representation. Approaching the modern-day context where knowledge related to "that the computer can read" has to be represented by the "machine language" translation of the information already present on the web page as well as the additional information that helps the computer respond to queries [1, 31].

The solution offered by the Semantic Web is to create a minimal, high-level, shared schema and one of the ontology languages defined for it. Relying on the idealization of technology when it comes to existing content on the Internet tagged with ontology-based "metadata" and published in one of the standard ontology languages, it is considered [32, 33].

The Semantic Web aims to connect knowledge of databases across the Internet hence the need to create such languages derived from the fact that unlike other "machine-readable" knowledge representation applications. The definition of ontology is "models of a domain" with two special characteristics that must be constructed from community-shared understandings. As well as modeled according to standard syntaxes and logic-based formal semantics, where the concept of ontology does not define the level of detail for modeling a domain ontology. Thus must be "a formal and shared conceptualization of a domain, i.e., a description of concepts and their relationships" [1, 4, 34].

3 Semantic Web Technologies

The Semantic Web at its core is the creation and implementation of technology standards that enable and facilitate the exchange of information between personal agents (devices). As well as establish a clear language for the most meaningful sharing of data between information systems in general and these devices. In this sense, a standardization of languages and technologies, as well as descriptive metadata is required. So, all web users obey certain shared and common rules regarding the storage of data and the description of this stored information that can be consumed in an appropriate manner automatic and unambiguous by other human or machine users. Thus, the creation and development of standards for data description and a language that allows the construction and codification of shared meanings that understand these standards is necessary, considering the current existence of the common technological infrastructure of the Internet [35].

Description Logics (DLs) are useful for to define, ascertain, integrate, prescribe and maintain ontologies, which are related with Semantic Web as one of the tools that can sustain the technology, which provides a common understanding of the basic semantic concepts employed to annotate Web pages [35, 36].

This tool provides an essential notion of the domain described by concept descriptions, with various inference capabilities that deduce implicit knowledge from the explicitly represented knowledge. In other words, they are expressions employing the concept and role constructors provided that are constructed from atomic concepts (unary predicates) and atomic roles (binary predicates) [35, 36].

Description Logics, in general, are grounded on concepts (classes) that are supported as sets of objects, and roles supported as binary relations on objects.

reflecting on the aspect of that particular DLs characterized by a set of constructors which provide building of complex concepts and roles from simpler ones. Still evaluating the Semantic Web's point of view provide well-defined semantics (are logics) still evaluating the provision of inference services [35, 36].

In summary, web ontology language is based on Description Logic, which are logic-based KR (Knowledge Representation) formalisms with emphasis on reasoning, and provide ontology design and expressivity. Description Logic is equipped with a terminological and an assertional formalism, and the large knowledge bases that ontologies needed, integration, and deployment supported by reasoning [35, 36].

A web document is a mixture of data and metadata, the term "meta" being relative to a self-referencing prefix, so that "metadata" is "data over data". Meaning the function of specifying characteristics of data that describe its meaning in a context, or even the way it will be displayed, used [35, 36].

3.1 HTML

The language still used today for the construction of most web pages is **HTML (HyperText Markup Language)** containing a set of tags, which make syntactic markers with descriptive power of data and commands for manipulating a document. This is derived from the standard **SGML (Standard Generalized Markup Language),** what in turn, is based on the idea that documents contain structure and other semantic elements that can be described without reference to how these elements will be displayed. That is, it is a metalanguage, respective to a language for describing other languages. Also considering that HTML tags are usable by a language derived from SGML, called **DTD (Document Type Definition)**. Thus, the HTML language is a defined set of tags, or an SGML-specific DTD, by having the purpose of constructing documents for display on computer devices (on the web) [37–39].

A web browser reading an HTML document interprets these tags by deciding how the data will be displayed. So, today's browsers interpret HTML because the DTD for defining HTML is fixed, and is known a priori by the browser interpreter. Although different browsers can interpret some display definitions in a particular way, the structure of HTML is rigid, i.e., without a language. In this sense, consequently, a browser update to interpret new tags, there is no possibility of adding new tag commands [37, 40].

3.2 XML

From the limitations of HTML and the needs of a language that describes the structure and display of documents, as well as semantic content and contextual meanings,

eXtensible Markup Language (XML), was created. Which is similar to HTML, also derived from SGML and containing tags for describing the content of a document. XML allows data to be described more meaningfully, allowing the semantics to be inserted into web documents and intranets. As long as it focuses on describing the data contained on documents and is flexible in that new tags can be added to the data as needed. As long as they are described in a specific DTD; that is, any developer community can create their specific tags that serve the purpose of describing their data, unlike HTML that is intended to control how the data will be displayed [41].

The XML standard is accepted as the emerging standard for web data exchange which, while allowing authors to create their own tags from a computational perspective, there is very little difference between certain tags. Which in some ways makes semantic markups created difficult unambiguously used by larger communities. Still, companies use XML and SGML-compliant standards in their databases and document bases, which enables the interoperability of the company's internal systems. Still, it is not enough just to have a flexible language like XML to construct metadata, but a way to share meaning that is unambiguously intelligible and consensual among all participants in a community. This can solve the naming explosion problem in different situations in which a univocal interpretation of data is not possible. Wherein the Semantic Web scope metadata patterns through the construction of XML code and a new meaning for the term ontologies have been developed. As the Dublin Core standard, which is for creating a controlled, albeit limited, vocabulary for use on the web, based on the assumption that the search for information resources must be independent of the medium in which they are stored [41–43].

The HTML along with XML and the web has made all online documents a huge book, inference, and data representation models that transform all data around the world into a single database that is needed [41, 43].

3.3 RDF

Just as **RDF**, which is a declarative language written in XML, became a standard recommended by W3C in 2004. It is a general method of breaking down any kind of knowledge into parts, with some rules about semantics or the meaning of those parts. It is also a method of expressing knowledge in a decentralized world pointed one of the foundations of the Semantic Web. In which computer applications use distributed and structured information spread across the Web, with a focus on the simplicity of being able to express any fact yet structured that a Computer application can do useful things with it [15, 44].

RDF encompasses an ontology standard for describing any type of Internet resources, such as a website and its content, establishing a metadata standard to be embedded in XML encoding. RDF has the idea of describing data and metadata through a resource-property-value "triple" scheme, and a coherent way to access web-published metadata standards through Dublin Core, or another shared namespace [45].

What distinguishes RDF technology from XML is that RDF technology is designed to represent knowledge in a distributed world, with focus and particular concern for meaning. In this sense, everything that is mentioned in RDF means something, either an abstract concept or a fact, whether a reference to something concrete in the world. What makes RDF well suited to distributed knowledge is that RDF applications have the ability to merge files placed by different people on the Internet. Still evaluating that it is easily learning from them new things that no document by itself asserts by associating documents through common vocabulary used, and then by allowing any document to use any vocabulary. Since standards designed for RDF describe logical inferences between facts and how to search for facts in large databases of RDF knowledge. This flexibility is unique to RDF, along with the ease with which new data can be added when compared to traditional data models [22, 32, 35].

In summary, the benefits of the RDF standard are over providing a consistent environment for publishing and using web metadata using the XML infrastructure. That enables applications to be able to act intelligently and automatically on information published on the Web, as their meanings are more intelligible. As well as to provide standardized syntax for describing the features and properties of documents on the Web. Still considering the RDF standard is still evolving, and solutions are still being developed and studied. So, the description of namespaces is done non-repetitively and more intelligently within the scope of a document and can further understand more properties [18, 35].

3.4 Linked Data and Open Data

Linked data within the semantic web environment is a network of data that needs formatting into standards accessible and manageable by the tools and end users of this content to make it "palatable" and "digestible". As long as it's data set by itself does not characterize a semantic network, but rather the relationships between these data, and this relationship is given this term linked data [46].

As depicted in Fig. 3, **Linked Data** consists of one of the main pillars of Semantic Web technology, which links between datasets that are understandable to both humans and machines. Also called Web of Data, and Linked Data provides the best practices that make these Possible links [46].

Open data is data that can be freely distributed and employed by anyone (user and machine) regarding the requirements to assign and share. However, Open Data is not equivalent to Linked Data, the open data can be made available to everyone (user and machine) without links to other data, where they can be linked without being freely available for reuse and distribution. And concerning Linked Data, is a powerful combination of linked data and open data [47].

The evolution of the Document Web towards a Data Web with technologies that aggregate semantics to help applications manipulate this information, seeks to reach a horizon in which there will be a global database. Provided that the data are published

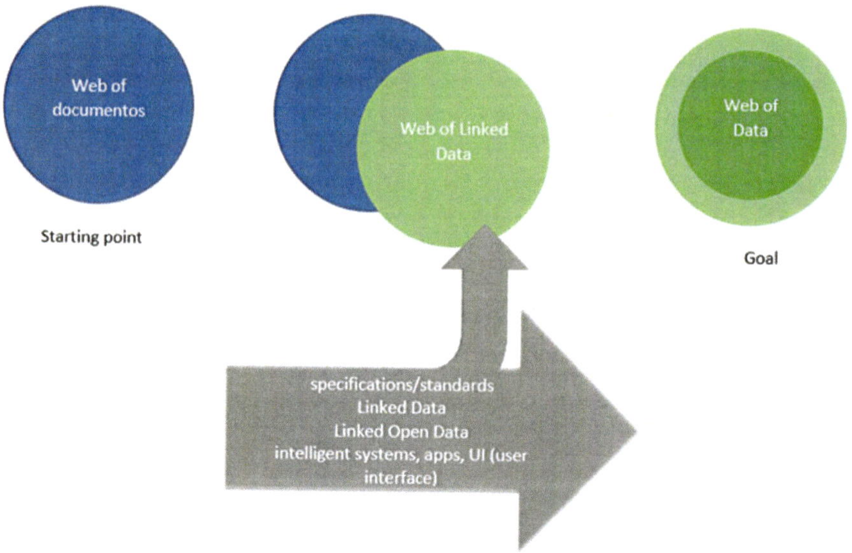

Fig. 3 Linked data

on the Web by different people being stored in different repositories around the world, where a growing set of information can be accessed by a diverse set of applications for different purposes [48].

4 Ontology

Ontologies allow digital access, interoperation and communication based on content by digitally describing through the web the respective classes to sets, collections or types of objects; individuals concerning basic objects; attributes regarding properties, characteristics or parameters that these objects can have and share; and relationships based on how objects can relate to other objects. That is, it allows intelligent access to documents on the Web by inferring or deducing the implicit knowledge of the explicitly stated rules and facts. Also related to the formal representation of these and the solution of inconsistencies related to the vocabulary of natural language such as homonymy (equal words in writing), synonyms (identical or similar meaning to other words), metonymy (using one term in place of another, with close affinity or sense relationship between them) [49–51].

An **Ontology** is a document or file that formally defines the relationships between terms and concepts, it is a specification of a conceptualization, designed to enable knowledge sharing and reuse. In order to create definitions necessary for the creation of common vocabulary or ontological commitments. Thus, they are the vocabularies chosen to define the level of relationship between data. Which are like an entity that

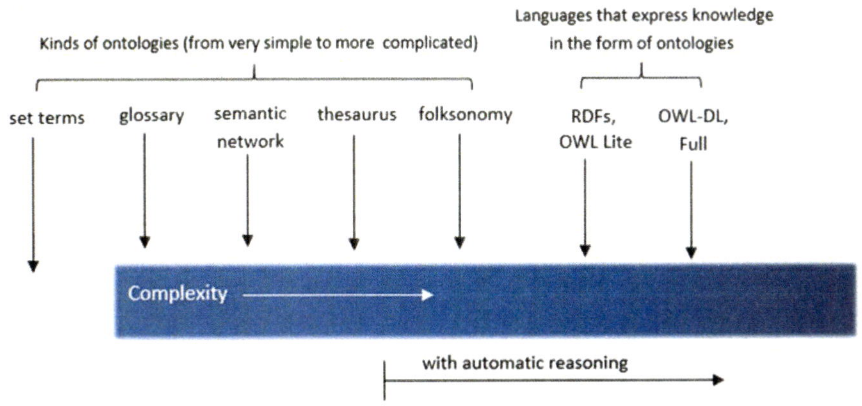

Fig. 4 Ontology

acts as a potentiator and integrator of meanings by eliminating the ambiguity of terms when data lacks greater semantic power [49–51].

It is through ontologies that it is possible to develop data models that provide the basis for applications that make use of the Semantic Web, consisting of a data model that represents a set of concepts within a domain and the relationships between them. An ontology is a formal and explicit specification of a shared conceptualization, this is used to make inferences about the objects in the domain. This defines terms for a computational agent to be able to extract as much information as possible from a document, which will provide the subsidies so that computational agents can search for information that has semantic meaning within the Web [49–51].

As depicted in Fig. 4, it is possible to notice the difference between **thesaurus** and ontologies which are knowledge representation models based on specific domain terminological control. A **thesaurus** is a documentary language characterized by the complexity of the relationship between the terms that communicate specialized knowledge, in contrast to ontology. It is a knowledge representation model, sometimes used as a documentary language, is used to represent and retrieve information through conceptual structures. It contains a central **glossary** that allows the description of terms in multiple domains, containing terms including all domain terms (concepts, instances, attributes, verbs, and their descriptions. Which must be clear and concise, containing independent general notions for the problem [49–51].

4.1 OWL

OWL (Web Ontology Language) is a semantic markup language for defining and instantiating Web ontologies, for publishing and sharing ontologies. This was designed to provide and be used to describe, in a natural way, classes and relationships, among them in documents and Web applications. Still evaluating their specification properties to derive logical consequences, i.e., facts that are linked by semantics, but not are present in an ontology.

An **OWL** makes use of constructors present in the description logic, which is equivalent to a logic-based knowledge representation formalism. It having been the remnant of **semantic networks**, being a collection of thousands of documents and definitions [5].

Folksonomy is a way of indexing information, having as its main feature the creation of tags based on the language of the people who use it. It is a relational way of categorizing and classifying information available on the Web, whether it is represented through texts, images, audio, video or any other format, its purpose is to sort out the chaos on the web.

OWL is a language that defines and instantiates Web ontologies, it can formalize a domain, define individuals and statements about them, define classes and properties of these classes, and using formal semantics. OWL has the ability to specify how to derive facts that are not present in the ontology, but are bound by semantics, it is the most widely used language of ontology developed [5].

OWL has better facilities for expressing semantics and meanings than XML, RDF, and RDF Schema, although it is based on RDF and RDF Schema and uses XML syntax. It is designed for applications that need to process information rather than just presenting it to humans, is designed for use by applications that need to process information content, rather than just displaying that information [51].

4.1.1 Lite, DL, and Full

OWL is divided into three sub-languages, **OWL Lite**, which supports users who need a simple classification hierarchy and constraints, using only a few features of the OWL language, and has more limitations than **OWL-DL** or **OWL-Full**. **OWL-DL** corresponds to description logics, which is a research area that studies a particular fragment of first-order logic that supports users who need maximum language expressiveness without losing computational fullness and decidability language reasoning. Is used by users who want maximum decidability, concerning all computations will end in a computational finite time just as expressivity, with completeness, which all conclusions are guaranteed to be computable. It includes all OWL language constructs but may be used under certain restrictions concerning the use of RDF and requires disjunction of classes, properties, individuals, and data values [52].

OWL-Full supports users who need maximum language expressiveness, with greater syntactic freedom but no computational guarantees, used by users who want maximum RDF expressiveness and syntax independence. Full and DL versions support the same set of OWL language constructs, although with different constraints. The Full version allows OWL and RDF Schema to be mixed without requiring the disjunction of properties, individuals, classes, and data values, i.e., a class can be both class and individual at the same time [5, 52, 53].

The use of each depends on the application in which it will be used, depending on the needs of the ontology. OWL-DL allows a greater level of expressiveness in the constructions, as the Full version brings a better detail [53].

In the field of health systems can be developed that monitor the effectiveness of medicines, as well as decision support applications for possible treatments and side effects. As well as the creation of assistive tools in epidemiological research, due to the combination of knowledge contained in the medical and pharmaceutical communities. With patient data, a wide range of intelligent applications are possible, as long as medical professionals describe or represent knowledge about symptoms, illnesses and treatments similarly to pharmaceutical companies describing information about medications, dosages, and allergies [12–14].

In short, ontologies are presented as a model of the relationship between entities and their interactions. In a particular domain of knowledge or specific to some activity, to build a shared vocabulary to exchange information between members of a community, be they humans or intelligent agents. In this sense, several standards and languages for sharing and building ontologies on the Web were created based on XML, with some differences in tag syntax. Among them can be exemplified as SHOE, Ontology Exchange Language (XOL), Resource Description Framework, Schema Language (RDFS). As well as Ontology Markup Language (OML and CKML), Ontology Interchange Language (OIL), which is a proposed extension of RDF and RDFS [49, 50].

4.2 *FOAF and SPARQL*

The success of the Semantic Web is considered proportional to how much knowledge available on the Web inter-referencing ontologies represents and interrelates. Since it is possible to create HTML pages and link them to existing ones, creating and publishing ontologies that reference other ontologies, in this knowledge-building model, there is no need for a centralized database [22, 32].

FOAF (Friend of a Friend) Ontology is a "machine-readable" ontology consisting of a specific descriptive vocabulary for people, their activities, and relationships with other people and even entities. It makes use of the RDF and OWL languages, enabling the description of social networks without the need for a centralized database. Allowing Semantic Web applications to list all the people a particular user and friend know in common or all Brazilians who have a FOAF profile, whereas the defined online social web. The interconnection of FOAF profiles is already complex and for many purposes, it is also representative, since this ontology is one of the most used on the Internet [54].

SPARQL (SPARQL Protocol and RDF Query Language) is a semantic web data query length and RDF data access protocol, these queries are quite a complex concept within programming languages. However, which are simply interpreted as queries, are technologies that enable the retrieval of embedded information on the network. In the context that the semantic web is a structured organization in meaning relationships of an infinite database, with the possibility of using specific languages and tools to query and filter all this valuable content [55].

The concept of artificial intelligence is related to inference, which generates positive and negative expectations for future scenarios, enabling devices and systems, from the information provided, to establish meaningful relationships. Which can help, facilitating, and automating decisions that are exhausting and complex for people. Within semantic web studies, the inference is the possibility of discovering new relationships between the terms used and their meanings, allowing new relationships and automatic processes to establish new rulesets [56].

Finally, in the context of the Semantic Web, the most commonly used languages are RDF, OWL, and SPARQL. Assessing that RFD was briefly created for modeling and describing information in real-world entities and Web resources consisting of making sentences about features in expression format subject-predicate-object type. In this context, RFD enables making sentences about sentences and extensions that allow working with classes, subclasses, and collections of instances, still providing the term that assigns a type to a resource (rdf: type). SPARQL is a standardized RDF query language, and concerning the OWL language. It was developed for descriptive logical addition to the RDF language, consisting of a set of Knowledge Representation languages with a formal semantics based on their mapping in the First Order Logic, extending the expressiveness of the RDF structure in the characterization of classes and properties [22].

In turn, the Semantic Web content pages will move to meaningful (semantic) pages allowing computers to perform more useful services through systems that offer smarter relationships. The basis of this technology will be the ontologies that allow nurturing meaning to content pages as well as relating them to each other. In this scenario, computers will be able to execute queries through virtual agents that find the desired information more quickly and accurately, together with providing the possibility of inference about them and their relationships [32].

To give meaning to the traditional web that is based on static content pages (HTML), it is necessary to adopt technologies such as RDF, or even the Resource Description Framework Schema (RDF-S), or yet the Simple Knowledge Organization System Reference (SKOS). Which are descriptive languages of the content of a page and work in conjunction with ontological languages like OWL, bringing out structured knowledge allowing the use of search and inference agents [48].

5 Architecture Semantic Web

The architecture that synthesizes the Semantic Web conception focuses on understanding the technology that allows users to visualize its role in the semantic construction of the Web. Which over the years and with the evolution of studies, the architecture has undergone several changes and conceptions since the original design. By aiming to highlight the reflexes of the new technological approaches of representation and retrieval of information resources, as illustrated in Fig. 5 [48].

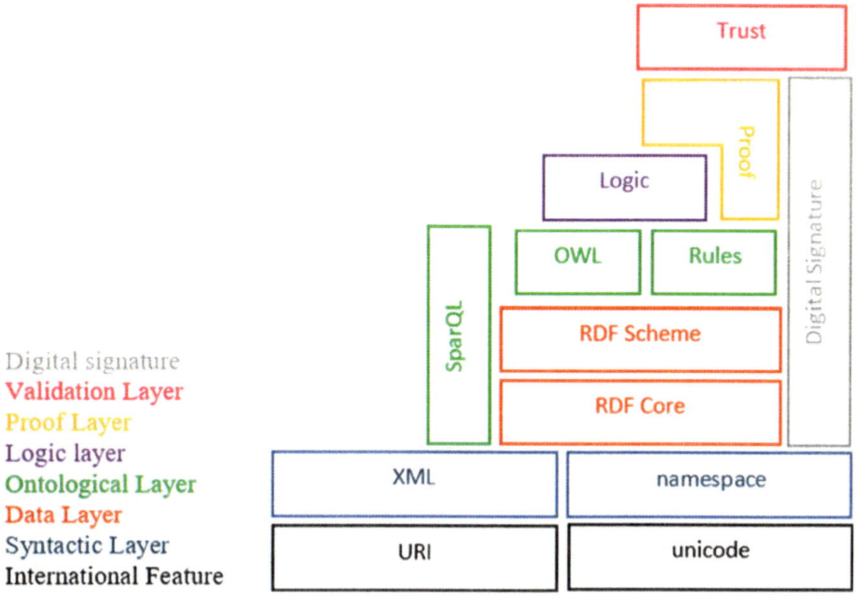

Fig. 5 Semantic web design

As depicted in Fig. 5, the **International** layer represents the bottom level with a focus on ease of data exchange through a concrete serialization syntax. In it, **Unicode** is a standard that allows computers to consistently represent and manipulate text from any existing writing system with an international character encoding where each character is given a unique identifier. So, a top-layer XML document can be represented regardless of platform or language that manages it. Concerning Universal Resource Identifier (**URI**), allow unambiguous identification of resources within the Semantic Web, not unambiguously identifying a Web resource. The **Syntactic** layer has **XML** and XML Schema, which is a markup language for data exchange used by the upper layers, however, it is not required as other languages could be used. The **Data** layer has **RDF** and **RDF Schema** providing flexibility in the representation of concepts and logical rules related to data that generate added value. Allowing the representation of object-oriented semantic models in the form of concept networks. Key concepts included include those that allow being defined: classes, properties, sub-properties, category and domain constraints, tags, and comments [57].

In the **Ontological** layer, formal and explicit specifications of concepts expressed in the lower layer are contemplated, which through ontological **rules** and **SPARQL** technologies capture and explain the necessary vocabulary in semantic applications, ensuring unambiguous communication. The **Logic** layer allows the definition of the logical rules that infer new knowledge through a rational process, proving that the results are correct concerning the rationality internal to a proofing language [17, 18].

In the **Proof** layer and the **Trust** layer, there are mechanisms for proving that the information received is logically coherent. Where it is structured in the set of rules of the lower layers, together with the need to validate it from a reliability point of view. Therefore, the reliability of the information must be verifiable under a logical validity based on requirements that prove unchanged be of correct source, and cannot be denied by its issuer. The **Digital Signature** is inserted in the structure with the function of incorporating security mechanisms that guarantee the reliability of the information, together to obtain a reliable web where the exchanged information is reliable through certification [18, 32].

However, despite the benefits and advantages of the Semantic Web, some obstacles are still present for its full adoption related to the little existing semantic content which makes its evolution difficult. Still considering that no ontological language is accepted as the ideal for the Semantic Web, besides not being fully standardized with each other. This also hinders their integration, where there must be efforts of adoption and standardization in order to reap the rewards that the meaning next to search automation will bring about the Semantic Web [58–60].

Where in an increasingly expanding virtual world flooded with data that can be converted into useful information, knowledge is a living, organic being. Which has the potential to branch into newspapers, books, libraries, research centers, companies, government institutions, social networks, among others. Forming a universe of specificities and peculiarities intrinsic to each of these segments requiring a broad portfolio of vocabularies to standardize these inter-relations of your data with men and machines [1].

6 Discussion

The Web is flooded with data in which this volume only grows by the day, but it is also a fact that this data does not have a clear and established meaning. This makes it impossible to use in an integrated manner without conflict, to determine this meaning and convert this data into Information usable by any agent (human or computer) is the ultimate goal of the Semantic Web. In a way, the Semantic Web is a vision of what the web will be like in the future, where computer agents can finally understand the meaning of data in the same way that people understand and act upon it, performing repetitive tasks and helping users of the most diverse ways.

The semantic search comes from the idea of the semantic web, making search results more accurate for users, because for those who perform a search, the expectation is a simple answer, and thus the semantic search is much more accurate. Considering there are also advantages for search engines, which offer less spammy search results, more knowledge of user intent and more conversation, gives users a better experience.

Impacting also in the production of content focused on semantic search, since quality content must be produced, where search engines need to know what your material is about in order to recommend it to people. Assessing that good content produced makes authority in the digital market and is much more likely to be referenced.

The goal of semantic search is to improve the accuracy of results by understanding the searcher's intent using factors other than the keyword. It can be viewed as an optimization for search engine searches given that semantic search has more to do with it. With the meaning of the search and not with the expression itself, it appears as an answer to a question based on information cataloged on the Internet, even considering the use of metadata and entity-based search are factors that help to rank the answers.

Importantly, the Semantic Web is not an industry standard today, but there are patterns, formats, and languages being used and developed creating a large-scale virtual environment. Still reflecting on applications revolutionizing the way the "real world" is viewed and understood through the "digital world", making the Semantic Web the promising future version of the Web.

It is possible also to see the maturity of these concepts and technology over the 3 decades since the web was created, and then is possible to see that the web is evolving and needs to take its next step to something closer to that of the semantic web. The term Semantic Web is like an "umbrella" of techniques, concepts, and patterns and not an inseparable set of languages and frameworks that should be used in a mandatory way, which is why most existing applications today use their principles making the web more flexible and decentralized.

7 Trends

HTML5 arrives to increase the semantic capacity of the code, increasing its power of representation and meaning, provided with new elements that give more meaning not only to the syntax of the code but also to human interpretation. It has a meaning that transcends the language of the machine by establishing a direct relationship with the human way of organizing content for the web.

The structural-semantic changes that technology gives shape the new framework that will influence design, communication, and information architecture professionals. Allowing the content produced to increasingly dispense with the fluidity and adaptability of the structural elements that should fit a new universe of information content distribution.

Search engines are also impacted when information is tagged and distributed through elements that carry a richer level of meaning because they can establish new relationships of relevance and hierarchy in content. Making an intelligent system with trillions of facts must be equipped with a shorter list of essential truths and a set of rules that allow users to deduce their implications.

8 Conclusions

The growth of the internet was huge and in less than two decades and what was a project to share information from around the world created unimaginable proportions. It is very useful because of the breadth and quantity of sources with diverse content, but eventually caused real chaos of information. Because anyone publishes their own information, and the volume is absurdly large that it pollutes the results and makes browsing increasingly confusing and scattered.

The term semantic refers to the meaning or essence of something, when applied to the search, the concept is related to the study of words and their logic. So, semantic search aims to improve the accuracy of results by understanding the intention of the search user. As long as there is a semantic search system using data such as localization, word variation, synonyms, among other characteristics.

Currently, there are still problems related to the non-understanding of search by current search systems. That is, when a web search is performed, the system returns the information not based on an understanding, but based on keywords that may or may not be related to what was searched.

From this principle emerges the Semantic Web, based on the Web that can bring to users more accurate information, where both people and computers are able to understand this data and, in this paradigm, the Web comes to be seen as a large database. And so, the Semantic web through its technologies is a way for computers to understand what the user wants, and from this view, the problem is now considered the need to create technologies capable of making computers understand more precisely the meaning of words and terms.

This type of scenario is very positive since it deals with existing technologies that only need to be standardized, and there is no need to simply redo everything that already exists and bring new technologies. Some adjustments are needed, evolving the web languages already in use for a fully implemented semantic point of view.

References

1. Workman, M.: Introduction to This Book. In Semantic Web, pp. 1–5. Springer, Cham
2. Andročec, D., Novak, M., Oreški, D.: Using semantic web for internet of things interoperability: a systematic review. Int. J. Semant. Web Inf. Syst. (IJSWIS) **14**(4), 147–171 (2018)
3. Bernstein, A., Hendler, J., Noy, N.: A new look at the semantic web. Commun. ACM **59**(9), 35–37 (2016)
4. Machado, L.M.O., Souza, R.R., Graça Simões, M.: Semantic web or web of data? A diachronic study (1999 to 2017) of the publications of tim berners-lee and the world wide web consortium. J. Assoc. Inf. Sci. Technol. **70**(7), 701–714 (2019)
5. Nagvenkar, A., Pawar, J.D., Bhattacharyya, P.: Indowordnet conversion to web ontology language (OWL). In: Mititelu, V.B., Forascu, C., Fellbaum, C. Vossen, P. Bucharest (eds.) Proc. 8. Global WordNet Conference, Romania, 27-30 Jan 2016. 2016; 255–258. http://irgu.unigoa.ac.in/drs/handle/unigoa/4419 (2016)

6. França, R.P., Iano, Y., Monteiro, A.C.B., Arthur, R., Estrela, V.V. Betterment Proposal to Multipath Fading Channels Potential to MIMO Systems. In: Brazilian technology symposium, pp. 115–130. Springer, Cham (2018, October)
7. França, R.P., Iano, Y., Monteiro, A.C.B., Arthur, R.: Improvement for channels with multipath fading (MF) through the methodology CBEDE. In: Fundamental and Supportive Technologies for 5G Mobile Networks, pp. 25–43. IGI Global (2020)
8. França, R.P., Iano, Y., Monteiro, A.C.B., Arthur, R.: A proposal of improvement for transmission channels in cloud environments using the CBEDE methodology. In: Modern Principles, Practices, and Algorithms for Cloud Security, pp. 184–202. IGI Global (2020)
9. França, R.P., Iano, Y., Monteiro, A.C.B., Arthur, R.: Improvement of the transmission of information for ICT techniques through CBEDE methodology. In: Utilizing Educational Data Mining Techniques for Improved Learning: Emerging Research and Opportunities, pp. 13–34. IGI Global (2020)
10. França, R.P., Peluso, M., Monteiro, A.C.B., Iano, Y., Arthur, R., Estrela, V.V.: Development of a kernel: a deeper look at the architecture of an operating system. In: Brazilian technology symposium, pp. 103–114. Springer, Cham (2018, October)
11. Monteiro, A.C.B., Iano, Y., França, R.P., Arthur, R., Estrela, V.V.: A comparative study between methodologies based on the hough transform and watershed transform on the blood cell count. In: Brazilian technology symposium, pp. 65–78. Springer, Cham (2018, October)
12. Monteiro, A.C.B., Iano, Y., França, R.P., Arthur, R.: Methodology of high accuracy, sensitivity and specificity in the counts of erythrocytes and Leukocytes in blood smear images. In: Brazilian technology symposium, pp. 79–90. Springer, Cham (2018, October)
13. Monteiro, A.C.B., Iano, Y., França, R.P.: Detecting and counting of blood cells using watershed transform: an improved methodology. In: Brazilian technology symposium, pp. 301–310. Springer, Cham (2017, December)
14. Monteiro, A.C.B., Iano, Y., França, R.P.: An improved and fast methodology for automatic detecting and counting of red and white blood cells using watershed transform. VIII Simpósio de Instrumentação e Imagens Médicas (SIIM)/VII Simpósio de Processamento de Sinais da UNICAMP (2017)
15. McIlraith, S.A., Son, T.C., Zeng, H.: Semantic web services. IEEE Intell. Syst. **16**(2), 46–53 (2001)
16. Vandenbussche, P.Y., Atemezing, G.A., Poveda-Villalón, M., Vatant, B.: Linked Open Vocabularies (LOV): a gateway to reusable semantic vocabularies on the Web. Seman. Web **8**(3), 437–452 (2017)
17. Klusch, M., Kapahnke, P., Schulte, S., Lecue, F., Bernstein, A.: Semantic web service search: a brief survey. KI-Künstliche Intelligenz **30**(2), 139–147 (2016)
18. ud@@@ Din, I., Khusro, S., Ullah, I., Rauf, A.: Semantic history: ontology-based modeling of users' web browsing behaviors for improved web page revisitation. In: Proceedings of the Computational Methods in Systems and Software, pp. 204–215. Springer, Cham (2018, September)
19. Krause, S., Hennig, L., Moro, A., Weissenborn, D., Xu, F., Uszkoreit, H., Navigli, R.: Sargraphs: a language resource connecting linguistic knowledge with semantic relations from knowledge graphs. J. Web Seman. **37**, 112–131 (2016)
20. Klimek, B.: Proposing an OntoLex-MMoOn alignment: towards an interconnection of two linguistic domain models. In: LDK workshops, pp. 68–73 (2017)
21. Hitzler, P., Krotzsch, M., Rudolph, S.: Foundations of Semantic Web Technologies. Chapman and Hall/CRC (2009)
22. Poirier, L.: Making the Web Meaningful: A History of Web Semantics. The SAGE Handbook of Web History, pp. 256–269. SAGE, London (2018)
23. Berners-Lee, T., Fielding, R., Masinter, L.: Uniform Resource Identifiers (URI): Generic Syntax (1998)
24. Seltzer, W.: World Wide Web Consortium (W3C) standards for the open web platform. In: Open Source, Open Standards, Open Minds Conference Proceedings, Washington DC (2016, April)

25. Abou-Zahra, S., Brewer, J., Cooper, M.: Web standards to enable an accessible and inclusive internet of things (IoT). In: Proceedings of the 14th Web for All Conference on the Future of Accessible Work. 2017. W4A '17: Proceedings of the 14th Web for All Conference on The Future of Accessible Work April 2017, Article No. 9, pp. 1–4. https://doi.org/10.1145/3058555.3058568
26. Peng, P., Zou, L.: Survey on federated RDF systems. Front. Data Comput. **1**(1), 73–81 (2019)
27. Ristoski, P., Paulheim, H.: Semantic Web in data mining and knowledge discovery: a comprehensive survey. Web Seman. Sci. Serv. Agents World Wide Web **36**, 1–22 (2016)
28. Pereira, F.A., Krzyzanowski, R.F., de Morais Imperatriz, I.M.: Técnicas de Search Engine Optimization (SEO) aplicadas no site da Biblioteca Virtual da FAPESP. Cadernos BAD **1**, 251–265 (2019)
29. Bala, M., D. Verma.: A critical review of digital marketing. Int. J. Manag. IT Eng. **8**(10), 321–339 (2018)
30. Kasemsap, K.: Software as a service, semantic web, and big data: theories and applications. In: The Resource Management and Efficiency in Cloud Computing Environments, pp. 264–285. IGI Global (2017)
31. Zhong, N., Liu, J., Yao, Y.: Web intelligence. Springer Science & Business Media (2003). https://books.google.com.br/books?hl=ptBR&lr=&id=9ovNrElSkKEC&oi=fnd&pg=PA1&dq=Web+Intelligence.+Springer+Science+%26+Business+Media&ots=0eQcmbjUyJ&sig=mC91HVXN_i7xzrsgWK0ed02azw#v=onepage&q=WebIntelligence.SpringerScience%26BusinessMedia&f=false
32. Calbimonte, J.P., Jeung, H., Corcho, O., Aberer, K.: Enabling query technologies for the semantic sensor web. Int. J. Seman. Web Inf. Syst. (IJSWIS) **8**(1), 43–63 (2012)
33. Souza, E.G., Bezerra, D.A., Costa, W.F.C.: Description of resources in a metadata structure inspired in the FRBR model: Brapci 2.0. Em Questão **22**(1), 113z136 (2016, Jan/Apr); **24**(2), 136–113 (2018)
34. Palmirani, M., Rossi, A., Martoni, M., Margaret, H.: A methodological framework to design a machine-readable privacy icon set. In: Data Protection/LegalTech Proceedings of the 21st International Legal Informatics Symposium IRIS 2018 (2018)
35. Pauwels, P., Zhang, S., Lee, Y.C.: Semantic web technologies in AEC industry: a literature overview. Autom. Constr. **73**, 145–165 (2017)
36. Baader, F., et al. (eds.): The Description Logic Handbook: Theory, Implementation and Applications. Cambridge University Press, Cambridge (2003)
37. Singhal, A., Pandey, D., Nagpal, R., Mehrotra, D.: Measuring informativeness of a web document. In: 2016 6th International Conference-Cloud System and Big Data Engineering (Confluence), pp. 654–657. IEEE (2016, January)
38. Krause, J.: HTML: hypertext markup language. In: Introducing Web Development, pp. 39–63. Apress, Berkeley, CA (2016)
39. Cover, R., Duncan N., Barnard, D.T.: The progress of SGML (Standard Generalized Markup Language): extracts from a comprehensive bibliography. Literary Linguist. Comput. **6**(3), 197–209 (1991)
40. Frohme, M., Steffen, B.: Active mining of document type definitions. In: The international workshop on formal methods for industrial critical systems, pp. 147–161. Springer, Cham (2018, September)
41. Gao, C.F., Qian, Y., Zhou, D.P.: U.S. Patent No. 10,318,616. U.S. Patent and Trademark Office, Washington, DC (2019)
42. O'farrell, W.G., Consens, M.: U.S. Patent No. 10,394,685. U.S. Patent and Trademark Office, Washington, DC (2019)
43. Wang, M., Wang, J., Guo, K.: Extensible markup language keywords search based on security access control. Int. J. Grid Util. Comput. **9**(1), 43–50 (2018)
44. Creamer, T.E., Hrischuk, C.E. U.S. Patent No. 9,817,914. U.S. Patent and Trademark Office, Washington, DC (2017)
45. Lefrançois, M., Zimmermann, A., Bakerally, N.: A SPARQL extension for generating RDF from heterogeneous formats. In: The European Semantic Web Conference, pp. 35–50. Springer, Cham (2017, May)

46. Adjallah-Kondo, G.C.., Ma, Z.: A survey on JSON mapping with XML/RDF. In: Emerging Technologies and Applications in Data Processing and Management, pp. 92–113. IGI Global (2019)
47. Zaveri, A., Rula, A., Maurino, A., Pietrobon, R., Lehmann, J., Auer, S.: Quality assessment for linked data: a survey. Seman. Web **7**(1), 63–93 (2016)
48. Tammisto, Y., Lindman, J.: Definition of open data services in software business. In International Conference of Software Business, pp. 297–303. Springer, Berlin, Heidelberg (2012, June)
49. Gottschalk, K., Graham, S., Kreger, H., Snell, J.: Introduction to web services architecture. IBM Syst. J. **41**(2), 170–177 (2002)
50. Guarino, N., Oberle, D., Staab, S.: What is an ontology?. In Handbook on Ontologies, pp. 1–17. Springer, Berlin, Heidelberg (2009)
51. Steffen, S., Studer, R. (eds.) Handbook on Ontologies. Springer Science & Business Media (2010)
52. Gruber, T. (2018). Ontology. In: Liu, L., Özsu, M.T. (eds.) Encyclopedia of Database Systems. Springer, New York, NY. https://doi.org/10.1007/978-1-4614-8265-9_1318
53. Silva, A.M.D.D.: Folksonomies in archives: controlled collaboration for specific documents. Ariadne **77** (2017)
54. Rebstadt, J., Brinkschulte, L., Enders, A., Mertens, R.: A visual language for OWL lite editing. In: SEMANTiCS. Posters, Demos, SuCCESS (2016, September)
55. Pauwels, P., Terkaj, W.: EXPRESS to OWL for construction industry: towards a recommendable and usable ifcOWL ontology. Autom. Constr. **63**, 100–133 (2016)
56. Mandal, S., Roy, S.K.: Linked open data: FOAF-enabled graph visualization. Chin. Librarianship 45 (2018)
57. Chiba, H., Uchiyama, I.: SPANG: a SPARQL client supporting generation and reuse of queries for distributed RDF databases. BMC Bioinf. **18**(1), 93 (2017)
58. Patel, P., Ali, M. I., Sheth, A.: From raw data to smart manufacturing: AI and semantic web of things for industry 4.0. IEEE Intell. Syst. **33**(4), 79–86 (2018)
59. Erl, T.: Service-Oriented Architecture: Analysis and Design for Services and Microservices. Prentice-Hall Press (2016)
60. Sabou, M., Aroyo, L., Bontcheva, K., Bozzon, A., Qarout, R.K.: Semantic web and human computation: the status of an emerging field. Seman. Web **9**(3), 291–302 (2018)

Semantic IoT: The Key to Realizing IoT Value

Hemanta Kumar Palo

Abstract The virtual representation and integration of the internet with the physical objects, devices or things have been growing exponentially in recent years. This has motivated the community to design and develop new Internet of Things (IoT) platforms to cater, capture, access, store, share, and communicate data for information retrieval and intelligent applications. However, the associated dynamism, resource-constrain, cost and the nature of the IoT warrants special design obligations for its effectiveness in the days ahead, hence pose a challenge to the community. The understanding of web data from machines according to the subject of terminology in different fields is a complex task. It opens up new challenges to researchers as such an effort mandates the provision of semantically structured, appropriate information sources in this information age. The advent of numerous smart devices, operators, and IoT service providers subject to time-consuming and complex operations, inadequate research and innovations give rise to design complexity. For efficient functioning and effective implementation of the domain requires the inclusion of semantics and the desired interoperability among these factors. This motivates the authors to review and emphasizes a few of the emerging trends of the semantic technology impacting the IoT. Particularly, the work focuses on different aspects as information modeling, ontology design, machine learning, network tools, security policy and processing of semantic data—and discuss the issues and challenges in the current scenario.

Keywords Internet of things · Semantic IoT · Interoperability · Ontology · Semantic web

H. K. Palo (✉)
Department of ECE, Institute of Technical Education and Research, Siksha 'O' Anusandhan (Deemed To Be University), Bhubaneswar, Odisha, India
e-mail: hemantapalo@soa.ac.in

R. Pandey et al. (eds.), *Semantic IoT: Theory and Applications*, Studies in Computational Intelligence 941, https://doi.org/10.1007/978-3-030-64619-6_4

1 Introduction

A numerous range of new services and products including smart city, e-healthcare, green energy, automobiles, retail, logistics, educational institutes, and environmental monitoring have been surfaced with the emergence of diverse and efficient IoT devices, sensors, connectivity, and communication networks [1]. The support and collaboration industries, academia, and standardization bodies concerned with telecommunication, informatics, and Semantic Web (SW) have set the desired momentum in this field. There are three different aspects concerned with an IoT paradigm that are intermixed and contribute to its definition [2]. These are, things-oriented, Internet-oriented, and semantic-oriented. The things oriented IoT, refers to elements such as the RFID tags with emphasizes on the related procedures that enhance an object's visibility like its tractability and status. The internet-oriented IoT envisages promoting the Internet Protocol and its numerous versions (i.e., the network technology to communicate different smart objects across the universe). The semantic-oriented IoT provides and finds the requisite solutions for modeling, describing, presenting, interconnecting, and processing things intelligently with limited or no human intervention.

Among these three visions of IoT, the SW aims to combine Artificial intelligence and knowledge engineering to represent, share, and integrate objects and related information besides inferring new and up-date knowledge today. It helps to generate machine-interpretable data which can be self-descriptive in the IoT domain. The impact of Semantic IoT (SIoT) on this world is many and ever-growing. A recent study shows an estimated 25 billion equipment and services will be connected to the Internet by 2020 [3]. Such an astounding number of heterogeneous services and devices will require automatic interconnections, interoperability and intelligent communications to facilitate the end-users to track, locate, discover, represent, store, and exchange huge amounts of information. This has led to the design and development of many technologies integrated into IoT such as the SW (i.e., ontologies, semantic, annotation, etc.), Linked Data, and SW services [4–6]. The critical area of concern is to integrate smart home appliances usually stuck between the early adoption phase and the mass-market phase owing to fragmentation [7]. The complexity increases due to the involvement of numerous smart devices, operators, and IoT service providers subject to time-consuming and complex operations, inadequate research and innovations, vendor lock-in, glitches in overall performances, etc. To be efficient, the IoT domain requires interoperability among these factors.

Semantics refers to the mutual understanding of the relevance of the meaning of shared knowledge, information or data. The ever-increasing demand of heterogeneous sensors and smart devices to observe and measure surrounding variables to gather valuable information makes incorporation of semantics in IoT. The obvious effect of heterogeneity is a lack of interoperability among IoT devices and services. This has made the SW system a widely accepted technology for modeling and integrating data collected from many sources. Nevertheless, such systems require a huge amount of computing devices or resources due to the involvement of a large number of sensors to build a smart environment. Thus, developing a cost-effective, semantically annotation IoT system with a limited number of resources and sensors has an area of concern [8]. To tackle the issue, the authors have followed a lightweight semantic annotation approach. The objective is to develop a resource-constrained IoT platform with fewer sensors, lower response time and data size.

The SW provides a future vision or scope to quickly search and interpret data or information by machines to accomplish numerous complex tasks concerned with discovering, retrieving, blending, and executing the available Web information. It depicts a set of data or information so connected that it is possible to process these easily by machine rather than human operators. It can be presumed to be an extension of the state-of-art World Wide Web (WWW) and can represent the data effectively by linking the entire database available globally [9, 10]. As compared to the SW, the SIoT operates in a more dynamic environment wherein there is a frequent change in the meaning of data or annotations over time and space. It allows semantic functionalities such as object recommendations or search and facilitates massive IoT data management. It plays a major role in the model, describe, integrate, interconnect, and process IoT intelligently or to scale-up or down the IoT infrastructure with limited or no human intervention [11]. A semantic-based discovery service *QoDisco* to address the device location, capabilities, context data type, quality, and contextual situations has been proposed by the authors to develop an effective SIoT system. It involves repositories storing resource descriptions based on an ontology-oriented information model with multi-attribute and range querying abilities [12]. The use of different approaches such as publish-subscribe interactions and the parallel interactions with multiple repositories has reduced the cost of semantic search as claimed by the authors.

To summarize the SIoT provides the essential framework for interoperability, consistency, discovery, scalability, composability, and reusability. It improves human–machine interaction, analyses, processes and activities concerning IoT resources, data, services, and automatic operations. It is more complex than SW, demands continuous pre-processing, monitoring, filtering, annotation, aggregation, and integration.

The SIoT aims to simulate the evolution of the existing Web by enabling the interaction of the available unstructured documents or information or data with the desired accessibility. Its objective is to provide a well-defined meaning, adequately allowing the people to work in coordination and co-operation, and enabling computers to be more efficient [4]. The SW is driven and built based on the World Wide Web Consortium (W3C)'s Resource Description Framework (RDF). It is coined with Uniform Resource Identifiers (URIs) syntaxes known as RDF syntaxes to represent data. The inclusion of data to RDF files facilitates discover, search, assess, add or collect, and process the web data by Web spiders or computer programs or web users with less effort. Such an effort assists humans in multi-tasking operations such as information retrieval, online bookings, online accessibility to dictionaries, etc.

The aims to achieve a few major objectives are summarized as

- Adequate data interoperability and communications between the applications and devices without any prior agreement.
- Maintain generic interworking with automated management of resources.
- Provide motivations for semantic discovery/data querying.
- Provision of semantic matching and binding of applications and devices.
- Application of Semantic reasoning to acquire new facts and knowledge from the set of asserted facts.
- Efficient understanding and monitoring of the surrounding environment.
- Facilitating smart and strategic decision making with adaptability to environment changes dynamically.

The evolution of SIoT over the years has been in Fig. 1.

The collection of data and conversion into meaningful information for strategic decision making gives an edge to the user. Collecting data is not sufficient, only

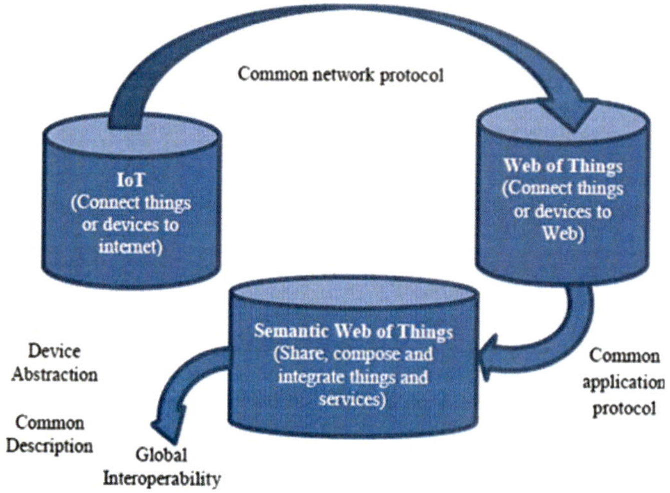

Fig. 1 The Evolution of SIoT

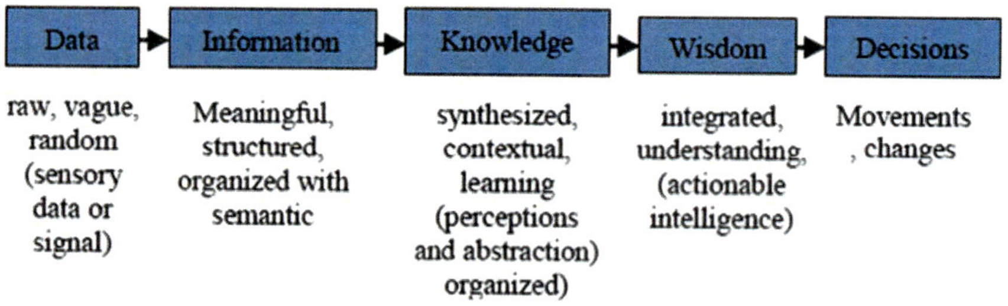

Fig. 2 The knowledge hierarchy in SIoT as a conversion of raw data into decisions

your ability to convert data into decisions that give you the edge. Figure 2 shows the knowledge hierarchy in SIoT that relies on the conversion of raw data into decisions.

2 Semantic Internet of Things (SIoT)

There are three paradigms to realize the IoT domain [2]. These are

- internet-oriented represented by middleware
- things oriented representation by the sensors
- Semantic-oriented described by the knowledge.

These paradigms are interdisciplinary nature, hence they are essential in realizing the IoT concept effectively in a platform where these three domains converge as shown in Fig. 3.

The semantically oriented architectures and execution environments can accommodate future IoT requirements with storing, scalable, and communication facilities for the development of suitable IoT applications such as smart cities or homes, weather prediction, smart health care, inhabitant monitoring, environments' abnormity, etc. utilize the SIoT system [13–18]. Nevertheless, the ubiquity, diversity, heterogeneity, and volatility or sensor data pose many challenges to both machines and men to comprehend and process. A generalized architecture of including semantics is provided in Fig. 4.

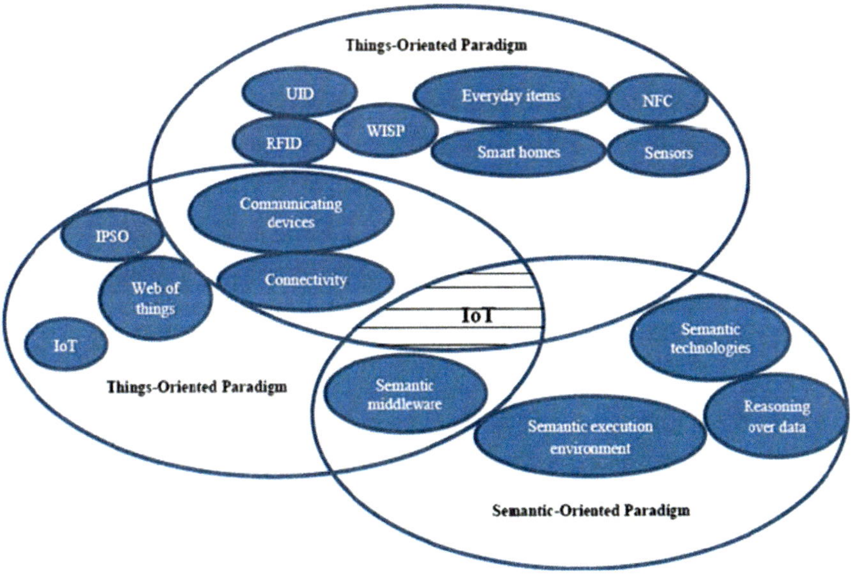

Fig. 3 Intersection of different IoT visions [2]

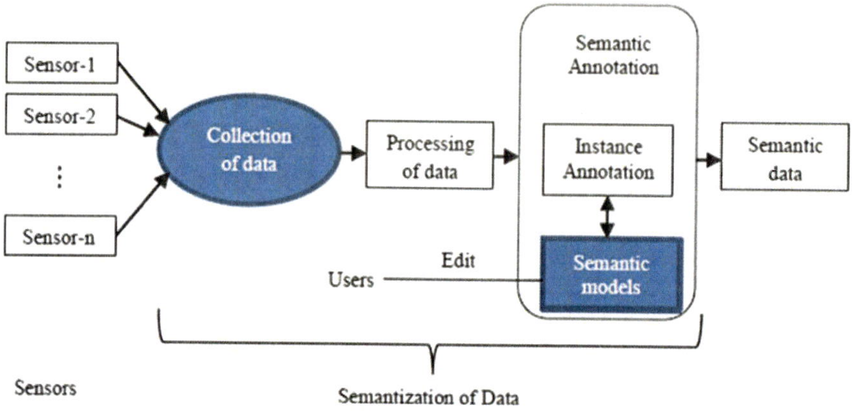

Fig. 4 A generalized architecture of including semantics

Figure 5 provides the various steps and processes involved to deal with heterogeneous data sources for effective SIoT based on SEG 3.0 methodology which supports the SI from the data sources to the end-user applications.

Fig. 5 The steps and processes of SIoT (SEG 3.0 methodology) [17, 18]

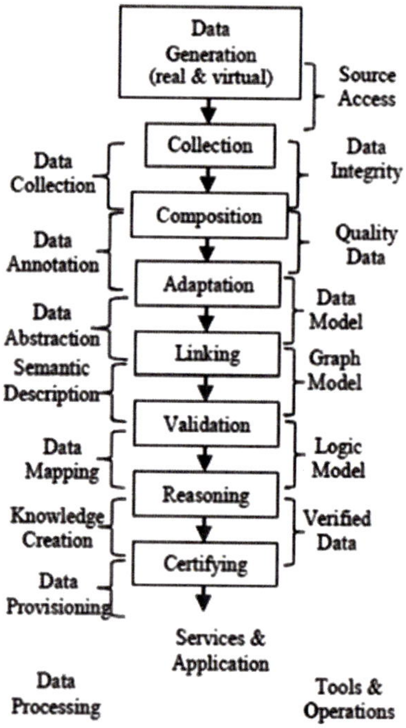

3 Semantic Interoperability (SI)

Due to an exponential increase in heterogeneous sensors, actuators, devices, and demand for better and smarter IoT platforms or services, it becomes essential to distribute, support, monitor, control, coordinate, and communicate with these things that require interoperability. The interoperability can be segmented into (a) technical Interoperability (b) Syntactic Interoperability (3) Semantic Interoperability as shown in Fig. 6 based on their dependability. Each layer has the prerequisite corresponding to its upper layer.

The technical interoperability can be defined as the ability of an object or thing to communicate with other things using software and hardware. On establishing the desired communication, the syntactic interoperability handles the data models, data encoding, data formats, communication protocols, and serialization techniques based on certain agreed standards. Finally, the SI provides a meaning to the content and facilitates to comprehend of the unambiguous and shared meaning of data. To understand the significance of interoperability the taxonomy has to be studied for five major perspectives as shown in Table 1

The SI can be viewed as the ability of various services, agents, devices and applications to communicate and exchange shared data information, and knowledge [22]. The mechanism is essential to facilitate flexible interoperability in today's IoT environments surrounded by distributed, dynamic, and heterogeneous data for the systems to collect and understand valuable information. For such purposes, the SI utilizes the metadata and ontologies to facilitate the business, industries, city planners, consumers, etc. to share meaningful information for understanding and reuse. Thus, multiple standalone systems, devices or applications used for different tasks need not include redundant or similar data. For instance, the temperature in Celsius recorded by sensors for weather warnings may not be valuable when the consumer

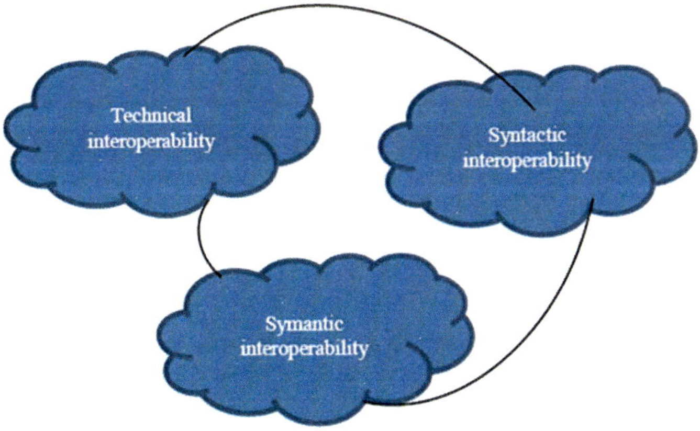

Fig. 6 Interoperability layers

Table 1 Interoperability taxonomy for five major perspective

Interoperability taxonomy	Attributes
Device interoperability [19]	• Consists of high-end and low-end devices • High-end devices (smartphones, Raspberry Pi, etc.) have enough computational capabilities and resources • Low-end devices (low-cost actuators, sensors, RFID tags, OpenMote, Arduino, etc.) are resource-constrained, low energy, communication, and processing capabilities • The objective is to communicate among the heterogeneous devices and integrate these with new IoT domains
Network interoperability [20]	• The network is multi-vendor, multi-service, heterogeneous, and largely distributed • Allows efficient exchange of message among different smart systems using different networks for end-to-end communication • Able to tackle issues like resource optimization, addressing, routing, QoS, security, mobility support, etc
Syntactical interoperability [21]	• Deals with interoperation of the structure and format of the data structure during the exchange of service and information among heterogeneous IoT domains or entities or systems • Incorporates a set of syntactic rules in the same or some different grammar • It is required when there is a mismatch in the encoding and decoding rules of the sender and receiver respectively
Semantic interoperability [22]	• To enable various services, agents, and applications for a meaningful exchange of data, information, and knowledge on and off the web • The need arises when the information and data models of any IoT systems cannot inter-operate automatically and dynamically due to different understandings and descriptions of operational procedures and resources
Platform interoperability [21]	• It is essential due to the existence of widely diverse operating systems (OSs), data structures, programming languages, access mechanisms, and architectures in IoT systems • Designers and developers must find mechanisms to access data from different IoT platforms or integrate the data in a cross-platform structure efficiently • Similarly, cross-domain within the heterogeneous domain of the different IoT platform needs to be addressed

cannot interpret these values or figures. The use of meta-tagged data and its sharing with its other apps in SI can alleviate this issue.

Interoperability has been a major burden to the developers of SIoT Systems due to many heterogeneous domains involving data formats, communication protocols, and technologies. With the limited availability of worldwide acceptable interoperability standards, tools are limited. An SI Model among heterogeneous IoT systems in healthcare has been proposed to monitor and communicate the current health status of patients with the physicians [23]. The health-related information between the patient and the medicos is semantically annotated and communicated in an effective way using the RDF. The SI is helping to unleash a new age of digitalization across industries. From manufacturing to the home, the potential to increase connectivity and productivity across a range of sectors is massive. It sets to bring about new ways in which businesses operate and engage with their products and customers. All this is driving the fourth industrial revolution, Industry 4.0. The word "smart," from smart cities with smart factories, derives from this very revelation. These smart environments encompass a variety of devices and technologies, and we're already seeing them today. Even in the manufacturing and industrial sectors, where automation and computerized control systems have been commonplace for many years, digitalization is driving a massive level of change. The challenge of integrating these legacies, mostly proprietary, systems that were not designed to communicate across product lines and functional areas, means the journey toward Industry 4.0 should not be underestimated. While digitalization is already well advanced in industrial environments, it is constrained by a lack of standards. To make the most of this digital revolution, it is desired to harness the full power of Industry 4.0. It envisages capitalizing on potential savings that automation will bring, companies need to federate IoT platforms and operating systems across multiple production lines and subsystems.

3.1 Issues of Semantic Interoperability (SI)

The followings are a few issues that need to be considered to make the SI effective.

- The words need to be used to represent a chosen set of concepts.
- A common vocabulary for IoT to tackle the semantic gaps between machines.
- The requirement to attach a multiple-level of semantics to raw data or a piece of information.
- The possibility of interpreting different levels of meaningfulness to represent a device or data.
- A strategic decision needs to include ontologies that facilitate the highest level of semantic clarity and transparency, but are inexpensive both in money and time. A trade-off has to be made between the cost and the amount of data sources while including the ontology with SI for big and open environments.
- The addition of value on the available data or its representatives is apps or purpose dependent. It poses difficulties for its adaptability by small IoT markets. In such a case, city planners can usually access only restricted data without SI.
- The initial costs incurred to achieve the apps with speed may increase exponentially with the number of devices, apps, and their integrations. It will restrict the use of available information for multiple purposes.
- With the increase in the number of IoT services, equipment, and the desire of cities to become smarter, providing the SI service at an affordable price and flexibility remains a challenge.

- Although SI adopts shared vocabularies and ontologies for a better understanding of meanings, it still requires maintaining and updating the intrinsic information from various data sources. It needs to preserve the existing domain knowledge and supporting data maintenance.
- For better maintenance and utilization of the shared ontologies and vocabularies, it is essential to develop a periodic and centralized updating mechanism.
- The software facilities and infrastructure relying on the shared ontologies have still issues related to scalability, an open and well-known issue of SW.

A potential solution to aforesaid issues can be using new ontology models or restructuring the existing ontologies aligning or integrating these semantically with various vocabularies using equivalences among their properties and classes [22]. However, a unified semantic model for annotation of IoT data, intelligent integration of distributed data sources, and the proper management and utilization of suitable sensors, still poses a challenge to this domain. The fusion and composition of different data sources, searching and discovery of actuators and data sources, fitting to an application based on their capabilities, reasoning, and analysis on semantic resources via visualization tools or reasoners needs further consideration [24, 25]. To realize the full potential of IoT, "things" need to talk to and understand each other. An even greater obstacle to attaining this outcome is the different data models employed within the countless different proprietary implementations.

4 Semantic IoT Versus Machine Learning

There have been many machine learning (ML) algorithms found useful in pattern recognition, speech, emotion identification, etc. [26–28]. The integration of SIoT with Machine Learning (ML) algorithms has been explored by many researchers in the field of pervasive computing, ubiquitous computing, wireless sensor networks, ambient intelligence, human activity recognition, etc. The state-of-art Artificial Neural Network (ANN) with a backpropagation algorithm has been explored to detect human movements such as sitting, walking, running, and for smart home activities [29, 30]. To process the sensor data in IoT, the trends of Convolution NN and Deep NN have been utilized effectively [31]. Similarly, various pattern recognition tools such as the Bayesian networks, Naive Base Classifiers, Support Vector Machines, K-Nearest Neighbor, Hidden Markov Model have been intelligently utilized for home automation, context-aware search systems, navigation systems in IoT based applications [32, 33].

Integrating SIoT and machine learning undoubtedly will increase the revenue in any business or industry. It is essential to choose certain words to describe a certain set of concepts suitably. The choice of a common vocabulary for SIoT to bridge the semantic gap existing between machines need to be defined. However, existing organizations use the concept superficially by not transforming it into a true profit-generating engine in reality. For example, a fast-food industry wants to introduce AI

(Artificial Intelligence) enabled diet charts to recommend or suggest add-on items based on the restaurant traffic, current selection, or environmental conditions like time or weather of a day. Although it is a lucrative upsell opportunity, it is not followed seriously as a critical-mission system. The success of the industry requires proper integration of SIoT, machine learning, and AI as the top revenue-generating engines. It must consider modern analytic models to be embedded with a customer lifecycle completely. For example, displaying a client's name and his or her choice or preferences on the menu chart enhances his or her satisfaction level and makes the customer feel special. It further provides both the customer and the organization many inputs such as his interests, behaviors, or future intention in real-life environments. Similarly, the SIoT powered AI chat-bouts assist a client to tackle simple issues, make him pleased with the user-friendly experience. The information on recommendations or offers based on customer choice will boost revenue and drives his intentions due to personalization. A survey shows 91% of consumers are inclined to buy a product based on offers and recommendations from companies which, remember, recognize, and value his or her association with a product [34]. Similarly, 83% of customers are willing to share their information to enable a personalized experience that helps SI in IoT. Thus, it is possible for a business to forecast a customer's next move, by consistently feeding and upgrading SI information collected from the different IoT framework. An intelligent embedded IoT, AI, and SI in business models using customers' profiles and information facilitate the recommendation of the next-suitable-action on every stage of a customer life-cycle. In this regard, the simulating engines or machine learning algorithms can provide the desired feedback for further new developments for better outcomes. Ultimately, it boosts revenue with increased productivity, reduced operational expenses, improved personalization at scale and customer satisfaction. The SI enabled automated IoT embedded with intelligence can allow thousands of such models to function concurrently for the benefit of both the service or product provider and the customer. A few of the state-of-art ML algorithms with their attributes, advantages, and disadvantages have been summarized in Table 2 [35].

5 Semantic Ontology

A few of the popular approaches applied to envisage the SI are the proxy gateway, Unified Data Models, and Frameworks, Ontologies.

Proxy gateway is an intermediate communication that sends a request and delivers the corresponding responses for another node. It communicates with other nodes in the other native data model as well as encoding scheme hence remains transparent. For example, the data model translator is an entity in the proxy gateway which translates between two data models, although for the heterogeneous network, it is not suitable. Due to this disadvantage, the proxy gateways apply to small networks and can facilitate technical interoperability. The unified data models such as the LWM2M objects, IPSO Smart Objects, Cluster Library (Zigbee Alliance), and ETRI

Table 2 Various machine learning algorithms and their attributes

ML algorithms	Attributes	Pros.	Cons.
Decision tree	• Uses a greedy top-down tree approach • The information gain is estimated at each node • The highest score of attributes is chosen	• A comprehensive analysis • Easy to use, simple to understand and interpret • Assign specific values to each problem reduces uncertainty, clears up the ambiguity, A white-box model	• Unstable • Relatively inaccurate • The information gain is biased • Calculations can get very complex, particularly if many values are uncertain
Random forest	• Based on the bagging algorithm • Uses an Ensemble learning technique	• Longer Training Period • Complex	• It can handle large data • Robust and scalable • Solve both classifications as well as regression problems • Works well with both categorical and continuous variables • It can automatically handle missing values • No feature scaling required • Handles non-linear parameters efficiently • Robust to outliers • Less impacted by noise
K-nearest neighbors (KNN)	• Instance-based learning approach • The most frequent class is determined based on K-nearest neighbors of an input feature	• Accurate and insensitivity to outliers • Suitable for both numerical and nominal features	• Noise intolerance • Memory insensitive • Curse of dimensionality • Subject to boundary complexity

(continued)

Table 2 (continued)

ML algorithms	Attributes	Pros.	Cons.
Artificial neural network (ANN)	• Requires fewer parameter • Data-driven and self-adaptive learning • Can perform tasks not possible by linear programming techniques • Requires to be reprogrammed • Parallel structured • Can model real-world complex problems	• No assumption of data distribution is required • Eligible for multivariate non-linear tasks • Universal function approximation • Noise tolerant	• Slower in the presence of large data set • Large processing time • A black-box approach • Does not converge to a stable version like SVM
MLP (multilayer perceptron)	• An NN-based approach • Uses back-propagation algorithm • Consists of more than one hidden layer • Distributed learning • Class separation using hyper-planes	• Simple • The stochastic nature of the learning process reduces the possibility of getting stuck in local minima • Easily takes advantage of redundant data • Easy to implement	• Slower to train • More hidden layers
Probabilistic NN (PNN)	• ANN-based approach • Single parameter variation • Bayes' optimized solution guaranteed	• Faster than MLP and RBFN • Noise robust • Easy to train • Outlier insensitive	• Large storage requirement • Requires a representative training feature set
Radial basis function network (RBFN)	• Class separation using hyper-spheres • Faster than MLP as no back-propagation algorithm is used • Good generalization capability and universal approximation	• Multi-parameter adjustment	• Slower than PNN

(continued)

Table 2 (continued)

ML algorithms	Attributes	Pros.	Cons.
Support vector machine (SVM)	• Linear discriminant learner • Use either RM (Risk Minimization) or ERM (Empirical RM) to enhance accuracy in a sample set • Accuracy can vary with different kernel functions	• Can handle large feature set • Guaranteed convergence with a unique solution	• Space complexity • Not accurate for a very large feature set • Kernel dependent • Unsuitable for variable-length data
Gaussian mixture model (GMM)	• Probabilistic approach • Uses Expectation Maximization to speed up response • A stochastic classifier that uses statistical parameters	• Robust • Computationally efficient • Faster • Suitable for a large feature set • Ease of implementation • Suitable for global features	• Text-dependent • Can't avoid exponential functions • Not suitable for low dimensional feature sets
Hidden Markov model (HMM)	• Text-dependent	• Large HMM structures possible using individual HMMs	• complex • Requires initialization of the suitable model parameter before training • Computationally slow

(continued)

Table 2 (continued)

ML algorithms	Attributes	Pros.	Cons.
Deep neural network (DNN)	• Best-in-class performance in multiple domains • Suitable for data-mining and large feature sets • Reduce the need for feature engineering, selection, and optimization which consume much time • Can adapt to new problems easily	• Maximum utilization of unstructured data • Elimination of the need for feature engineering • Ability to deliver high-quality results • Elimination of unnecessary costs • Elimination of the need for data labeling	• Not suitable for small feature sets • Needs thousands of samples to perform satisfactorily • Computationally expensive to train (requires GPU in most cases due to its requirement of large data handling) • No strong theory or properly defined mathematical model to guide for determination of DNN topology, training method, favor, and hyper-parameters. Learning is spread over the hidden layers and hence is a black-box approach • It is not possible to comprehend what is being learned

Data Model provide semantic and syntactic information. Nevertheless, it cannot find a general framework for any large-scale. Ontology can be viewed as the formal specification of any concept. It provides the working model on the properties, relationships, types, and interactions among different entities in an IoT system. It helps the SIoT by facilitating the context or reference on a wide range of vocabularies. Its major components are (a) class (b) individuals (c) attributes (d) relations [24] and may not be limited to physical objects only. A few popular ontologies are SSN (Semantic Sensor Network) Ontology, IoT-O Ontology, Thing Description, etc. Semantic interoperability allows the exchange of unambiguous and machine-understandable data. The use of semantic level facilitates intelligent data federation, data inference, and true machine intelligence. However, it is difficult to provide the desired semantic understanding without modeling them uniformly with ontologies. Ontologies utilize the vocabularies as well as the taxonomies or the terminologies to represent the properties, concepts, and interrelations among data meaning. As ontologies are machine-understandable these are crucial for the inter-machine automation to provide a high value to the IoT domain.

Ontology can capture reusable information in a domain and is a specification of a conceptualization. Web Ontology Language (OWL) is an ontology language widely applied in the field of Semantic IoT. With the growing demand for efficient SIoT, much large-scale ontology is developed in real-time to occupy and integrate information and knowledge. It is essential to focus on the complexity of these ontologies for clear understanding, reusability, and integration. Based on the software metric concept, a set of ontologies at the class-level and ontology-level have been investigated to study the design complexities [10]. The ontology metrics are evaluated against Weyuker's criteria with empirical analysis of the characteristics and applicability. Thus, SW ontologies can facilitate the integration and management of knowledge for autonomous processing of the data by software professionals. There have been many ontology languages such as OWL, RDF, RDF Schema, etc. that provide the much-needed vocabularies to represent the essential domain knowledge, data aggregation and integration from different sources. These languages can only provide the conceptual data rules and models, but not the specific serialization format. Some other specific languages such as Turtle, N3 (Notation3), JSON-LD (Javascript Object Notation for Linked Data), extensible Markup Language (XML), N-Triples can describe semantic data [36]. The SIoT has a standard query language known as SPARQL.

The ontologies in SIoT provide a common language to describe objects or things and their interrelationships. In this regard, the SSN Ontology designed by the W3C SSN Incubator Group helps to represent the properties of sensors for desired observations and can be extended to generate new features in SIoT. The NCI Thesaurus Ontology is an OWL ontology that applies SW to develop mutually agreeable and consistent vocabularies to contain and integrate domain knowledge from many sources. It has nearly 60,000+ named classes, an approximately similar number of anonymous classes, and 100,000+ properties (connections) among these classes. This way, the ontology can represent information on approximately 8000 therapies and 10,000 cancers. Similarly, using the Linked Data project, it is possible to generate and integrate several RDF datasets, such as US census data, DBLP4, DBPedia3, FOAF5, etc. Two major constraints in the design and development of semantic ontologies are the quantity and the complexity that directly affects the maintainability, reusability, understanding, integration, and applicability [37]. Ontology complexity, in general, is the difficulty in designing, developing, modifying, and reusing the ontology which can be measured using some software metric standard such as Weyuker's criteria [38]. There are other ontologies measuring matrices namely the number of the leaf class, the number of root classes, and the average depth of inheritance tree used to measure the ontology cohesiveness [39], entropy-based metric to measure ontology's structural complexity described as a UML diagram [40], etc. These metrics focus on either one or two factors of ontology such as the structural complexity or empirical validations. In this regard, a meta-ontology O2 matrix characterizing the semiotic objects to identify three ontology measuring parameters such as structural, functional and usability-profiling has been proposed [41]. Similarly, more than two measuring metrics have been proposed by other researchers as well with some of them have a limited utility [42, 43]. Nevertheless, these ontology metrics need to follow a few guidelines such as name anonymous classes, name anonymous individuals, materialize the subsumption hierarchy and unify names, etc. They need to propagate instances to the deepest possible class or property within the hierarchy and normalize properties while creating the metrics. These analyses show the absence of systematic methods and measuring parameters hence need to be addressed by future researchers.

6 Network Tools for Efficient SIoT

There are several semantic modeling tools have been emerged to facilitate the SIoT domain. The 'Hyper Thing' is considered to be a semantic web URI validator to distinguish between the URI identities a RealWorld Object and a web document resource. It checks whether the method of publishing of URIs follows 303 URI or the W3C hash practices. It also attempts to verify the redirection between the Document URIs and the Real-World Object URIs to get rid of the data publisher mistakenly redirect. The 'NeON' is a network tool that provides the methodology for Ontology while the 'OWL' tool allows project for semantic validation with ontologies written in OWL/XML, RDF/XML, OWL Functional Syntax, OBO Syntax, Manchester OWL Syntax, and KRSS Syntax. On the other hand, the 'OQuaRE' happens to be a square-based approach to evaluate an ontology quality based on software quality evaluation and software quality requirement specifications. The 'OntoClean' tool in a problem domain defines the meta-properties for the construction of ontology language descriptions. Other similar tools find useful in the field of SIoT are OnToology, Oops Ontocheck, OntoAPI, OntoMetric, Prefix, etc. Table 3 provides describes these tools in brief.

Table 3 A brief description of the ontology tools

Web page	Tools	Descriptions
https://www.hyperthing.org	HyperThing	Distinguish between URI identities in a real-world object and a web document resource
https://neon-toolkit.org/wiki/Main Page	NeON	The methodology for ontology
https://miuras.inf.um.es/oquarewiki/index.php5/MainPage	OWL validator	Allows project for semantic validation with ontologies written in OWL/XML, RDF/XML, OWL functional syntax, OBO syntax, manchester OWL syntax, and KRSS syntax
https://miuras.inf.um.es/oquarewiki/index.php5/MainPage	OQuaRE	Evaluates an ontology quality based on software quality evaluation and software quality requirement specifications
https://www.ontoclean.org/	OntoClean	Defines the metaproperties for the construction of ontology language descriptions
https://ontoology.linkeddata.es	OnToology	It automates part of the collaborative ontology design process. It Surveys the repository with the OWL file to create, documents for validation
https://oops.linkeddata.es	Oops	Detects the most common ontology development pitfalls when (1) the relationship with a range or domain resides in the intersection different classes (2) Absence of any naming convention in an identifier of the ontology elements (3) the presence of a cycle between the hierarchy of two classes in ontology
https://www2.imbi.uni-freiburg.de/ontology/OntoCheck/	Ontocheck	It verifies the metadata completeness and the naming conventions in an ontology
https://sourceforge.net/projects/drontoapi/	OntoAPI	It is a consumer API to evaluate the ontology web service provider response
https://oa.upm.es/6467/	OntoMetric	A method to select an appropriate ontology
https://prefix.cc/	Prefix	It helps to simplify the development of an RDF using the URI prefixes

(continued)

Table 3 (continued)

Web page	Tools	Descriptions
https://linkeddata.uriburner.com:8000/vapour	Vapour	It uses a scripting approach to denote the linked data validator for debugging. It tests the content negotiation on a vocabulary
https://vocab.cc	Vocab	It is an open-source project which helps the RDF designers to find linked data vocabularies
https://www.w3.org/RDF/Validator/	The W3C RDF validator	It is an online service to check and visualize a customer's RDF document

7 Security Policy in SIoT

SIoT can be assumed to be a black hole and is a nebulous term. Hence, the security policy must be comprehensive and has expansive visibility. Further, there are many decades-old technologies included in the IoT world that need to segment or managed effectively in addition to the identifiable emerging devices. For example, one of the oldest iterations of IoT is the involvement of Multi-function printers with scanning and copying facilities. However, often these are neglected from observed security strategies since any employee can easily use them from their workplace or route it anywhere in an organization. The security proliferation of a smart building can put an entire city, its infrastructure, including the traffic and power grid systems comprising millions of IoT endpoints under threat. However, as all these systems are connected to the cloud, an organization can observe the usage patterns for efficient environments. Efficient integration of IoT enabled smart devices such as cameras, wired or unwired devices, biometric access keypad, face or gesture recognition models with interoperability will make way to the right people at the right place. For example, in a smart city, it remains a challenge to secure the traffic systems, parking meters, other wired or wireless sensors from proliferating as the city is covered with a mesh of networked IoT endpoints. As visibility is an obvious requirement for robust security and policy management, interoperability or every conceivable device, cameras, a meter is an essential requirement of IoT.

8 Conclusions

The real barrier of SIoT is poor or improper semantic modeling or descriptors of the data. The data has a certain meaning, name, nomenclature or value, but the way it can be repetitive or normalized or easily understandable to everybody is of great concern. It is often based on analogies by the person involved. The proliferation of the types and the number of connected devices, the fragmentation of the IoT market, the web-based threat, security vulnerabilities, etc. pose a challenge for the effective implementation of SIoT. To protect the systems for remote attacks, it is advisable to use secret keys to provide hardware protection. It also requires development time, extensive security expertise, and costs to configure and the desired provisioning for each device. With industries developing millions of connected devices each year, the scalability of the device architecture is another major issue in this field. The producers can provide and configure high-volume orders as ease. Thus, the low or mid-sized deployments have been mostly occurring in small companies with low performing options. The SIoT adoption and innovation have the issues concerned with the data interoperability. The cross-domain interoperability is likely to remain the core element for a promising future generation IoT platform that is going to be coined as the Internet of Everything. The absence of adequate interoperability leading to the integration of heterogeneous IoT devices into platforms results in the technological and economical drawbacks. Further, the problems associated with the design of applications of exploiting data collected from multiple IoT domains, hindrances of IoT enabled technology explorations at a large-scale, increase in the cost of implementation and provisioning of services and reluctance in adopting advanced IoT technologies hamper the overall user satisfaction.

References

1. Van Kranenburg, R., Bassi, A.: IoT challenges. Commun. Mob. Comput. **1**(9), 1–5 (2012)
2. Atzori, L., Iera, A., Morabito, G.: The internet of things: a survey. Comput. Net. **54**(15), 2787–2805 (2010)
3. Evans, D.: The internet of things: how the next evolution of the internet is changing everything. CISCO White Pap. **1**, 1–11 (2011)
4. Berners-Lee, T., Hendler, J., Lassila, O.: The semantic web. SciAm **284**(5), 28–37 (2001)
5. Berners-Lee, T.: Linked data. Int. J. Semant. Web. Inf. Syst. **4**(2) (2006)
6. McIlraith, A., Son, T.C., Zeng, H.: Semantic web services. IEEE Intell. Syst. **16**, 46–53 (2001)
7. Greenough, J.: The US smart home market has been struggling-here's how and why the market will take off. Business insider. Available online https://www.businessinsider.com/the-us-smart-home-marketreport-adoption-forecasts-top-products-and-the-cost-and-fragmentation-problems-that-could-hindergrowth-2015-9 (2016)
8. Al-Osta, M., Ahmed, B., Abdelouahed, G.: A lightweight semantic web-based approach for data annotation on IoT gateways. Proc. Comp. Sci. **113**, 186–193 (2017)
9. Zhang, H., Li, Y.F., Tan, H.B.K.: Measuring design complexity of semantic web ontologies. J. Syst. Softw. **83**(5), 803–814 (2010)
10. Amarilli, F., Amigoni, F., Fugini, M.G., Zarri, G.P.: A semantic-rich approach to IoT using the generalized world entities paradigm. In: Managing the Web of Things. Morgan Kaufmann, pp. 105–147 (2017)
11. Gomes, P., et al.: A semantic-based discovery service for the internet of things. J. Internet Serv. Appl. **10**(10), 2–14 (2019)
12. INFSO D.4 Networked Enterprise and RFID INFSO G.2 Micro and Nanosystems. In Co-operation with the Working Group RFID of the ETP EPOSS, Internet of Things in 2020, Roadmap for the Future, Version 1.1, 27 May 2008
13. Toma, I., Simperl, E., Hench, G.: A joint roadmap for semantic technologies and the internet of things. In: Proceedings of the Third STI Road Mapping Workshop, Crete, Greece (2009)

14. Vyas, D.A., Bhatt, D., Jha, D.: IoT: trends, challenges and future scope. Int. J. Comput. Commun. **7**(1), 186–197 (2015)
15. Shi, F., Li, Q., Zhu, T., Ning, H.: A survey of data somatization in internet of things. Sensors **18**(1), 2–20 (2018)
16. Serrano, M., Barnaghi, P., Carrez, F., Cousin, P., Vermesan, O., Friess, P.: Internet of things IoT semantic interoperability: research challenges, best practices, recommendations and next steps. In: IERC: European Research Cluster on the Internet of Things, Tech. Rep (2015)
17. Gyrard, A., Serrano, M.: Connected smart cities: interoperability with SEG 3.0 for the internet of things. In: Proceedings of 30th IEEE International Conference on Advanced Information Networking and Applications Workshops, pp. 796–802 (2016)
18. Hahm, O., Baccelli, E., Petersen, H., Tsiftes, N.: Operating systems for low-end devices in the internet of things: a survey. IEEE Internet Things J. **3**(5), 720–734 (2016)
19. Bello, O., Zeadally, S., Badra, M.: Network Layer Inter-Operation of Device-to-Device communication technologies in Internet of Things (IoT). Ad Hoc Networks, pp. 1–11 (2016)
20. Noura, M., Atiquzzaman, M., Gaedke, M.: Interoperability in internet of things: taxonomies and open challenges. Mob. Netw. Appl. **24**, 796809 (2019)
21. W3C: Semantic Integration and Interoperability Using RDF and OWL. www.w3.org/2001/sw/BestPractices/OEP/SemInt (2018)
22. Jabbar, S., Ullah, F., Khalid, S., Khan, M., Han, K.: Semantic interoperability in heterogeneous IoT infrastructure for healthcare. Wirel. Commun. Mob. Comput. **9731806**, 1–10 (2017)
23. Serrano, M., Gyrard, A.: A review of tools for IoT semantics and data streaming analytics. Build. Blocks IoT Anal. **6**, 139–163 (2015)
24. Swetina, J., Lu, G., Jacobs, P., Ennesser, F., Song, J.: Toward a standardized common M2M service layer platform: Introduction to oneM2M. IEEE Wirel. Commun. **21**(3), 20–26 (2014)
25. Mohanty, M.N., Palo, H.K.: Segment based emotion recognition using combined reduced features. Int. J. Speech Tech. **22**(4), 865–884 (2019)
26. Palo, H.K., Mohanty, M.N., Chandra, M.: Efficient feature combination techniques for emotional speech classification. Int. J .Speech Tech. **19**(1), 135–150 (2016)
27. Palo, H.K., Sagar, S.: Comparison of neural network models for speech emotion recognition. In: 2nd IEEE International Conference on Data Science and Business Analytics (ICDSBA), pp. 127–131 (2018)
28. Khan, A.M., Lee, Y.K., Lee, S.Y., Kim, T.S.: A triaxial accelerometer-based physical-activity recognition via augmented-signal features and a hierarchical recognizer. IEEE Trans. Inf. Technol. B **14**(5), 1166–1172 (2010)
29. Altun, K., Barshan, B.: Human activity recognition using inertial/magnetic sensor units. In: International Workshop on Human Behavior Understanding. Springer, Berlin, Heidelberg, pp. 38–51 (2010)
30. Lane, N.D., Bhattacharya, S., Georgiev, P., Forlivesi, C., Kawsar, F.: An early resource characterization of deep learning on wearables, smartphones and internet of things devices. In: International Workshop on Internet of Things towards Applications. ACM, pp. 7–12 (2015)
31. Chen, Y., Zhou, J., Guo, M.: A context-aware search system for internet of things based on hierarchical context model. Telecommun. Syst. **62**(1), 77–91 (2016)
32. Bhide, V.H., Wagh, S.: I-learning IoT: an intelligent self learning system for home automation using IoT. Int. Conf. Commun. Sig. Process. 1763–1767 (2015)
33. https://www.accenture.com/_acnmedia/pdf-77/accenture-pulse-survey.pdf (2018)
34. Ruta, M., Scioscia, F., Loseto, G., Pinto, A., Di Sciascio, E.: Machine Learning in the Internet of Things: a Semantic-enhanced Approach. Semantic Web, IOS Press, pp. 1–22 (2018)
35. Sezer, O.B., Dogdu, E., Ozbayoglu, M., Onal, A.: An extended IOT framework with semantics, big data, and analytics. In: IEEE International Conference on Big Data (Big Data), pp. 849–1856 (2016)
36. Koru, A.G., Tian, J.: An empirical comparison and characterization of high defect and high complexity modules. J. Syst. Softw. **67**(3), 153–163 (2003)
37. Weyuker, E.J.: Evaluating software complexity measures. IEEE Trans. Softw. Eng. **14**(9), 1357–1365 (1988)

38. Yao, H., Orme, A.M., Etzkorn, L.: Cohesion metrics for ontology design and application. J. Comput. Sci. **1**(1), 107–113 (2005)
39. Kang, D., Xu, B., Lu, J., Chu, W.C.: A complexity measure for ontology based on UML. In: Proceedings of 10th IEEE Int Workshop on Future Trends of Distributed Computing Systems (FTDCS'04), IEEE CS, Washington, DC, USA, pp. 222–228 (2004)
40. Gangemi, A., Catenacci, C., Ciaramita, M., Lehmann, J.: Modelling ontology evaluation and validation. In: Proceedings of 3rd European Semantic Web Conference (ESWC'06). Budva, Montenegro, pp. 140–154 (2006)
41. Wang, T.D., Parsia, B., Hendler, J.A.: A survey of the web ontology landscape. In: International Semantic Web Conference on Lecture Notes Computer Science, vol. 4273. Springer, pp. 682–694 (2006)
42. Vrandečić, D., Sure, Y.: How to design better ontology metrics. In: ESWC'07: Proceedings of 4th European Conference on the Semantic Web. Springer-Verlag, Innsbruck, Austria, pp. 311–325 (2007)
43. Das, S.K., Palo, H.K.: Internet of Things (IoT) Application in Green Computing: an Overview. Advances in Greener Energy Technologies. Springer, Singapore, pp. 85–102 (2020)

Provenance Data Models and Assertions: A Demonstrative Approach

Rajiv Pandey and Mrinal Pande

Abstract Provenance as perceived is a trail of a piece of a data item that helps in linking derived pieces of web resources or Internet of Everything (IoE) data to its creators. Provenance allows the software agents and developers to assert that devices across the IoE landscape can be made trustworthy if the same has a valid derivation path that is associated with it and is reliable/trustworthy. Provenance is considered as metadata that must be embedded into an OWL ontology, this metadata supports the semantic agents/reasoners to evaluate the data or workflow trail of the item in question. Contemporary researchers have proposed several models of trust that are based on mathematical calculations, however, the implementation of trust on semantically generated and modified documents i.e. ontologies at large is still evolving. This chapter thus aims to discuss, deliberate, and implement trust in an existing ontology using provenance assertions. This implementation of trust is based on the PROV-DM (Data Model) that has been suggested by the World Wide Web consortium. The chapter illustrates the implementation and inferencing of trust embedded in an OWL-based University Ontology. Provenance assertions using various scenarios and their inference have been highlighted to signify the validity and consistency of the ontology an XML serialized dataset.

Keywords Provenance · Scenario-based assertions · Semantic IoT · Interoperability · Ontology · Semantic web

1 Introduction

The Semantic Web since its inception has come a long way. Most of the ontologies and sematic models that have been proposed are limited to vocabulary confirmations. The

R. Pandey (✉)
Amity Institute of Information Technology,
Amity University Uttar Pradesh Lucknow Campus, Lucknow, India
e-mail: rpandey@lko.amity.edu

M. Pande
AIIT, Amity University Lucknow Campus, Lucknow, India
e-mail: mrinalkidak@gmail.com

R. Pandey et al. (eds.), *Semantic IoT: Theory and Applications*, Studies in Computational Intelligence 941, https://doi.org/10.1007/978-3-030-64619-6_5

major aspects that these ontologies or semantically generated datasets of connected devices and agents lack is the semantic interoperability of the provenance annotations and assertions.

Sharing ontologies with inbuilt provenance will to some degree ensure interoperability. Interoperability in semantic ecosystems is a concern and can also be achieved by embedding metadata in an accepted universal format to various agents and machines. It is further needed to implement an interface/agent that can comprehend and respond to semantic interoperable definitions. In case different data models are used, interoperability can be achieved by applying semantic translation. One such proposed architecture to achieve semantic interoperability is IoT PlatformSemanticMediator(IPSM) tool [1].

The next important aspect that an ontology lacks is provenance related metadata. Provenance [2] being a history of a part of data allows for the easy replication of data in an organization. Provenance is data or workflow trail that provides for easy traceability of the lineage of a given data item. Provenance related metadata in an ontology will ensure trustability. Most ontologies that exist do not have the feature of trustability associated with them. Though they allow for effective semantic searching, by specifically establishing a relationship between the subject and the object, yet they do not provide for mechanisms that facilitate the users to ensure or rely on the trustworthiness of data items. For example, an object book can be related to the author using a property *isAuthoredBy*, it may enhance semantic searchability but will not ensure provenance. It is thus required that we implement these using some procedures or data models that provide for effective search procedures across the IoE ecosystems. PROV-DM [3] data model is one such kind of data model that allows the users to effectively search for data present on the web and ensure that the same is trustworthy. This is a standard data model that has been suggested by the World Wide Web Consortium (W3C) and provides for constructs that can be incorporated into the ontology thereby facilitating for effective search results.

The provenance model proposes the ontology including concepts such as an Entity, Activity, and Agent. The description of Entity, Agent, Activity forms the basis for provenance integration and access. Numerous provenance models exist, the common ones are listed in Sect. 4. This chapter, however, describes provenance using the PROV-DM data model of W3C (Fig. 1).

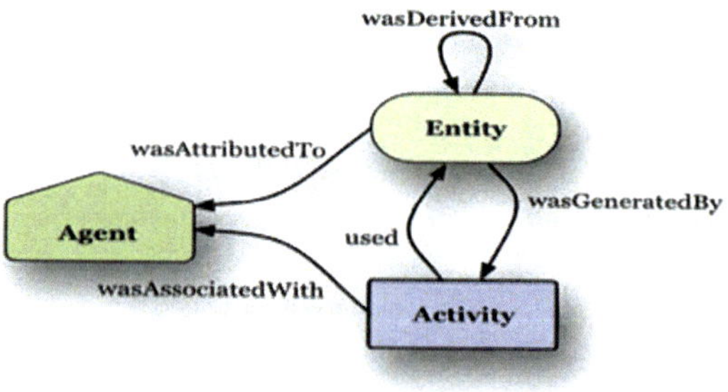

Fig. 1 The relationships between entity, agent and activity [3]

The PROV-DM concepts are defined as:

- Entity—any real, digital thing.
- Activity—procedures by which an Entity is created and comes to existence. It describes a method by which a new Entity is formed along with its attributes.
- Agent—an individual that participates in the creation of an Entity through an Activity.

This chapter aims to embed and highlight the relevance of provenance in an OWL-based University Ontology. It demonstrates the implementation of provenance through scenario assertions using Protégé editor and embedded tools developed by Stanford Center for Biomedical Informatics Research.

After introducing the concept and the need for provenance in the semantic systems we further proceed with the implementation aspects. The chapter explores various aspects relevant to the implementation of provenance, whereby the concept of trust is introduced in Sect. 2, and the types of provenance and common models are deliberated in Sects. 3 and 4 respectively. Section 5 deciphers the provenance layered cake and the subsequent sections demonstrate the incorporation of provenance through various assertions and their implementations.

2 Trust in Semantic Web

The Semantic Web is an interconnected ecosystem of knowledge that exhibits properties like heterogeneity, openness, and ubiquity. Semantic Web enables the user to make semantic based serach unlike the keyword based search.The Semantic Web allows users to contribute to an environment where humans as well as software agents possess the capability to affect a change. In such circumstances, it becomes very critical to evaluate the trustworthiness of the data items that are retrieved using the semantic agents and their ontologies. Trust allows users to leverage and exploit the full potential of the semantic systems [2].

A trust ecosystem can be a structure of metadata on individuals that comprises of class instances having annotations about the trustworthiness of the same. Trust is a crucial factor that needs to be incorporated into ontologies as it would help in creating a semantic network that would help in retrieving trustworthy data.

Provenance and trust are interrelated since assertion for provenance will ensure trust. Trust is a numerical measure that may depend on its context and use. Establishing trust in an object or an Entity involves analyzing its origination trail and its authenticity. Provenance by virtue of its process flow shall ensure trust. It can be thus concluded that provenance is a platform for the trust algorithm [2].

3 Provenance: Types and Models

Provenance understood as a lineage refers to the records describing the Activity, Entity, and Agent that helped or contributed to the creation, updating, usage, and enhancement of data items present on the web or IoE. Trust depends upon provenance to a great extent as the same help in answering the questions like:

- When, by whom and how was a data item created?
- Where and by whom the data item was updated?

Provenance describes those activities that contributed to its creation, and entities involved in the creation [3].

Various organizations are using provenance for ascertaining the quality of information available on the web, however, its integration with ontologies is not done. Provenance as a data trail of when, who, and where is being used by the American Food and Drug Administration to keep the record of drug ingredients. The medical history of patients is also being recorded in hospitals using provenance. The design of aircraft is also undertaken using provenance in various organizations [4, 5].

PROV [3], a specification that relates all the recommendations of provenance and is organized as a layered arrangement, Fig. 2 is a process-flow model. PROV helps to represent and interchange provenance information using widely available formats such as RDF and XML.The base class Activity denotes things that evolve, and act upon with entities. Activities are ordered within a provenance trace. Thus, alignment with OWL is natural [6].

Ongoing researches [8–11] prove that semantic web applications can successfully use provenance in healthcare, banking, finance, and life sciences. Thus, the given scenarios prove that provenance can be used in one way or the other in the semantic systems. Though the mathematical models of provenance are present, yet the domain-specific implementation of provenance needs to be enhanced in scale across domains. This chapter suggests one such domain-specific assertions for provenance on a University Ontology using the PROV-DM data model.

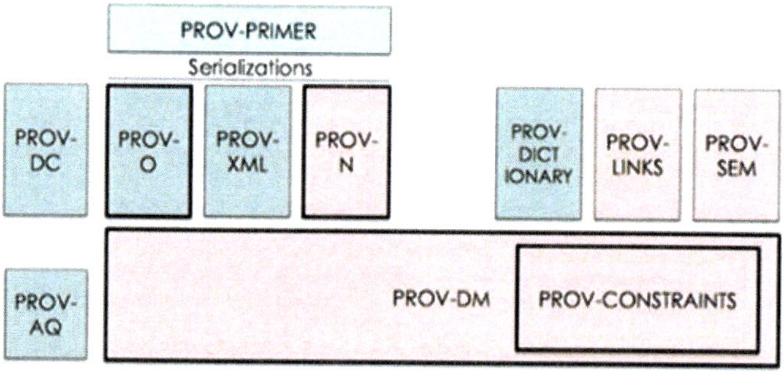

Fig. 2 The provenance architecture stack [7]

3.1 Data and Workflow Provenance

Since provenance signifies the origin or the source of something, a digital object's provenance is perceived as an 'audit trail' that exhibits details of both the data and the process used to derive the object [12].

Provenance provides indications on data quality and its trustworthiness. Thus, provenance is defined as the metadata about an Entity that informs the users about the Activity, Agent, and Entity, involved in its production [7].

Researchers have suggested these two types of provenance:

- Data provenance
- Workflow provenance

Data Provenance

Provenance is used in many fields of applications like e- science, data warehousing, etc. In such cases, provenance helps in developing the functional aspects and also contributes towards its improvement. Provenance is used in these domains to know about the origin of the data items. Data provenance in these cases helps in tracking the source of the data item. Thus, data provenance of the data item consists of the processes and the sources that led to the creation of the data item.

With the velocity at which the data is being used by processes and also consumed by the software agents, provenance becomes necessary as it allows the users to track the history of the data items that are residing in storage spaces and are available on the internet data stores.

Scientists in the field of biology, chemistry, and physics use content that is amalgamated from various other data items. Researchers are thus interested in getting the source of the data item and the processes that are part of the creation of data items. It allows the researchers in determining the processes that led to the development and transformations that were applied to the data item. This information is critical as this helps in determining the quality of data and examining and reconsidering the processes by rerunning the specific processes.

Data warehouses allow the data to be integrated from various resources and representations and helps in analyzing data items. Provenance helps in the analysis of the data items by providing the information about the sources of data items and the changes made subsequently to that data item.

Data provenance has gained wide popularity in workflow management systems. Systems based on Service Oriented Architecture (SOA) are also using data provenance. Data provenance here helps in understanding how a complex computational task was accomplished. It helps in determining the procedures that led to the creation of a result from a computation [13].

Workflow Provenance

Workflow provenance is linked to preserving the track of the various processes that transform a raw data item into a finished product by subjecting it to various activities or processes. A workflow is signified as chaining of various activities together. An activity A2 may be taking input data from an activity A1. Thus in workflow provenance, the emphasis is on activity rather than the data that is being worked upon. Whereas in data provenance the emphasis is on the data. Computationally-driven data-intensive experiments using workflows enable automation, scaling, adaptation, and provenance support [14]. In workflow provenance, the emphasis is on the procedure/workflow rather than the data. Thus, workflow analysis can enable evaluating the degree of provenance.

4 Provenance Models

Models of provenance evolved since there was no formal way to represent data about provenance in the World Wide Web and IoE. Hence considering the above concern, the model of provenance was first proposed in 2006 when a common consensus emerged among the members of the W3C consortium [15]. A common model of provenance was prescribed for representing provenance due to the following reasons.

- need for a common vocabulary of provenance representation,
- uniformity in the agent structure and functionality,
- documentation of process related to provenance,
- how of data derivation, is data item the result of an activity or an agent,
- syntax of data annotation.

Subsequent to the consensus for the uniform model, a challenge was proposed by the W3C which led to the development of OPM or the open provenance model for prescribing provenance [16].

4.1 OPM Model Leading to PROV-DM

The OPM model based on the principle of causal relationships emphasizes, the capturing dependencies evolving out of the processes.

It is further stated that these dependencies can be captured by capturing the processes through which the artifacts were derived using these.

The OPM model studies causal relationships or dependencies on processes by which the artifacts were derived and the Agent that initiated these processes. For example, a book B was produced by the editing process E and by the help of agent A and subsequent modifications/additions to book B. It also allows us to assert the derivation dependency between different artifacts without specifying any processes that were involved in creating that artifact. It is thus inferred that agents can be human, machines, or organizations that might be involved in the creation of an artifact with the help of a process.

After a careful study of OPM, the provenance working group proposed a formal definition of provenance known as the PROV-DM data model which is considered the universal standard provenance model after the OPM, that would specify rules that would help in the creation of provenance [17].

4.2 PROV-DM Data Model

"Data provenance also known as pedigree or lineage is the description of a piece of information from its origins" [18]. It specifies the processes by which the given piece of data in a given database. We consider the example of a database to understand the above scenario and to demonstrate the need for why and where model of data provenance. Suppose a database D = Q(d) is constructed from a query Q over the data d, and the user wishes to evaluate how and by which source and process the given data is derived from the database or which are the tuples that contributed to the formation of the given piece data item. The users of these databases are not able to track the accuracy and the timeliness of data being delivered to them. The data in the field of scientific research is derived from various public databases, however only a handful of these databases provide experimental data, the remaining databases do not provide for good quality data and are in some way or the other the dataset views of the other databases. This is because these remaining databases carry additional values added to them by experts by the means of corrections and annotations. These databases although carry huge sources of information yet the users of these databases are oblivious to how the provenance information can be tracked by them. These problems are addressed in [18, 19] by stating their derivation in a relational database using the tuples that led to the development of the piece of a data item. However, questions like "from where did the piece of data of data item come" and "why is it in the database" remain unanswered. The who and how assertions, with help entity and activity of the provenance in a data model seek to find answers to such questions. Contemporary researches have demonstrated the why of provenance models and been reflected by authors in their articles [19, 20].

This chapter demonstrates a generic implementation of provenance over the PROV-DM data model that allows us to compute, derive, and understand both the why and where of a given piece of data item [21].

5 The Provenance Architecture

The PROV cake is the basic data model proposed to represent entities, human agents, and objects that may be associated with generating data pieces. The provenance stack is layered, and the details of each layer are elaborated in the following sections.

5.1 PROV-DM

Provenance architecture is built over the provenance data model layer (PROV-DM). PROV-DM is generic and allows applications/agents of the Semantic Web to transform provenance instances into the PROV-DM instances. A valid PROV instance corresponds to a consistent history of objects and interactions to which logical reasoning can be safely applied [22]. Semantic Web agents can deploy the provenance for specific purposes.

PROV-DM groups the entire ontology in a group of core and extended structures. Core structures are considered to map domain-specific vocabulary. An Entity a real-world, physical, or digital object that comes to existence by activities or any software or human agents that belongs to the class of a thing. The extended structures consist of the subtypes, plan, bundles, extended relations, revision, collections, etc. Six elements of PROV-DM are as follows [7, 23]:

- Element 1: entities along with their associated activities, displaying the time stamp of its evolution, utilization, and invalidation.
- Element 2: the generation of entities from other entities.
- Element 3: agent (human or software) that is responsible for performing various activities.
- Element 4: bundles, a metadata that displays provenance of provenance
- Element 5: includes attributes that link entities of similar type.
- Element 6: collections that enable users to frame a logical structure of the provenance instances.

These six elements address the needs through various layers on top of the PROV-DM.

Motivating Features to Prefer PROV-DM Over Counterparts

Backward or across serialization of multi-format systems can enable interoperability as is the case of XML to OWL serializations, where a XML syntax can be serialized to an OWL syntax. Most semantic agents deploy provenance models apart from PROV-DM, those applications prefer models such as OPM, Provenir [24], and PML [25] for an interchange of semantic information. These semantic applications need not be tailored to use PROV-DM, but the serialization feature of PROV-DM can be considered for information interchange across such applications.

To ensure the interoperability of system on the level of provenance and its analytics, we need to evaluate that the provenance data interchange across the systems syntactically conforms to PROV-DM syntax. XML serialization of PROV-DM can be used for provenance information interchange across systems and it allows validation of these serializations using XML schema validators [26, 27].

The backward and across serialization is thus the primary reason for PROV-DM to take preference over other models and be proposed by the W3C. The PROV-N, PROV-XML are various serializations of the PROV-DM data model.

5.2 *PROV-N, PROV-O, PROV-XML*

PROV-N stands for the provenance notation. PROV-N is a language that enables ease of learning, discussing, illustrating, and formalizing all issues concerning provenance. Simple and easy to use syntax, interoperability with parsers, ease of understanding, and implementing.

PROV-O and PROV-XML are various serializations of the PROV-DM thus enable easy transfer of provenance information across all syntaxes. PROV-XML has been proposed to support users and applications that deploy XML for data interchange and PROV-O is used to express PROV-DM using the OWL-2 Web Ontology Language [26]. PROV-N, PROV-O, PROV-XML serializations permit sets of classes, attributes, and constraints that support the creation and transfer of information created by various semantic agents across multiple domains.

5.3 *PROV-CONSTRAINTS and PROV-SEM*

A concerning issue with PROV-DM is that it permits users to define different entities that refer to different aspects of the same thing. These entities are termed as alternate entities. PROV-CONSTRAINTS layer concentrates on developing valid PROV instances by incorporating constraints and thus limiting the problem of alternate entities. PROV instances are validated to ensure that they refer to the same objects at a given time period and can be used for logical reasoning and drawing inferences [22].

A document changed multiple times by numerous authors can be represented using different entities for the various versions which are distinguished based on their generation derivation and invalidation events. This creates ambiguity hence the PROV-CONSTRAINTS layer can be used to place constraints on replications/alternate entities. Provenance instances can be validated through the process of normalization.

PROV-SEM the semantic layer assists the PROV-CONSTRAINTS layer. PROV-SEM layer interprets the PROV-DM statements as atomic formulas and maintains that instances can be validated by applying the principles of definition, inferences, and constraints on the atomic formulas specified by the layer itself. PROV-SEM layer ensures that an atomic formula can be further divided into Entity formula, relationship formula, and auxiliary formulas for easily establishing and understanding relationships that exist between entities, agents, and objects [28].

5.4 *PROV-DICTIONARY and PROV-LINKS*

PROV–DICTIONARY layer allows users to consolidate provenance information into a specialized structure that is a dictionary. A dictionary is a specialized form of collection that allows users to give a better structure to a collection and consists of key Entity pairs [29]. Similarly, PROV-DICTIONARY is used to provide a structure to collections on the other hand PROV-LINKS is used to link across bundles.

To evaluate trust in information on the web, we monitor its provenance, and for trusting the provenance, we evaluate the provenance of provenance. PROV-LINKS layer is meant for applications where provenance information is generated and updated by multiple agents over a period. Bundles are a concept suitable for such applications.

A bundle is defined as a named set of provenance descriptions and is a mechanism by which provenance of provenance can be expressed [28]. Bundles provide a feature to annotate provenance descriptions, Bundles are considered as valid entities whose provenance can be represented by PROV-DM. The interconnected systems and devices contain numerous agents and applications that tend to create and modify entities. An entity created by one and modified by the other changes the provenance description. Considering that the provenance descriptions are present in the form of bundles, a need is felt to link these bundles generated by various entities. This relationship is ensured by the PROV-LINKS layer. The layer ensures the fact that the provenance created by the originator is consumed efficiently. Thus, by using the PROV-LINKS layer a consumer has to mention a relation/link to the entity in the bundles to ensure the provenance trail. The mention of an entity in a bundle (containing a description of this entity) is another entity that is a specialization of the former and that presents at least the bundle as a further additional aspect [28].

5.5 *PROV-AQ*

Key consideration while designing the PROV-AQ (Provenance Access and Query layer) is the fact that it is designed to be independent of any serialization format. Since it is aimed at retrieval of provenance records from provenance dataset using the HTTP protocol. Since most semantic parsers/HTTP applications will need to evaluate the provenance record. PROV-AQ allows for additional pingback mechanisms by the consumer of the provenance to the creator of the provenance. Pingback is a mechanism that notifies the web authors about their document being linked to another document. In such a scenario, a web publishing software automatically informs the authors thereby allowing for the possibility of automatically creating links to the original documents [30]. The process of pingback comes into being when the consumer of provenance defines a new resource based on the producer of the resource. Presumably, a piece of news in a newspaper portal changes frequently. In this scenario, if the provenance of a resource is considered, i.e. the web page of the newspaper's website, the web page will have multiple target URI referring to the same web page. Therefore, it is obvious that the same resource can be represented by multiple entities and can thus be referenced by multiple URI.

5.6 ROV-DC

PROV-DC (Dublin Core architecture) enables the conversion of provenance from the DC vocabulary [31] to the PROV-DM vocabulary. This signifies that an application which supports PROV can easily consume provenance already exposed as Dublin Core. DC is used for multiple reasons ranging from a simple description of a resource to combining metadata vocabularies besides providing interoperability for metadata in the Semantic Web as many of its terms are used for expressing provenance.

5.7 ProvONE

Provenance models proposed so far have been the ones that provide for database provenance. These models aim to support the authenticity and attribution of a data item. Thus, such kind of models is suitable for tracking the lineage of material objects like artworks, manuscripts, database flows, etc. In scientific experiments, however, we require shreds of evidence in the support of experimental results. The data in these scientific workflows is obtained by the means of computational results. That is in the form of structured graphs specific to each computational step. The provenance information thus provided is encoded by the means of machine processes and is used for query and analysis. Workflow tracking systems like Taverna [32], Kepler [33], and VisTrails [34] collect this information using the proprietary models. These systems also adopt varied models to specify workflows. This creates a system of heterogeneity which makes it difficult for scientists to analyze and compare provenance information generated from different workflows that are generated or enacted using different systems. This in turn results in missed opportunities for stitching the traces provided by different workflows [35]. ProvONE data model was thus proposed to fulfill the above-mentioned requirements and provide a standard format for modeling provenance with regards to scientific workflow systems. ProvONE is a model that will work on the top of the existing PROV-DM model and provides for procedures that easily amalgamate with the existing PROV-DM data model [36].

6 Tools for Capturing Provenance

6.1 Karma

Karma [37] is a tool that is widely used in cyberinfrastructure for collecting and maintaining provenance records. Karma's modular architecture allows for the multiple instrumentation plug-ins that allow for its use in different architectural settings.

Karma provenance management tool allows users to easily capture provenance information inside scientific workflows.

6.2 *Taverna*

Taverna [32] is another tool that is used for capturing workflow provenance however it cannot be used to capture the provenance of the workflow definition. Taverna considers that scientists manage the evolution of workflow definitions by annotating the same using the attribution attribute.

6.3 *Protégé*

Protégé [33] an open-source editor provides a graphical interface to define ontology. This tool for its integration for visualization, reasoning, and extended support for the development of ontologies using OWL and its serializations, has been the reason for our preference to demonstrate provenance in the subsequent sections.

6.4 *CamFlow*

CamFlow [38] is a tool that is used for capturing data provenance which can be easily integrated into business processes and will generate good provenance records for the same.

6.5 *Kepler*

Kepler is free software that serves multiple roles. Kepler can process scientific workflows, semantic workflows, sharing workflows and maintains provenance descriptions regarding the same.

6.6 *Linux Provenance Modules*

Linux Provenance modules [39] abbreviated as LPM are systems that are used for managing cybercrimes. The tool provides for fraud detection and data protection at an early stage.

6.7 *Open Provenance Model*

OPM or the open provenance model suggested by W3C provides various provenance collection and management tools like ProvStore [30], validators, and translators.

6.8 *ProvStore*

ProvStore is a repository that allows for the storage, browsing, and management of provenance documents using a web interface. It allows us to upload the data related to provenance on the cloud well.

7 Provenance Implementation in University Ontology Using Basic Constructs of PROV-DM

The deliberations that follow shall closely evaluate the implementation of provenance in a University Ontology (the entire ontology and its code is beyond the scope of the chapter and deliberately skipped). Protégé an open-source editor tool along with its Hermit reasoner shall form the basis of all assertions and inferences. The various aspects of Activity, Entity, and Agent are critically evaluated, and the results are supported with the relevant screenshots for the comprehend-ability and readability of the chapter.

An OWL ontology is deemed as a valid dataset, any inconsistencies in the representation of the ontology with respect to defining data and object properties, cardinality and description logic aspects render the ontology as invalid when exposed to a semantic agent or a reasoner.

The chapter provides inferred results to signify that the creation and implementation of provenance keep the validity of the ontology intact.

In the screenshots and codes provided below, we have shown the implementation of the PROV-DM [7, 22, 40, 41] data model in the University Ontology.

Figure 3 shows a hierarchy of classes that are part of the ontology under discussion and implementation. The classes and their instances shall form the bases to implement the Entity, Activity, and Agent relationship in the latter part of this chapter.

The initial implementation is aimed to render the features exhibited in the creation of the Entity, Activity, and Agent, the relevant code snippets in OWL are provided to corroborate the creation.

The subsequent sections after the creation shall demonstrate the reasoning and inference derivation based on the created University Ontology.

Figure 4 signifies that the class Activity, Agent, and Activity belong to a super class Thing. All individuals in the provenance assertions shall belong to either of the three core classes namely Entity, Activity, and Agent.

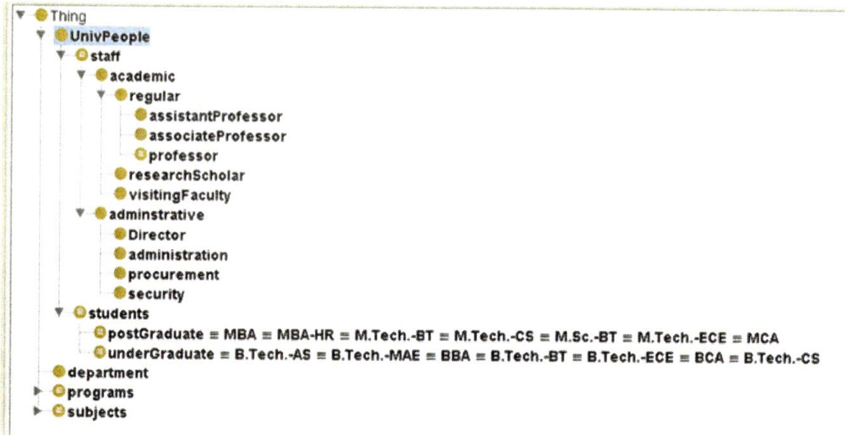

Fig. 3 A class hierarchy of university ontology

Fig. 4 Entity, agent and activity relationship to super class thing

Figure 5 provided below states that a university publishes a hard-bound journal and publishes an e-journal both of which are an Entity. The e-journal is an alteration of the university journal. The e-journal has been created by the Activity- creation of university e-journal by Agent Vijay.

The relationship between Agent and Activity has been mentioned using the *wasAssociatedby* property and the relationship between the university journal and the *e-journal* has been mentioned by the provenance assertion named *alternateOf*. Further, the relationship that specifies that the Activity creation of a university journal is an individual/instance of class Entity is specified by the relationship has individual. The figure provided below shows the asserted and inferred model of the ontology.

The assertion in Fig. 6 states that e-journal is an alternate of university journal. The relationship between the two is implicitly defined by the assertion *alternateOf* and the reasoner explicitly evaluates the validity of the property. Figure 7 demonstrates that a valid identifier in terms of URI is associated with every instance of Entity, Activity, and Agent, which enables it to be referenced on the semantic ecosystems.

Creation of an Entity the university e-journal takes place by an Agent instance Vijay through the Activity creation of the university e-journal therefore we are bound to specify the relationship through the assertion *wasAssociatedWith.*

Figure 8 highlights the aspects of a provenance model.

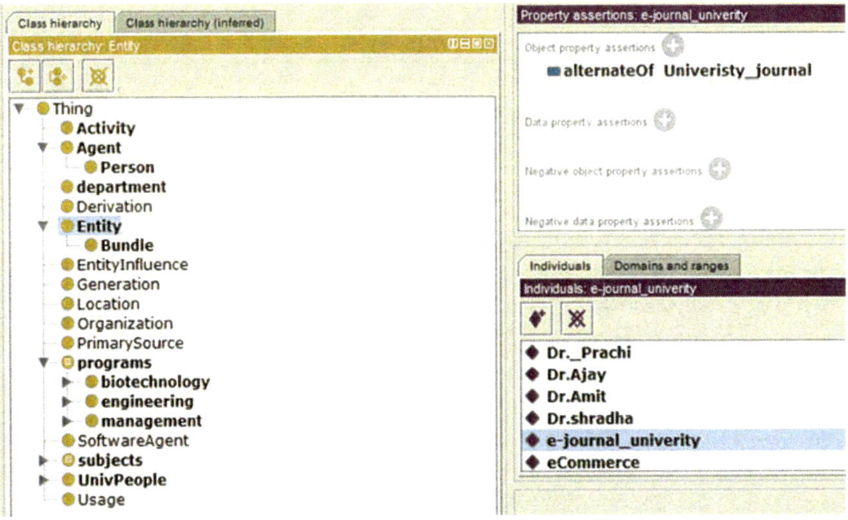

Fig. 5 Asserted model of the ontology for e-journal of university

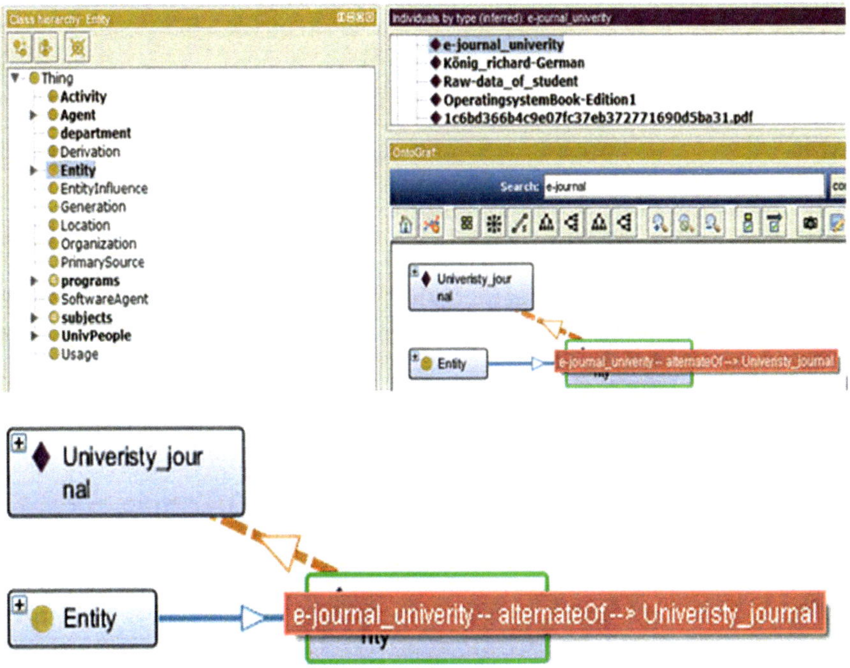

Fig. 6 The inferred model of the ontology highlighting the alternateOf provenance assertion

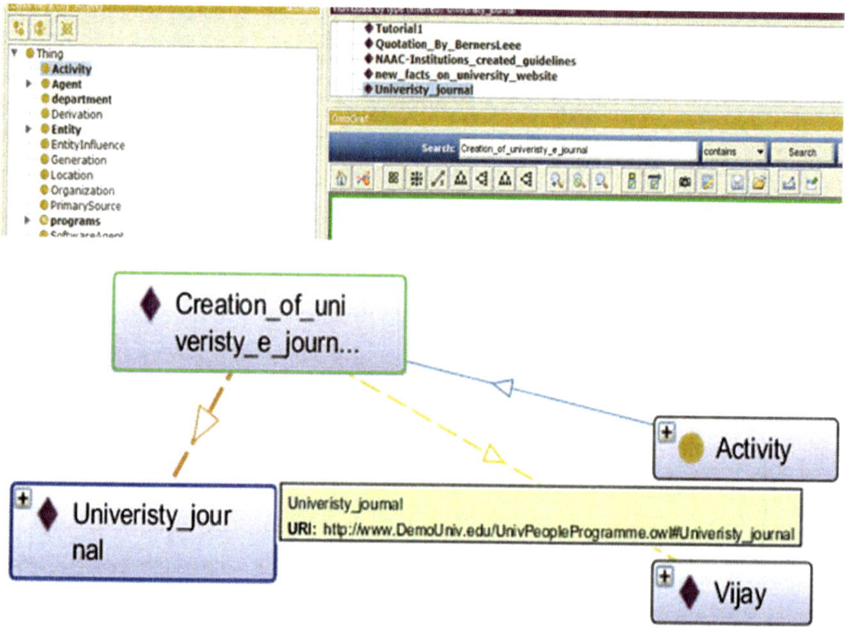

Fig. 7 Every individual-entity, agent, and activity has a valid URI

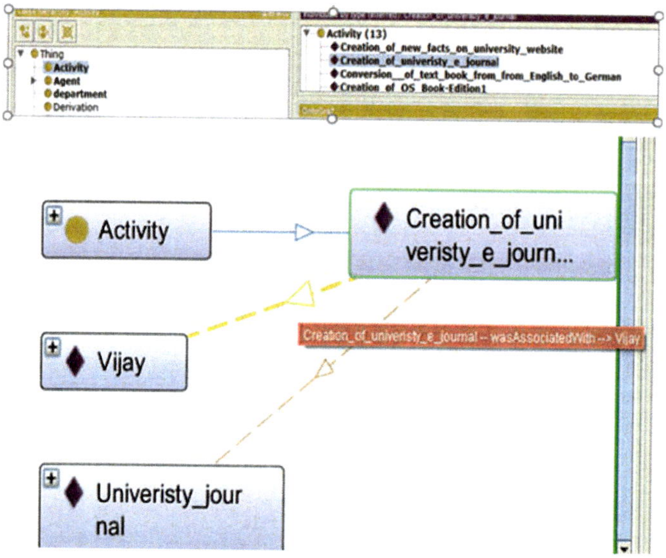

Fig. 8 An inferred model of the ontology demonstrating an activity-creation of university e-journal using the wasAssociatedWith assertion

8 Embedding Assertions: A Scenario-Based Approach

8.1 *Provenance Assertions for Entity*

University journal or IQAC report is the Entity that will undergo changes when Agents and Activities are subjected to them. The whole purpose of provenance is the evaluate the trail/lineage and determine the trust-ability of the resultant document.

Undermentioned is an OWL code of the Entity:

<Declaration>
<NamedIndividual IRI = "#IQAC_report"/>
</Declaration>
<ClassAssertion>
*<Class IRI = "*https://www.w3.org/ns/prov#Entity*"/>*
<NamedIndividual IRI = "#IQAC_report"/>
</ClassAssertion>

8.2 *Provenance Assertions for Agent*

OWL code of the Agent:

<Declaration>
*<Class IRI = "*https://www.w3.org/ns/prov#Agent*"/>*
<NamedIndividual IRI = "#Vijay"/>
</Declaration>

8.3 *Provenance Assertions for Activity*

OWL code of the Activity that creates a Entity:

<Declaration >
<NamedIndividual.
IRI = "# creation of university e-journal"/>
</Declaration>
<ClassAssertion>
*<Class IRI = "*https://www.w3.org/ns/prov#Activity*"/>*
<NamedIndividual IRI = "# creation of university e-journal "/>
</ClassAssertion>
<ObjectPropertyAssertion>
*<ObjectProperty IRI = "*https://www.w3.org/ns/prov#*AssociatedWith"/>*

9 Scenario Assertions for Provenance Representation

This section demonstrates the implementation aspects of provenance with the help of an ontology. The demonstrations are corroborated using a built-in reasoner named Hermit. In order to embedded trust in an ontology, we have discussed above that the entire model is divided into Entity, Agent, and Activity [42, 43].

An Entity being any real or hypothetical object has a URL/URI associated with it.

In the undermentioned demonstration, the *University_Journal* is an Entity on which the provenance trail of Activity in relation to an Agent is explored in the following sections.

The Agent involved in performing some Activity is usually one of three types-personal agent, human agent, and organizational agent.

Activity-Activity is a process through which the agent performs some task.

Figure 9 shows that *Rohit* and *Shivam* are types of human agents.

The screenshots have been created using Protege [43] and have been shown to be consistent using the Hermit reasoner. The screenshots further prove that the PROV-DM data model can be successfully implemented in ontologies and the same can be shown to be valid for various provenance assertions.

Figures 9 and 10 have been reproduced to illustrate the graphical inference and the textual inference respectively, that is rendered through the reasoner for an Agent.

Code for an asserted model of the ontology showing the provenance assertion for Agent:

```
<Declaration>

    <NamedIndividual IRI = “#Rohit”/>

</Declaration>
<ClassAssertion>

    <Class IRI = “https://www.w3.org/ns/prov#Person”/>
    <NamedIndividual IRI = “#Shivam”/>

</ClassAssertion>
<Class IRI = “https://www.w3.org/ns/prov#Person”/>

    <NamedIndividual IRI = “#Rohit”/>

</ClassAssertion>
```

It can further be asserted that the agent *Rohit* teaches Artificial_Intelligence and thus the same can be classified as a human agent. Thus, establishing a provenance assertion between Agent and Entity.

It is further stated that entity *Artificial_Intelligence* an individual of class Entity is related to Agent *Rohit* by the relationship teaches. Figure 12 shows the asserted model of the ontology while Fig. 11 represents the inferred model for the same (Fig. 13).

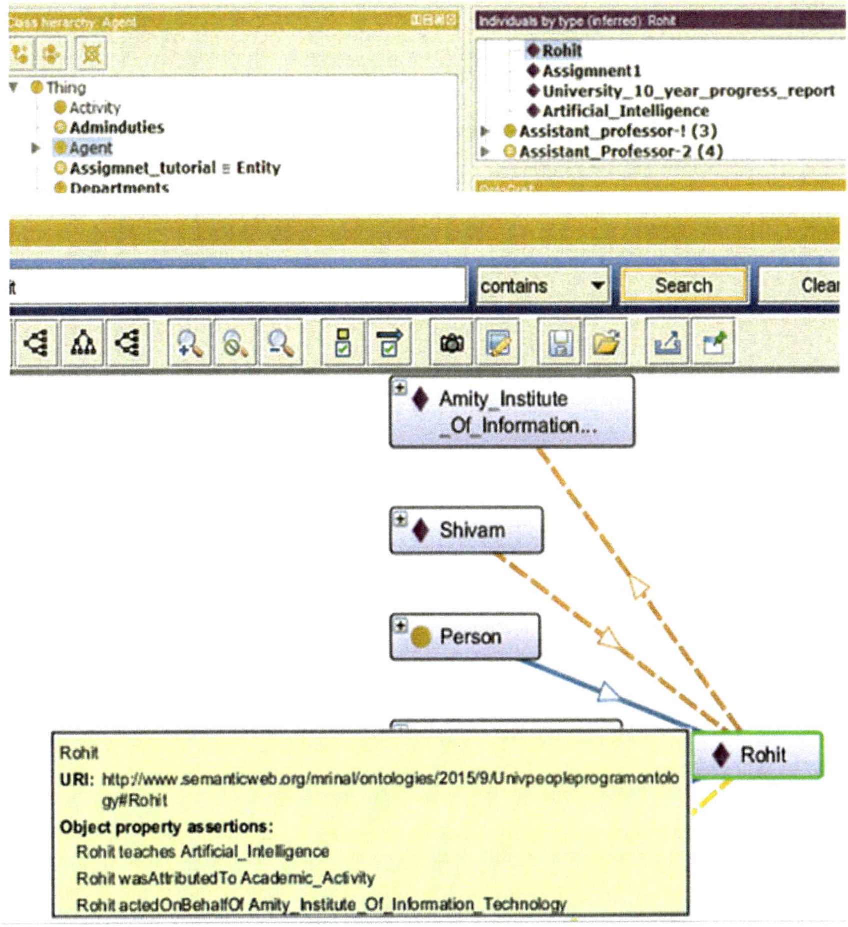

Fig. 9 Inference model of the university ontology using the Hermit reasoner for agent

The above assertion demonstrates that *e-journal_of_university* is an individual of entity *Assigment_Tutorial*, these assertions further justify the validity of the ontology and provenance, else the reasoner would have highlighted that as a data inconsistency.

Figure 14 below, on the other hand, exhibits the instances of the Activity in the left pane. The instance of class Activity, *Academic_Activity* is displayed on the top-right pane with the graphical representation in the bottom right pane.

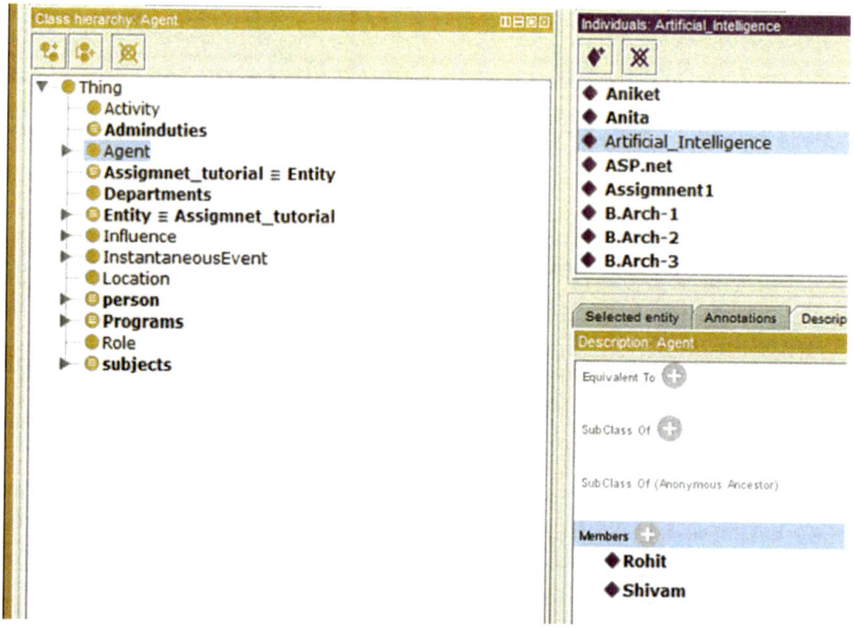

Fig. 10 The asserted model of the ontology using the Hermit reasoner for agent

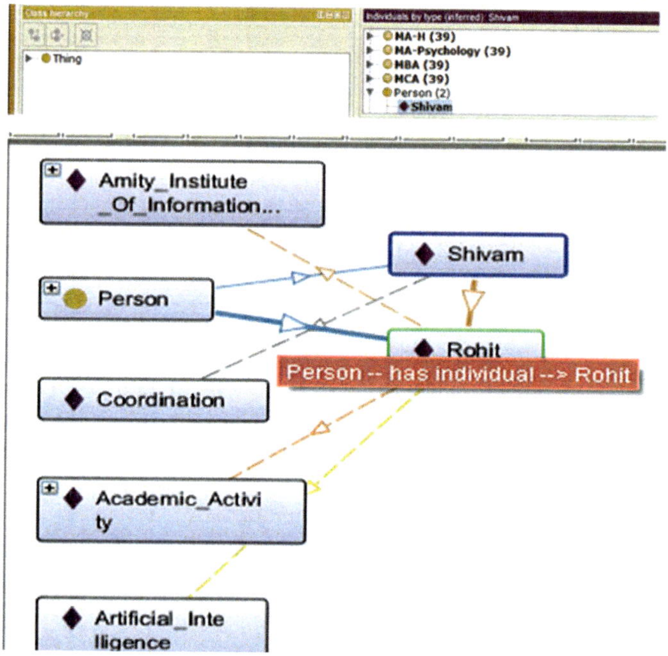

Fig. 11 Individual relationship relating stating that *Rohit* is a type of person agent

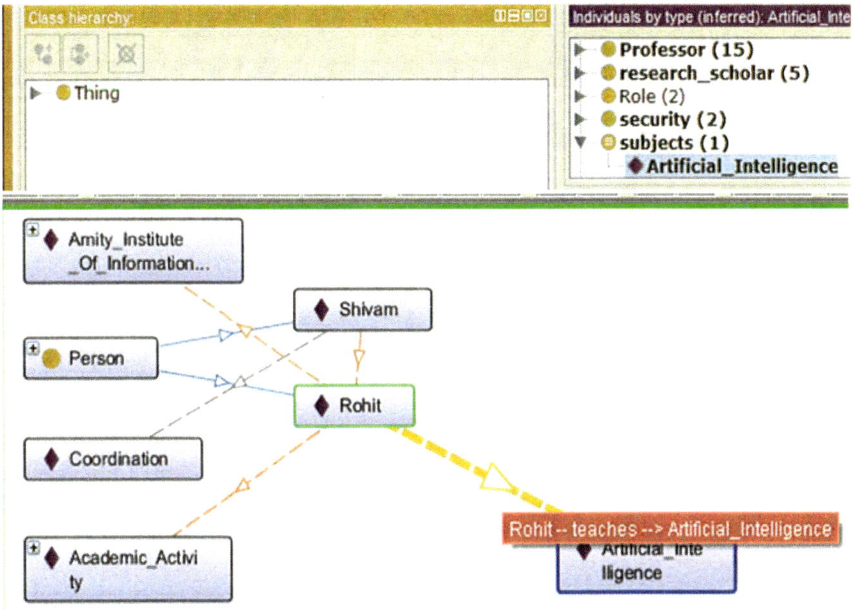

Fig. 12 Relationship between agent rohit and the entity *Artificial_Intelligence*

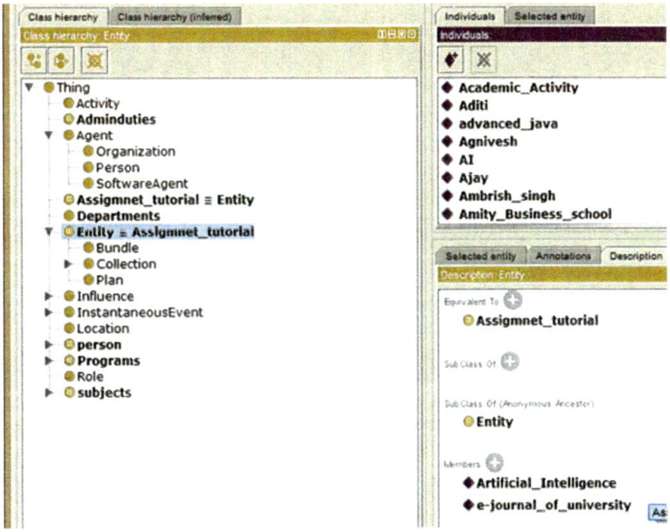

Fig. 13 Asserted model of the ontology showing the individuals that belong to class entity

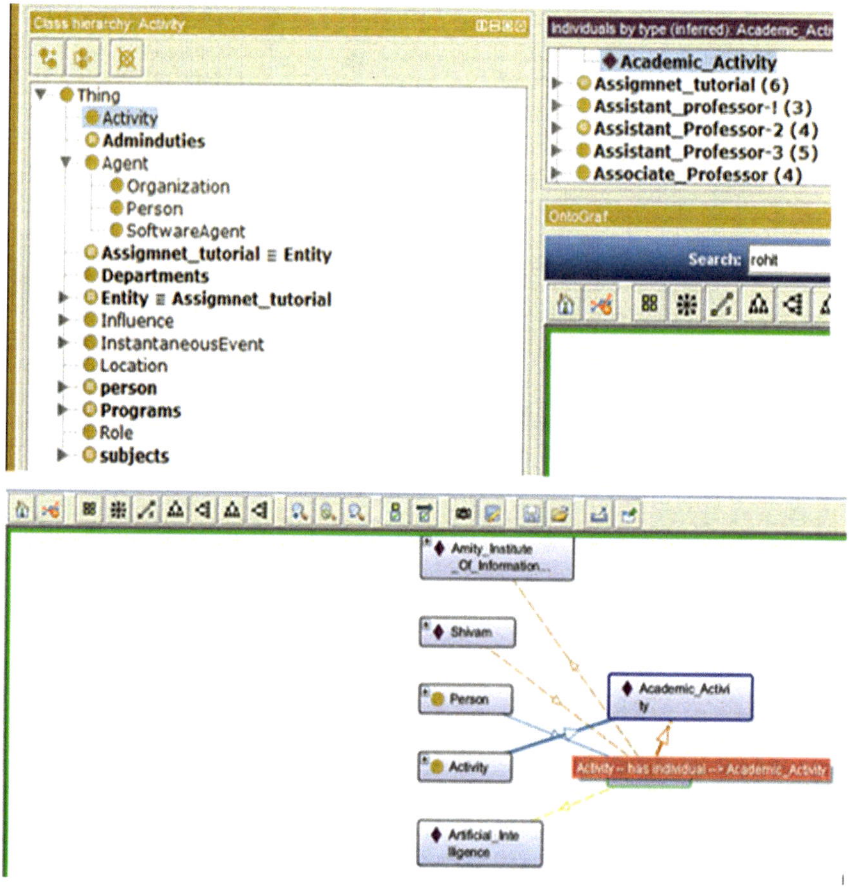

Fig. 14 An inferred model of the ontology showing the members of class activity and its individuals

9.1 Scenario: ActedOnBehalfOf *Provenance Assertion of an Agent*

The *actedOnBehalfOf* provenance assertion states that a given agent has performed an activity by acting as a substitute for another agent. In the given scenario the agent *Rohit* acts on behalf of another agent *Amity_Institute_of_Information_Technology*. Figure 15 and the code provided below show the same.

```
<Declaration>
    <NamedIndividual IRI = "#Rohit"/>
</Declaration>
<NamedIndividual IRI = #Amity_Institute_Of_Information_Technology"/>
</ClassAssertion>
```

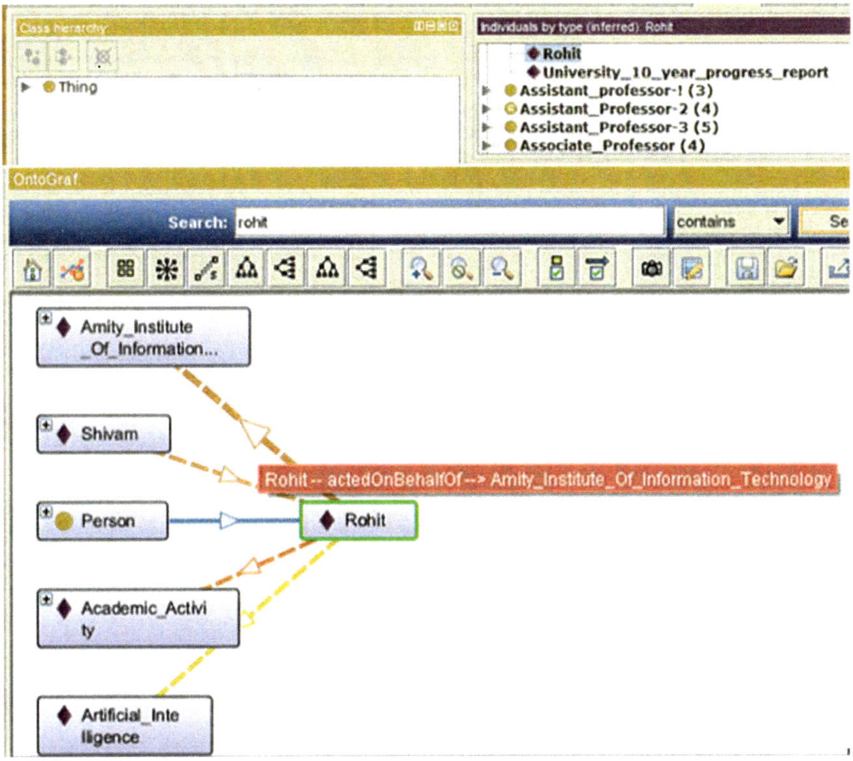

Fig. 15 The Inferred model of the ontology using the actedOnBehalfOf of clause

```
<ObjectPropertyAssertion>
    <ObjectProperty IRI = "#teaches"/>
    <NamedIndividual IRI = "#Rohit"/>
    <NamedIndividual IRI = "#Artificial_Intelligence"/>
</ObjectPropertyAssertion>
<ObjectPropertyAssertion>
    <ObjectProperty IRI = "https://www.w3.org/ns/prov#actedOnBehalfOf"/>
    <NamedIndividual IRI = "#Rohit"/>
    <NamedIndividual
    IRI = "#Amity_Institute_Of_Information_Technology"/>
</ObjectPropertyAssertion>
    <ObjectProperty IRI = "https://www.w3.org/ns/prov#actedOnBehalfOf"/ >
    <NamedIndividual IRI = "#Shivam"/>
    <NamedIndividual IRI = "#Rohit"/>
</ObjectPropertyAssertion>
```

9.2 *Scenario:* WasAttributedTo, *Provenance Assertion of an Activity*

The *wasAttributedTo* provenance assertion states that an activity was performed by an agent through a process. In our ontology, the *wasAttributedTo* provenance has been used to associate the agent *Rohit* to the activity named as an *Academic_Activity*.

The code and Fig. 16 reflect the same:

<ObjectPropertyAssertion>

ObjectProperty IRI = "http://www.w3.org/ns/prov#wasAttributedTo"/
<NamedIndividual IRI = "*#Rohit*"/>
<NamedIndividual IRI = "*#Academic_Activityc*>

</ObjectPropertyAssertion>
<ObjectPropertyAssertion>

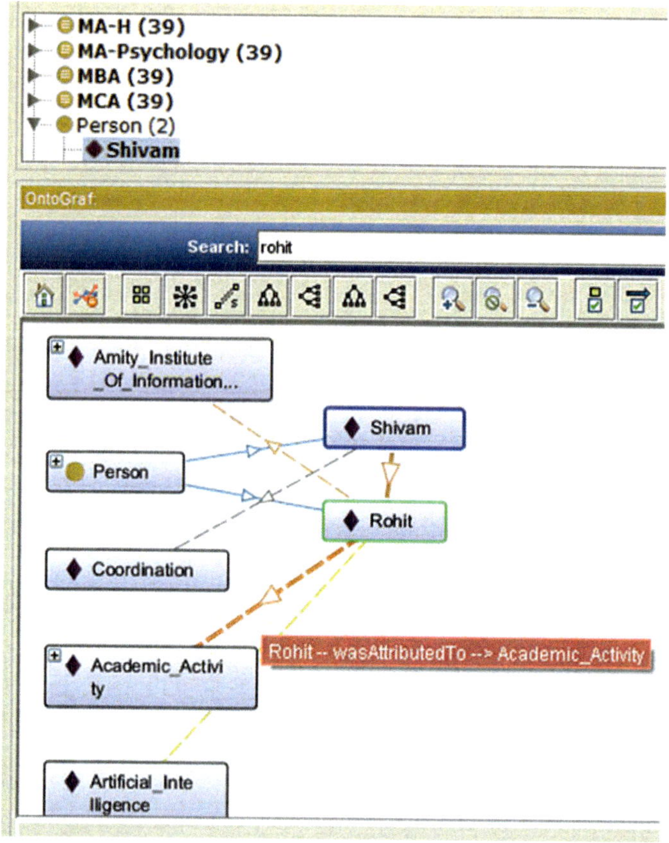

Fig. 16 An inferred model of the ontology using the *wasAttributedTo* clause

<ObjectPropertyAssertion>
<Declaration>

<NamedIndividual IRI = "*#Academic_Activity*"/>

</Declaration>
<ClassAssertion>

<Class IRI = "https://www.w3.org/ns/prov#Activity"/>
<NamedIndividual IRI = "*#Academic_Activity*"/>

</ClassAssertion>

10 Conclusions

The Semantic Web has come miles from its inception. It has envisaged technology and languages to describe data transfer through XML and RDF. It further evolved to embrace Description Logic and cardinality; however, the real transformation was visible when the researchers became able to develop Semantic Web documents in the form of ontologies through the various flavors of OWL recommended by W3C. The main concern with the semantic representation and information interchange in IoE ecosystems is the trust and reliability of the contents and origination of the said ontology/data item, i.e. can the user trust it, who developed it, who or which agent modified the contents further, are still the questions to be answered. This led to the incorporation of provenance through which any ontology can be embedded with aspects like the version of the entity, who was the agent that manipulated the said ontology, and through which activity of the agent was the ontology manipulated. The latest recommendation by W3C in Time Ontology [6] has also not exploited provenance to its full and is a work in progress. This chapter has explored the need for provenance, the primary components of provenance, its types, and other detailed implementations. The authors have presented the code snippets and implementational screenshots for ease of understandability. The deliberations on the detailed development of the ontology, discussing its properties, and demonstrating all aspects of implementation are beyond the scope of the chapter and thus have been intentionally avoided.

References

1. Ganzha, M., Paprzycki, M., Pawłowski, W., Szmeja, P., Wasielewska K.: "Towards Semantic Interoperability Between Internet of Things Platforms" Integration, Interconnection, and Interoperability of IoT Systems, Internet of Things (2018). https://doi.org/10.1007/978-3-319-61300-0_6
2. Effective Design of Trust Ontologies for Improvement in the Structure of Socio-Semantic Trust Networks

3. Pandey, M., Pandey, R.: Provenance constraints and attributes definition in OWL ontology to support machine learning. In: 2015 International Conference on Computational Intelligence and Communication Networks (CICN), pp. 1408–1414 (2015). Extracted from https://www.w3.org/TR/prov-primer/
4. Foster, I., Kesselman, C., Nick, J.M., Tuecke, S.: Grid services for distributed system integration. Computer **35**, 37–46 (2002)
5. Chen, L., Yang, X., Tao, F.: A semantic web service-based approach for augmented provenance. In: 2006 IEEE/WIC/ACM International Conference on Web Intelligence (WI 2006 Main Conference Proceedings) (WI'06), pp. 594–600 (2006)
6. w3.org/TR/2020/CR-owl-time-20200326/, Time Ontology in OWL W3C Candidate Recommendation 26 Mar 2020
7. PROV-N: The Provenance Notation [Online]. Available https://www.w3.org/TR/prov-n/ (2013)
8. Tan, W.C.: Provenance in databases: past, current, and future. IEEE Data Eng. Bull. **30**, 3–12 (2007)
9. Bose, R., Frew, J.: Lineage retrieval for scientific data processing: a survey. ACM Comput. Surveys (CSUR) **37**, 1–28 (2005)
10. Blount, M., Davis, J., Ebling, M., Kim, J.H., Kim, K.H., Lee, K., Misra, A., Park, S., Sow, D., Tak, Y.J.: Century: automated aspects of patient care. In: 13th IEEE International Conference on Embedded and Real-Time Computing Systems and Applications (RTCSA 2007), pp. 504–509 (2007)
11. Misra, A., Blount, M., Kementsietsidis, A., Sow, D., Wang, M.: Advances and challenges for scalable provenance in stream processing systems. In: International Provenance and Annotation Workshop, pp. 253–265 (2008)
12. Lakshmanan, G.T., Curbera, F.: Provenance in web applications. IEEE Internet Comput. **15**(1), 17–21 (2011)
13. Glavic, B., Dittrich, K.R., Kemper, A., Schöning, H., Rose, T., Jarke, M., Seidl, T., Quix, C., Brochhaus, C.: Data provenance: a cctegorization of existing approaches. In: BTW'07: Datenbanksysteme in Buisness, Technologie und Web, pp. 227–241 (2007)
14. Khan, F.Z., et al.: Sharing interoperable workflow provenance: a review of best practices and their practical application in CWLProv. https://doi.org/10.1093/gigascience/giz095
15. Ceolin, D., Groth, P.T., Van Hage, W.R., Nottamkandath, A., Fokkink, W.: Trust evaluation through user reputation and provenance analysis. URSW **900**, 15–26 (2012)
16. Missier, P., Belhajjame, K., Cheney, J.: The W3C PROV family of specifications for modelling provenance metadata. In: EDBT'13 (2013)
17. Groth, P., Moreau, L.: PROV-Overview. An Overview of the PROV Family of Documents (2013)
18. Simmhan, Y.L., Plale, B., Gannon, D.: A survey of data provenance in e-science. ACM SIGMOD Record **34**, 31–36 (2005)
19. Cui, Y., Widom, J.: Practical lineage tracing in data warehouses. In: Proceedings of 16th International Conference on Data Engineering (Cat. No. 00CB37073), pp. 367–378 (2000)
20. Woodruff, A., Stonebraker, M.: Supporting fine-grained data lineage in a database visualization environment. In: Proceedings 13th International Conference on Data Engineering, pp. 91–102 (1997)
21. Buneman, P., Khanna, S., Wang-Chiew, T.: Why and where: a characterization of data provenance. In: International Conference on Database Theory, pp. 316–330 (2001)
22. Constraints of the PROV Data Model [Online]. Available https://www.w3.org/TR/2013/REC-prov-constraints-20130430/ (2013)
23. PROV-DM: The PROV Data Model [Online]. Available https://www.w3.org/TR/prov-dm/ (2013)
24. https://www.provenir.com/decision-engine-software/decisioning-platform/
25. https://dvcs.w3.org/hg/prov/raw-file/default/ontology/working-dir/pml-3/Overview.html
26. The W3C Working Charter [Online]. Available https://www.w3.org/2011/prov/wiki/Interoperability
27. OWL Web Ontology Language XML Presentation Syntax (2003)

28. Linking Across Provenance Bundles [Online]. Available https://www.w3.org/TR/2013/NOTE-prov-links-20130430/ (2013)
29. PROV-Dictionary: Modeling Provenance for Dictionary Data Structures [Online]. https://www.w3.org/TR/2013/NOTE-prov-dictionary-20130430/ (2013)
30. https://openprovenance.org/store/
31. PROV-AQ: Provenance Access and Query [Online]. Available https://www.w3.org/TR/2012/WD-prov-aq-20120619/ (2013)
32. https://taverna.incubator.apache.org/introduction/
33. https://kepler-project.org/
34. https://www.vistrails.org/index.php/Main_Page
35. W3C. ProvONE: A PROV Extension Data Model for Scientific Workflow Provenance. Available Projects/job/ProvONE-Documentation-trunk/ws/provenance/ProvONE/v1/provon
36. Víctor Cuevas-Vicenttín, B.L., Ludäscher, B., Missier, P., Khalid Belhajjame, P., Fernando Chirigati, Y.W., Saumen Dey, U.D., Parisa Kianmajd, U.D., David Koop, S.B., Altintas, I., San Diego, U.C., Christopher Jones, M.B.J., Walker, L., Peter Slaughter, B.L., Yang Cao, U.: ProvONE: A PROV Extension Data Model for Scientific Workflow Provenance.
37. https://pti.iu.edu/impact/open-source/karma.html
38. https://camflow.org/
39. https://linuxprovenance.org/
40. https://www.dublincore.org/specifications/dublin-core/dcmi-type-vocabulary/
41. Kravari, K., Bassiliades, N.: ORDAIN: An Ontology for Trust Management in Theinternet of Things. https://www.researchgate.net/publication/320520680_ORDAIN_An_Ontology_for_Trust_Management_in_the_Internet_of_Things
42. Semantics of the PROV Data Model [Online]. Available https://www.w3.org/TR/prov-sem (2013)
43. https://protege.stanford.edu/

IoT Data and Interoperability

Need and Relevance of Common Vocabularies and Ontologies in IoT Domain

Arunima Sharma and Ramesh Babu Battula

Abstract Data plays a foremost part in Future Communication. Internet of Things encompasses various sources of data of different type, by different name. The complexity and variance in data increase the criticalness of the model. Same data has different meaning in different applications. These variances make it difficult to apply any query on data. To resolve the issues in naming of data values and to create consistency Vocabulary and Ontologies are must. Data gathered on a big platform e.g. for IoT via cloud need to manage data in proper manner, which is spread all over the internet on different servers. The features and related research is explained in this chapter, having a brief introduction with application areas.

Keywords Semantic web · Ontology · Vocabulary · Internet of thing · IoT · Smart city

1 Introduction

Internet of Things (IoT) brings a heterogeneus of technologies and devices together to simplify the basic life of human beings. Smart Home, Traffic, Healthcare etc. are some of the major areas using different IoT components. On basis of applications we can consider IoT as a complex group of several devices, sensors and people who interacts with each other and controlled by a centre control station. This interaction or communication generates a large amount of data gathered from several source in different formats. This data has various formats of data related to each other. Due to complex relationship and variation this data is considered as heterogeneous data. Heterogeneous data is complex with perspective of management, access and collection. On web, data is linked together on basis of some relationship and similarity. The extension of World Wide Web which deal with related web pages, and extensions of similar data is known as Semantic Web.

A. Sharma (✉) · R. B. Battula
Department of Computer Science and Engineering, Malviya National Institute of Technology, Jaipur, India
e-mail: aru.92@rediffmail.com

R. Pandey et al. (eds.), *Semantic IoT: Theory and Applications*, Studies in Computational Intelligence 941, https://doi.org/10.1007/978-3-030-64619-6_6

Complex linking in words on internet in interlinked with each other and creates a web of connections between words related to each other. Vocabulary and Ontology are two main characteristics of words which play most important role. The IoT assumes an ever-expanding job in empowering brilliant city applications. A philosophy based semantic methodology can help improve interoperability between an assortment of IoT-created just as reciprocal information expected to drive these applications. While various metaphysics indexes exist, utilizing them for IoT and brilliant city applications require critical measure of work. Semantic advancements propose an appropriate methodology for interoperability by sharing normal vocabularies, and furthermore empowering interoperable portrayal of construed information. IoT testbed suppliers have as of late began to add semantics to their structures permitting the creation of the semantic Sensor Web, which is an expansion of the current Web in which data is given all around characterized which means, empowering machine-to-machine interchanges and cooperations between items, gadgets and individuals.

Semantics generally model the space ideas in incredible detail. Despite the fact that they can be applied for questioning nearly anything about articles, these unpredictable models are regularly hard to execute and utilize, particularly by non-specialists. They request extensive handling assets and along these lines they are viewed as unacceptable for compelled situations. Rather, IoT models ought to think about the limitations and dynamicity of the IoT situations, particularly with the new pattern towards incorporating semantics in obliged gadgets for example, M2M doors or cell phones. Simultaneously they have to show the connections and ideas that speak to and permit interoperability between IoT elements.

In this manner, expressiveness versus intricacy is a test. It is essential to take note of that semantic models are most certainly not final results. They are ordinarily just piece of an answer also, ought to be straightforward to the end client. The semantic comment models ought to be offered with compelling strategies, Programming interface's and instruments to process the semantics so as to extricate significant data from crude information. Question techniques, AI, thinking and information investigation methods ought to have the option to successfully utilize these semantics. Semantic demonstrating is just the underlying piece of the entire plan, and it needs to consider how the models will be utilized; how the explained information will be recorded and questioned with continuous information; and how to make the distribution reasonable for obliged conditions and enormous scope organizations when applications frequently require low inactivity and handling time. To permit a typical jargon to interoperate between various frameworks a scientific classification is required to portray the estimations of the gadgets as far as the amount sorts and units, for example, temperature and degrees Celsius.

2 Vocabulary

Vocabulary is the knowledge regarding word. It is the basic information required about a word.

Controlled vocabularies provide a way to arrange data for successive information retrieval. Indexing, Heading, Thesauri, and other are collection of controlled data. This data is well defined selected for a category to belong on basis of knowledge, natural language and some restrictions. It is well formed set of information so that any query on data will give corrects outcome. It uses tagging system for words to get connected and retrieval becomes easy.

2.1 Vocabulary Tools

1. Subject Heading

Subject heading helps to create the catalogue of data in categories and sub-categories easily. It considers the whole document in pre-coordinated order.

2. Thesauri

It is collection of words well organized in a sequence. It covers a specific portion of document in direct order.

3 Ontology

Ontology is derived from Greek words means being logical discourse, in general like a dictionary. Ontology consist all the relations, rules, similarity and dissimilarity. It is the formal representation, categorization, and domain of data. All the ontologies are used to resolve the complex relationships. These ontologies are limited in number till there is no new requirement raised for a different sub category.

Ontologies, for example, OWL-S and low-level determinations, for example, the TD or the oneM2M Base Ontology can be utilized together to depict IoT/WoT frameworks, cultivating interoperability. OWL-S helps programming operators to find the web administration that will satisfy a particular need. When found, OWL-S gives the vital language builds to depict how to conjure the administration. It permits portraying inputs what's more, yields. Because of the elevated level portrayal of the administration, it is conceivable to form numerous administrations to perform increasingly complex errands. In OWL-S, administration depiction is composed into four regions: the procedure model, the profile, the establishing and the administration. In particular, the procedure model depicts how a help plays out its undertakings. It incorporates data about sources of info, yields, preconditions and results.

Additionally, the oneM2M Global Initiative characterizes a norm for machine-to-machine correspondence interoperability at the semantic level, the one M2M Base Ontology, which is a significant level metaphysics intended to encourage interoperability among different ontologies utilizing equivalences and arrangements. The TD is a focal structure obstruct in the W3C Web of Things (WoT) and can be considered as the passage purpose of a Thing. The TD comprises of semantic metadata for the Thing itself, an association model dependent on WoT's Properties, Actions, and Events worldview, a semantic pattern to make information models machine-justifiable, and highlights for Web Linking to communicate relations among Things".

Both oneM2M Base Ontology and the metaphysics characterized in TD take a stab at interoperability among different IoT applications and stages, every one covering a huge arrangement of utilization cases, so there is additionally a work in progress to adjust the oneM2M philosophy to the TD cosmology. Symmetrical to these ontologies, the SOSA/SSN philosophy is a cosmology for depicting sensors and their perceptions. Among other ideas, it characterizes the class SOSA: Procedure, which "is a reusable work process, convention, plan, calculation, or then again computational strategy that can be utilized, among others, to indicate how a perception movement has been made, can be either a record of how the activation has been performed or then again a portrayal of how to associate with an actuator (i.e., the formula for performing incitations).

Things utilizing ontologies in a way that permits us to legitimately convey the conduct usage. Work planned for displaying the conduct of Things utilizing FSMs and Web Ontology Language (OWL) exists in the writing. Unified Modeling Language (UML) FSMs utilizing OWL, playing out a very nearly balanced interpretation between UML ideas and OWL classes. In spite of the fact that their planning from UML to OWL takes into account a more machine-lucid data structure, its intricacy makes it unconventional to use. UML is utilized to indicate stage free route guides in web applications. They use OWL to portray a model for FSMs which fills in as a meta-model for semantic web portrayals of the route guides on the Semantic Web. There likewise exists some logical writing gave not exclusively to make a model to communicate the conduct of a help yet in addition to decipher furthermore, execute the conduct. The point is to build up a FSM that a unique server can peruse and mean executable elements. These executable substances are executed later by robots. They effectively fulfill their goal, their OWL FSM is space explicit and it incorporates properties that illuminate the multifaceted nature of their utilization case yet make the FSM excessively complex.

3.1 Components of Ontology

1. Individuals

Initial object considered in starting phase of classification are known as individuals. It is the basic components of ontology which comprise real objects such

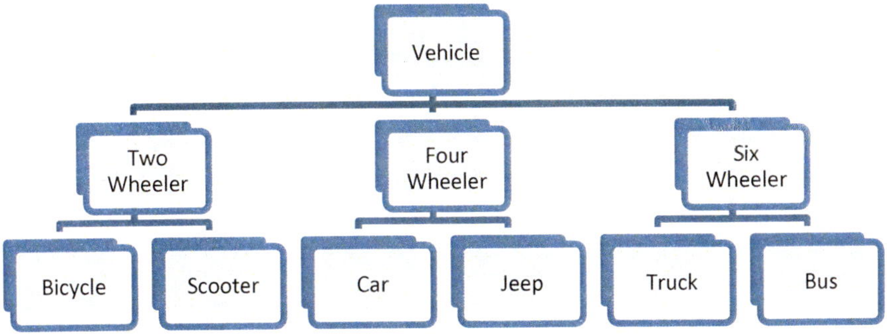

Fig. 1 Example of ontology

as individuals, amphibians, books, plants, along with nonconcrete objects such as numbers and words. The overall purpose of ontology is to offer a way to categorizing entities, in different classes grouped together or correlated with each other. In Fig. 1 car, bicycle, truck are individuals.

2. Classes

Categories in which Individuals are classified are known as classes. Classes are considered as type, category, sort, kind and extension. Classes are considered as abstract objects that are defined by aspect values or constraints to be member of a class. In Fig. 1 Vehicle is a class having two categories based on number of wheels. Classes have to fulfil criteria to belong to a specific group. Class can be a collection of other classes.

Ontologies differ on the basis of conditions related to some conditions like classes can encompass supplementary classes, or can be applicable to itself, and if there is a universal class, etc. Occasionally some restrictions are also applied on classes to avoid inconsistencies. A class can be extensional if it is categorized only by its association or class is identical else it is an intentional class. Extensional classes have a systematic format and try to avoid any inconsistency in data.

Most of the upper ontologies are considered as intentional classes having only necessary conditions for member association in a class. The amalgamation of compulsory and necessary conditions within a class is considered as fully defined.

3. Attributes

Features or Properties of object which help them to categorize in particular type of class are known as attributes. Individual objects in ontology can be related to other things, on basis of some features or portions. These connected properties are their attributes, even though there can be present independent objects also. Every single attribute can be reflected as a class or individual or a sub class. The relationship among individuals can be explained on basis of attributes easily with their similarity and differences. For example numbers of wheels are classifying the vehicles in different categories as shown in Fig. 1. An attribute can be present as a complex data form, i.e.

on basis of wheels, sources, colour many vehicles get classify in different categories but they are having some common features and inter relationships.

Ontologies are only strong if attributes are related to other attributes. If that is not the situation, then either taxonomy or controlled vocabulary is used.

4. Relations

The way in which objects or classes are connected with each other based on a property is known as Relation between objects or classes. Relationships (relations) among entities in ontology identify how entities are interrelated to each other. For example, in Fig. 1 both truck and bus are six wheeler vehicles. Set of associations defines the semantics of the area. An important category of relation is the sub-sumption relation. This relationship describes which object belongs to which class. This helps to create a tree structure of objects which represents relation among objects having common attributes and properties. Another one is mereology relation, which represents association in a class or classes of objects. It uses a directed acyclic graph for data representation. Relation can be categories for relations among classes, individuals, single individual and one class, a object and a collection of objects and collections

5. Function terms

Function terms are structure which is get developed on basis of a relation among individuals. It

6. Restrictions

A limiting condition which used to satisfied for development of range and domain for individuals is Restriction on Individual. It restricts objects in a certain type of concept.

7. Rules

Set of some conditions need to be fulfil by individual to qualify the minimum requirement to create a relation or to belong to a class. It is a logical statement

8. Axioms

Prior Knowledge about the individuals, help to create Axioms. Axioms are set of rules known before and based on learned data. These assumptions plays most important role. The prior knowledge is necessary about the application to decide the axioms in ontology. These will help to initialize basic known category for data. This will reduce the data processing time and complexity of the processing.

9. Events

The change of state of one individual to another based on some logic is known as event. It shows the transaction among the relationship or attributes of individuals on a input.

3.2 Types of Ontology

Ontologies help to represent information and theories in different forms at different levels of abstraction. These different domain ontologies are every so often incompatible with other ontologies. Different types of ontologies are as follow

1. Domain specific Ontology

Individuals belong to a particular type of class comes under this category. The Domain specific ontology is classified on such sets which have no intersection and have different meanings for different classes. For example, "name", can be "full name", "last name", "first name"... all these have different meanings when they get connected to different classes. Merging of different ontologies is complex and has to be done on basis of a common platform.

2. Hybrid Ontology

Hybrid is mix up of different ontologies in one which doesn't come in remaining categories. It is the most complex type of data which need variations in Axioms with changes of class for an individual.

3. Upper Ontology

The general relationships among individuals which are well known and accepted belong to upper ontology. It considers linguistic rules and learned domain for categorization of individuals.

Ontologies need to be designed for significant applications according to user requirements. It is required to be flexible enough for modification and reuse. Linked data, taxonomies, interoperability, axioms, vocabulary, need to be predefined and easy to use. Upper ontologies epitomize mutual notions and relations in range of domains and Domain ontologies characterize conceptions and relations of a specific.

3.3 IoT Ontologies

A cosmology is characterized as "a formal, unequivocal detail of a common conceptualization" and is utilized to speak to information inside a space as a lot of ideas identified with one another. There are four fundamental parts that form a cosmology: Classes, relations, qualities and people. Classes are the primary ideas to depict. Each class can have one or a few kids, known as subclasses, used to characterize increasingly explicit ideas. Classes and subclasses have qualities that speak to their properties and attributes. People are occasions of classes or their properties. At long last, relations are the edges that interface all the introduced segments.

IoT-based genuine world to be isolated into 3 layers: a physical layer, i.e., things; an data layer, i.e., information and metadata about information given by things; and a useful layer containing administrations gave by things. To coordinate our vision of the genuine world and its portrayal by the Internet of Things, fabricating a metaphysics that really models every one of the three layers. Indeed, the physical layer is spoken to by a Device Metaphysics. The data and administration layers are spoken to by a (Physics and Mathematics Domain Ontology and Estimation Models Ontology. To portray the ontologies all the more decisively:

1. Gadget Ontology

The Device Ontology models real equipment gadgets that may exist in the system. It tends to be viewed as the gadget depiction storehouse that can be gotten to for disclosure.

2. Space Ontology

The (Physics and Mathematics) Space Ontology models data about genuine world physical ideas and their relations among one another. It tends to be viewed as the fundamental archive to access for administration creation.

3. Estimation Ontology

The Estimation Ontology contains data about various estimation models ("straight introduction", "Kalman channel", "gullible Bayesian learning", and so on.), the conditions that drive them, the administrations that execute them, etc. It tends to be for the most part viewed as the archive portraying the gadget's nature of administration, and gives data expected to support structure. This cosmology to be utilized as a source of perspective by any middleware or application requiring IoT administrations, i.e., administrations gave by genuine world things. Those administrations, much of the time, produce inexact be that as it may, never 100% exact results.

Most existing philosophy work concentrated on demonstrating either gadgets as done, e.g., in MMI cosmology 2, or material science independently. The curiosity of our methodology is that it joins and takes favorable circumstances of the three ontologies by connecting, all together, the area of information for detecting, activating, and handling assignments and this present reality portrayal through IoT administrations, that know about their condition. A significant commitment is the degree of deliberation at which things, permit clients to portray gadgets in an expressive way while as yet staying away from complex subtleties. Truth be told, as target adaptability, consider straightforwardness in displaying information to be a basic models.

3.4 Advantages of Ontology

One of the main advantages of ontologies is that, by having information only regarding the necessary relationships between objects helps to identify information about data easily. Such relationship makes it is easy to implement for graphical data. Ontologies function helps to depict the human perspective about data and interlinked concepts among objects. In addition ontologies make available a extra rational and tranquil navigation for operators in the ontology organization.

Ontologies are tranquil to outspread as associations and notion equivalently easy to connect with the present ontologies. It also provides the meaningful significance of datum in different format; even data is unstructured, semi-structured or structured, making data amalgamation simpler with text quarrying, and information based analysis.

3.5 *Restrictions of Ontology*

Ontologies offer an opulent group of tools and techniques for information modelling, but it comes by way of some limitations also. One such restriction is the existing property paradigms. For instance, while providing dominant class, the most recent version of the Web Ontology Language—OWL2 has limited set of property paradigms.

Another limitation is that they specify how data has to be structured and avoid addition of extra data. Frequently, data introduced from a new source would be operationally inconsistent with the constraints set. Therefore, this new data need to be modified before being incorporated with existing data.

4 Need of Vocabulary and Ontology

Efficient data storage is must needed to keep relevant multimedia data connected and organized. This data is having lots of variance with large volume and velocity. IoT is going to gather data from different applications developed and used by different people is going to have inconsistency in naming methodology. If a similar vocabulary is used it become easy to gather relevant data. Query processing and decision making processes will also become an easy task for end users to gather important information and used it further.

5 Ontology Quality Methodology

For evaluation and validation ontologies, ontology quality methodology [1] is used. It has following categories to measure the quality of data

(1) Serialization

Serialization is the sequential process to convert data of an individual object into form of binary values. Bit data is easy to save and send during communication. Its main aim is to save all different states of data, which make its serial retrieval easy. The inverse process of serialization is called deserialization. Serialization allows user to save sequential states of an individual object and re-construct it also by retracing. It provide data storage and data exchange simple among distributed applications. With flexibility of serialization, user can perform different actions like sending the object by using a web service, object passing from one domain to another or through a firewall etc. Serialization helps to maintain security or user-specific shared information across different heterogeneous applications.

(2) Syntactic validation

Syntax validation is the method of to check the syntax of a program is free of any error. There are so many tools available to check the syntax of language. These validators work both online and offline. There are syntax validators, know as linters, for programming language

Syntax validators can check syntax, fading common errors such as dividing by zero. Syntax validators highlight code style. Most of the search engines on web use online code checkers.

(3) Interlinking

Linked data is structured data which is inter-related with each other so it becomes more valuable through semantic queries. It is based on standard Web technologies but rather than using those to use as web pages only it used to share information which is automatically accessed by digital devices. One part of the visualization of linked data is Internet which is source of global database.

(4) Documentation

Data documentation helps to keeping data organised easy to collect. The resulting data libraries include information and method how to process it. Brief and clear documentation help users to understand the framework of data. It increases the probability of data usage.

(5) Visualization

Ontology Visualization is representation of data in form of graph as node link diagram having edges and vertices for visual representation of data. Usually vertices contain semantic information and attributes. Sometimes external features are also used, like colour shape and codes to support user visualization flexibility.

(6) Availability of resources

The resource availability describe how ontologies availability changes over time and how user and developer observe those changes in ontologies. These attributes include longevity and observability. Attributes of individuals captures whether the availability of ontology is directly observable, partially observable, or unobservable to various entities.

(7) Discoverability

Ontologies can be used to determine information from Big Data. Based on these distinct projections, efforts can be on the way to exploration of the likelihood to use ontological knowledge for the requirement of Big data and meta data as a underpinning to proficiently determine useful statistics for exploration.

(8) Ontology Design

What's, why's and where's organise the ontology in data set to identify the best place for data. Ontology designers need to find correct place to classify data in ontology with its relationship with rest. Matching problems doesn't provide optimal solution which makes need of ontology design. Domain and task are two major entities in designing of ontologies. Domain establishes a relationship and information of individuals and task makes them identify able in the set.

Selection of Ontology has to be based on some parameters are as follow:

1. Goal of ontology

Sharing common understanding of the structure of data is one of the major goals of ontology. For example, if different Web sites contain medical data it becomes difficult to share it or correlate it with other sites data. If these Web domains share and distribute the similar ontology, data extraction and aggregation becomes easy. The users can practice this combined statistics to reply enquiries or as feedback into other applications.

Permitting reprocess of domain information was unique and powerful services after recent outpouring in ontology exploration. For instance, models used for representation of dissimilar time in different domains. This comprises the concepts of time pauses among domains, points in time, comparative time measurement, and much more. If we requires a large ontology, integration of several present ontologies help to create better and enormous province.

Production of unambiguous domain norms creates it easy on the way to alter if there is any changes occurred. Separating the domain describe mechanisms rendering to the mandatory measurement and implement it independently.

2. Size

Large size of ontology increases its efficiency as well as complexity. The small size ontologies are not enough to perform any operation, so it gets integrated with parent class but still remain as an individual class. The big ontology requires optimal search and update algorithms whereas classification in such data sets is more accurate. Interrelationship makes it difficult to manage and find a relation individually among objects or classes.

3. Documentation

Proper documentation of ontology makes it easy to use and access. Usually for developer it is easy to extend and integrate different data values with the help of documentation. The proper management of ontology make easy to identify an error.

4. Avability on Web

Avability of ontology on web make it easy to update frequently having more precise and updated knowledge. The linking of data also becomes easy to categorized data because of it. If an unknown user having less or no information will edit data it will decrease its performance and quality. Identification of such faults is too difficult in such cases. Avability on web make ontologies to easily available anytime anywhere. The queries and correlation of data become easy in this case. It will remain updated in real time with multi user accessibility.

5. Popularity

Popularity is the most important aspect to improve the quality of words and it plays an important role in data set. Unpopular words get vanished from daily conversation and rarely used. The elimination of this data is not a good decision because it helps to identify miscellaneous hidden information. But it has to give low priority for optimal identification. Most of the least used words represent unauthorized data with respective of forensics. Arrangement of ontologies on basis of popularity helps to make frequently used data easily available. It makes operations less time consuming. System optimality gets increased with this feature.

6. Maintenance with time

Maintenance of data allows speeding up query process by illuminating the inherent relations which are authenticated and approved through the semantics of ontology. The complication of reasoning within ontology, is in this manner moved from demand time to apprise time.

7. Meta data Vocabulary

"Data about Data" or set of words denoting the features of resources is known as Meta data. Ontology is a metadata vocabulary. Generally, metadata vocabularies are domain precise. Controlled vocabularies are homogeneous and ordered words provide a dependable way to define data. Metadata designers allocate terms from vocabularies to simplify information recovery.

Commonly vocabularies contain subject headings, lists, files, and thesauri. Controlled vocabularies can be organised in alphabetical lists with a hierarchical arrangement of data. Thesauri also comprise synonyms, associated data, scope and notes, data antiquity, alternative idioms, or numerical data. Ontologies include even more specification, such as description, concepts, hierarchy and relationships with other values.

6 Application

Smart City [2] requires interoperable system, to process data securely, and manages services. Secure data access will reduce the cost of operation. Some famous Ontologies for Smart Cities [1] are KM4City (Knowledge Model for City), Semantic Traffic Analytics and Reasoning for CITY(STAR-CITY), SCOnt (Smart City Ontology), and CityPulse

Agriculture and Food supply [3] is also needed a dictionary o track entities in system and to manage information from seeding to selling of crops. It helps to manage data of farm items and to keep transaction easy. Smart City, Smart Home,

And Smart Weather [4] have dependency among each other for various tasks and management.

Health Sector [5] shows how LOV4IoT will help to manage complex distributed data having different type of media contents but having same vocabulary can be easily manages. the data access will be easy due to inter linking from different departments. Transportation and Logistics [6] will also become easy to manage in real time. Time required to share information will get reduced and different representations of outcome can be done easily.

7 Background

Gyrard et al. [2] projected a semantic web search application for IoT based cities with case study of three procedure cases FIESTA-IOT EU, Machine-to-Machine Measurement (M3), and VITAL EU scheme. This project is combination of web based data from IoT to Semantic based but not suitable for real time interoperability practically. Kamilaris et al. [3] proposed Agri-IoT, an IoTbased smart farming applications over web which supports big data analysis, event detection, interoperability, online information and linked datasets accessible to end uses always. But it doesn't standardized data specifically used for agriculture and semantic web axioms. Noura et al. [4] created a corpus and discussed application of semantic web in Smart City, Smart Home, and Smart Weather. It created Knowledge Extraction for the Web of Things (KE4WoT) a set of ontologies based on some specific domain. It is efficient one if domain to which word belong is considered else it put word in unrelated category or discard it. This causes outlier data which can be useful in ignored data category. Gyrard et al. [1] suggest some techniques to make ontologies more effective with combined ontology sets for IoT and Smart city LOV, READY4SmartCities, Open Sensing City (OSC), in addition to LOV4IoT. Gyrard et al. [7] raise the requirement of interoperability of data needed for semantic web [8] and proposed a framework Machine-to-Machine Measurement (M3). But it still lacks combination of different domains. It is difficult to combine these frameworks together. Cross domain applications [9] can be useful to collect similar type of data from different applications but requires domain base knowledge. It provides a set of linked open rules which can be used generally for IoT applications having cross domain data. Bermudez-Edo et al. [10] states that semantic techniques upsurge the complexity and processing time which makes them unsuitable for IoT. To resolve this issue they proposed IoT-Lite for semantic sensor networks but it lacks interoperability. Linked Open Vocabularies for IoT (LOV4IoT) [11] overcome the challenges of classification, re-engineering and designing of interoperability. This shows evolved better results and easy to establish technique. It is up to the mark but doesn't consider previous established classification and classes for ontologies. IERC Cluster Semantic (IERC AC4) [12] resolve four interoperability issues Technical, Syntactical, Semantic and Organizational. IERC AC4 has some shortcomings also like reuse of methodologies, validation and evaluation of ontologies, and a well-designed structure. Machine-to-Machine Measurement

(M3) framework [13] tries to simplify domain specific issues. It highlighted uniform nomenclature requirement to improve performance of Ontology. In perspective of security M3 framework is not secure because it is accessible by third party easily. IoT-O [14] an extension of M2M is proposed to use known ontologies and defines some concepts relevant to IoT. It is based on principal of indexing of resources data; still semantic data interoperability is unachieved.

8 IoT-Related Ontologies

The space of ontologies is divided, paying little heed to the area of intrigue. The more extravagant a metaphysics is, the bigger zone it ranges. Subsequently, uniqueness and crossing points with different ontologies become increasingly many-sided and complex. Web of Things traverses tremendous number of areas, and extends with the developing prominence of "shrewd gadgets". Utilization of ontologies in the IoT mirrors this extensiveness. There are numerous ontologies that speak to models applicable to the IoT, including, in any case, not restricted to, gadgets, units of estimation, information streams, information preparing, geo-location, information provenance, PC equipment, techniques for correspondence, and so forth. Highlight of the IoT is a brilliant gadget fit for correspondence. From this point of view, ontologies that catch the possibility of a gadget, and are entrenched in the IoT space: SSN, SAREF, oneM2M Base Ontology, IoT-Lite, and OpenIoT. Every one of them takes an alternate way to deal with demonstrating the IoT space at the same time, in spite of the distinctions in conceptualization, they spread converging sections of the IoT scene. Beneath, we talk about disparity, oppositeness and covers between these ontologies.

SSN, or "Semantic Sensor Network" [15] is a metaphysics based on sensors and perceptions. It is an accepted augmentation of the SensorML language. SSN centers around estimations and perceptions, ignoring equipment data about the gadget. In particular, it portrays sensors as far as abilities, execution, use conditions, perceptions, estimation forms, and organizations. It is profoundly measured and extendable. Truth be told, it relies upon other ontologies in key territories (for example time, area, units) and, for every pragmatic reason, should be stretched out before genuine execution of a SSN-based IoT framework. SSN, detailed on head of DUL, is an ontological reason for the IoT, as it attempts to cover any utilization of sensors in the IoT.

SAREF or "The Smart Appliances REFerence" [16], metaphysics covers the zone of shrewd gadgets in houses, workplaces, open spots, and so forth. It doesn't concentrate on any mechanical or logical usage. The gadgets are described overwhelmingly by the function(s) they perform, orders they acknowledge, and states they can be in. Those three classifications fill in as building squares of the semantic depiction in SAREF. Components from each can be consolidated to deliver complex depictions of multi-practical gadgets. The portrayal is supplemented by gadget benefits that offer capacities. A critical module of SAREF is the vitality and force profile that got significant consideration, not long after its inception. SAREF utilizes WGS84 for geolocation and characterizes its own estimation units. oneM2M Base Ontology (oneM2M BO) is an as of late made philosophy, with first non-draft discharge in August 2016. It is generally little, readied for the discharge 2.0 of oneM2M details, and planned with the expectation of giving a common ontological base, to which different ontologies would adjust. It is like the SSN, since any solid framework fundamentally needs to expand it before execution. It depicts gadgets in an expansive degree, empowering (in an overall sense) particular of gadget usefulness, organizing properties, activity and administrations. The way of thinking behind this methodology was to empower revelation of semantically differentiated assets utilizing an insignificant arrangement of ideas. It is a base cosmology, as it doesn't expand some other base models, (for example, DUL or Dublin Core). Notwithstanding, arrangements to different ontologies are known.

IoT-Lite [10] is a "launch" of the SSN, for example an immediate augmentation of a few of its modules. It is a negligible cosmology, to which a large portion of the provisos of the SSN apply. In particular: center around sensors and perceptions, dependence on other ontologies (for example time or units ontologies), high measured quality and extendability. The thought behind the IoT-Lite was to make a little/light semantic model that would be less burdening (than other, increasingly verbose and more extensive models) on gadgets that process it. Simultaneously, it expected to cover enough ideas to be valuable. The metaphysics portrays gadgets, articles, frameworks and administrations. The principle augmentation of the SSN, in the IoT-Lite, lies also of actuators (to supplement sensors, as a gadget type) and an inclusion property. It expressly utilizes ideas from a geo-location cosmology to divide gadget inclusion and arrangement area.

OpenIoT [17] philosophy was created inside the OpenIoT venture. In any case, here, we utilize the expression "OpenIoT" to allude to the metaphysics. It is a nearly enormous model that reuses and joins different ontologies. Those incorporate all modules of the SSN (the principle reason for the OpenIoT), SPITFIRE (counting sensor systems), Event Model-F, PROV-O, WGS84, CloudDomain, SIOC, Association Ontology and others, including littler ontologies created at the DERI (at present, Insight Center). It likewise utilizes ontologies that give premise to those listed before, for example DUL. Other than ideas from the SSN, OpenIoT, utilizes an enormous number of SPITFIRE ideas, for example system and sensor arrange depictions. While some referenced ontologies are definitely not imported by the OpenIoT unequivocally, they show up in all models, documentation, what's more, venture expectations. Consequently, one can treat OpenIoT as a blend of portions of those. Essentially to the SSN, OpenIoT doesn't characterize its own area ideas and doesn't unequivocally import geolocation ontologies. It depends on different ontologies for that yet, rather than the SSN, it plainly shows LinkedGeoData and WGS84 of geolocation portrayals. It characterizes a constrained set of units of measure (for example temperature, wind speed), however just when they were pertinent to the OpenIoT venture pilot usage.

The rich set-up of utilized ontologies implies that OpenIoT gives rich depiction of gadgets, their functionalities, capacities, provenance, estimations, arrangements and position, vitality, pertinent occasions, clients and numerous others. Strikingly enough, it doesn't unequivocally portray actuators or impelling properties/capacities.

It tends to be seen that the expansive extent of the cosmology makes it rather entangled. This is likewise in light of the fact that, it isn't recorded all around ok, for example the detail level and straightforward entry of the documentation don't coordinate the range of inclusion of ideas in the model. Besides, it isn't plainly and expressly modularized, notwithstanding being an expansion of the SSN.

Let us note that, while there are other IoT models of expected intrigue (such as OGC Sensor Things, FAN FPAI, UniversAAL ontologies, IoT Ontology3, M3 Vocabulary), we won't consider them here. This is a result of space constraint, and the way that they have produced substantially less "general intrigue". In any case, we intend to remember these ontologies for resulting work.

Let us presently think about the choose ontologies [18] one next to the other. Chosen key angles, or classifications, straightforwardly relating to the IoT; set the first segment of Table 1. Notwithstanding, due to complexities and different ways of thinking behind thought about ontologies, every classification should be further researched.

Every one of considered ontologies proposes an alternate way to deal with demonstrating the IoT space [19]. The greatest contrasts are in the subtleties.

(a) OneM2M BO proposes a little base cosmology, like upper ontologies that gives just an insignificant set of profoundly unique elements. This takes into consideration a wide arrangement of space ontologies to be effortlessly lined up with it. It likewise implies that the BO itself isn't sufficient to model any solid issue

Table 1 IoT ontologies comparison (a), (b)

(a)

Sub Domain	Thing	Gadget	Gadget sending	Gadget properties and capacities	Gadget vitality	Capacity and administration
SSN	✓	✓	✓	✓	✓	
SAREF	✓	✓	✓		✓	✓
OneM2M BO	✓	✓	✓			✓
IoT-Lite	✓	✓	✓			✓
OpenIoT	✓	✓	✓	✓	✓	

(b)

Sub Domain	Detecting and sensor properties	Perception	Impelling and actuator properties	Conditionals
SSN	✓	✓		✓
SAREF	✓	✓	✓	
OneM2M BO		✓		
IoT-Lite	✓		✓	
OpenIoT	✓	✓		

(or arrangement) in the IoT. Moreover, it doesn't catch a few angles that are normal in different ontologies.

(b) OpenIoT contrasts this way of thinking by giving a nitty gritty model to a particular issue (for example pilot usage from the OpenIoT venture) that can be additionally applied in an increasingly broad case, or in different arrangements. Its overwhelming utilization of outside ontologies gives high semantic interoperability by structure.

(c) SSN is a created model of the IoT when all is said in done, however with solid spotlight on sensors. It depends on DUL, what's more, is unmistakably modularized, which makes it a decent contender for augmentations into solid frameworks and executions. This is prove by the way that other ontologies, assessed here, utilize it. With regards to particularity, it places itself in the center between oneM2M BO and OpenIoT.

(d) IoT-Lite is an expansion of chose SSN modules, essentially to incorporate actuators. Or maybe than concentrating on giving a point by point portrayal of a delimited issue space inside the IoT, it moves toward the displaying issue from the point of view of a usage gadget. It plans to convey a little, yet complete, model in request to rearrange preparing of semantic data. This is additionally its particular attributes.

(e) SAREF is a model with a solid spotlight on its own region—of brilliant apparatuses. Despite the fact that mappings to different norms exist, SAREF was created without any preparation to speak to a particular zone of use of the IoT. In this region, it conveys a solid and point by point base, that is additionally clear and simple to comprehend. Simultaneously, it is sufficiently general to be utilized when stretched out to different areas, or arrangements.

Curiously, every one of these ontologies totally ignore equipment determinations. It appears that the "place" of a gadget in an IoT framework is significantly more imperative to philosophy engineers than its equipment detail and coming about abilities.

9 Semantic Web

The Semantic Web is an expansion of the World Wide Web [20] through norms set by the World Wide Web Consortium (W3C). The objective of the Semantic Web is to make Internet information machine-decipherable. To empower the encoding of semantics with the information, innovations, for example, Resource Description Framework (RDF) and Web Ontology Language (OWL)are utilized. These advancements are utilized to officially speak to metadata. For instance, metaphysics can portray ideas, connections among elements, and classifications of things. These implanted semantics offer critical points of interest, for example, thinking over information and working with heterogeneous information sources.

These principles advance regular information configurations and trade conventions on the Web, on a very basic level the RDF. As per the W3C, "The Semantic Web gives a typical structure that permits information to be shared and reused across application, undertaking, and network boundaries." The Semantic Web is in this way viewed as an integrator across various substance and data applications and frameworks.

The term was authored by Tim Berners-Lee for a snare of information (or information web) that can be prepared by machines that is, one in which a significant part of the importance is machine-intelligible. While its faultfinders have scrutinized its possibility, advocates contend that applications in library and data science, industry, science and human sciences research have just demonstrated the legitimacy of the first concept.

Berners-Lee initially communicated his vision of the Semantic Web in 1999 as follows:

I have a fantasy for the Web [in which computers] become fit for investigating all the information on the Web—the substance, connections, and exchanges among individuals and PCs. A "Semantic Web", which makes this potential, still can't seem to develop, yet when it does, the everyday instruments of exchange, administration and our day by day lives will be dealt with by machines conversing with machines. The "astute specialists" individuals have promoted for a long time will at last materialize.

9.1 Difficulties

A portion of the difficulties for the Semantic Web incorporate incomprehensibility, dubiousness, vulnerability, irregularity, and double dealing. Robotized thinking frameworks should manage these issues so as to convey on the guarantee of the Semantic Web.

1. **Incomprehensibility**: The World Wide Web contains a large number of pages. The SNOMED CT clinical wording philosophy alone contains 370,000 class names, and existing innovation has not yet had the option to kill all semantically copied terms. Any computerized thinking framework should manage really immense sources of info.
2. **Unclearness**: These are uncertain ideas like "youthful" or "tall". This emerges from the unclearness of client inquiries, of ideas spoke to by content suppliers, of coordinating inquiry terms to supplier terms and of attempting to join diverse information bases with covering yet quietly various ideas. Fluffy rationale is the most well-known method for managing dubiousness.
3. **Vulnerability**: These are exact ideas with unsure qualities. For instance, a patient may introduce a lot of indications that relate to various diverse particular judgments each with an alternate likelihood. Probabilistic thinking strategies are commonly utilized to address vulnerability.

4. **Irregularity**: These are consistent logical inconsistencies that will unavoidably emerge during the improvement of huge ontologies, and when ontologies from isolated sources are joined. Deductive thinking bombs disastrously when confronted with irregularity, since "anything follows from a logical inconsistency". Defeasible thinking and paraconsistent thinking are two procedures that can be utilized to manage irregularity.
5. **Trickery**: This is the point at which the maker of the data is deliberately deceptive the customer of the data. Cryptography methods are right now used to lighten this danger. By giving a way to decide the data's honesty, including what identifies with the personality of the substance that delivered or distributed the data, anyway validity gives despite everything must be tended to in instances of possible misleading.

This rundown of difficulties is illustrative as opposed to comprehensive, and it centers around the difficulties to the "binding together rationale" and "evidence" layers of the Semantic Web. The World Wide Web Consortium (W3C) Incubator Group for Uncertainty Reasoning for the World Wide Web (URW3-XG) last report protuberances these issues together under the single heading of "uncertainty" [21]. Many of the procedures referenced here will expect expansions to the Web Ontology Language (OWL) for instance to comment on restrictive probabilities. This is a region of dynamic research.

9.2 Principles

Normalization for Semantic Web with regards to Web 3.0 is under the consideration of W3C.

9.3 Parts

The expression "Semantic Web" [22] is frequently utilized all the more explicitly to allude to the organizations and advancements that empower it. The assortment, organizing and recuperation of connected information are empowered by advances that give a conventional depiction of ideas, terms, and connections inside a given information area. These innovations are determined as W3C guidelines and include:

- Resource Description Framework (RDF),an overall technique for depicting data
- RDF Schema (RDFS)
- Simple Knowledge Organization System (SKOS)
- SPARQL, a RDF question language
- Notation3 (N3), planned in view of human-coherence
- N-Triples, an arrangement for putting away and transmitting information
- Turtle (Terse RDF Triple Language)

- Web Ontology Language (OWL), a group of information portrayal dialects
- Rule Interchange Format (RIF), a structure of web rule language lingos supporting principle trade on the Web.

9.4 The Semantic Web Stack

The Semantic Web Stack [23] outlines the engineering of the Semantic Web. The capacities and connections of the parts can be summed up as follows:

XML gives an essential punctuation to content structure inside records, yet connects no semantics with the significance of the substance contained inside. XML isn't at present an important part of Semantic Web advancements as a rule, as option sentence structures exists, for example, Turtle. Turtle is an accepted norm, yet has not experienced a conventional normalization process.

XML Schema is a language for giving and limiting the structure and substance of components contained inside XML records.

RDF is a basic language for communicating information models, which allude to objects ("web assets") and their connections. A RDF-based model can be spoken to in an assortment of language structures, e.g., RDF/XML, N3, Turtle, and RDFa. RDF is a key norm of the Semantic Web.

RDF Schema expands RDF and is a jargon for portraying properties and classes of RDF-based assets, with semantics for summed up progressions of such properties and classes.

OWL includes more jargon for portraying properties and classes: among others, relations between classes (for example disjointness), cardinality (for example "precisely one"), fairness, more extravagant composing of properties, qualities of properties (for example balance), and listed classes.

SPARQL is a convention and inquiry language for semantic web information sources.

RIF is the W3C Rule Interchange Format. It's a XML language for communicating Web decides that PCs can execute. RIF gives different renditions, called tongues. It incorporates a RIF Basic Logic Dialect (RIF-BLD) and RIF Production Rules Dialect (RIF PRD).

9.5 Applications

The goal is to improve the ease of use and convenience of the Web and its interconnected assets by making Semantic Web Services, for example, Servers that uncover existing information frameworks utilizing the RDF and SPARQL norms. Numerous converters to RDF exist from various applications. Relational databases are a significant source. The semantic web server connects to the current framework without influencing its activity.

Records "increased" with semantic data. This could be machine-reasonable data about the human-justifiable substance of the archive, (for example, the designer, title, depiction, and so forth.) or it could be absolutely metadata speaking to a lot of realities, (for example, assets and administrations somewhere else on the site). Note that anything that can be related to a Uniform Resource Identifier (URI) can be portrayed, so the semantic web can reason about creatures, individuals, places, thoughts, and so forth. There are four semantic explanation arranges that can be utilized in HTML records; Microformat, RDFa, Microdata and JSON-LD. Semantic markup is frequently produced consequently, as opposed to physically.

Basic metadata vocabularies (ontologies) and guides between vocabularies that permit record makers to realize how to increase their archives so operators can utilize the data in the provided metadata. Computerized specialists to perform undertakings for clients of the semantic web utilizing this information.

Electronic administrations (frequently with operators of their own) to gracefully data explicitly to specialists, for instance, a Trust administration that an operator could inquire as to whether some online store has a background marked by helpless assistance or spamming.

Such administrations could be helpful to open web indexes, or could be utilized for information the executives inside an association. Business applications include:

- Encouraging the joining of data from blended sources
- Dissolving ambiguities in corporate phrasing
- Improving data recovery subsequently decreasing data over-burden and expanding the refinement and accuracy of the information retrieved
- Distinguishing applicable data as for a given domain
- Giving dynamic help.

In a partnership, there is a shut gathering of clients and the administration can uphold organization rules like the selection of explicit ontologies and utilization of semantic explanation. Contrasted with the open Semantic Web there are lesser prerequisites on adaptability and the data circling inside an organization can be progressively confided when all is said in done; protection is less of an issue outside of treatment of client information.

9.6 *The Semantic Web Stack for the IoT*

The Semantic Web Technologies Stack presents the center Semantic Web Technologies utilized at various levels of an IoT framework [24]. The mix of Semantic Web innovations into IoT frameworks can be distinguished at three various levels. The "demonstrating level" gives a typical comprehension of Things' qualities and abilities. It utilizes shared and basic acknowledged vocabularies and ontologies to encourage the reconciliation of information created by various frameworks (for example sensor ontologies). The "information preparing level" utilizes depiction rationales and OWL semantics so as to empower thinking and deduction over the information. At last, the "IoT Administrations and Application" level uses particular portrayal furthermore, ontologies that empowers administration distribution, disclosure, piece and adjustment.

9.6.1 Models and Meta-models: Information Bases

The general nature of the last help or application depends of the nature of each included layer. In this specific circumstance, the primary layer is worried about information readiness. Deciphering furthermore, understanding the information is the primary essential in this process. This layer deals with the semantic coordination and accumulation of information from an assortment of sources. Semantically explained information can be changed and displayed by explicit necessities. The Model speaks to the Thing and structures the Assertions Box (ABox), while the Meta-model depicts the jargon used to portray the Thing and structures the Terminological Box (TBox). A Knowledge Base is made by these two parts. At last, the meta-metamodel gives the build jargon to the TBox.

9.6.2 Information Preparing

An IoT framework [18, 21] is, by its temperament, a conveyed framework and handling its information should be possible at various levels. While the restricted neighborhood data can give some essential translation furthermore, preparing in its area of intrigue, further understanding on the information is acquired at more elevated levels, when information from numerous sources are assembled, prepared and associated. We stress two distinct methodologies for preparing this information: (1) utilizing semantic reasoners and (2) utilizing Big Data explicit calculations (for example AI).

(1) Reasoning and Inferences.

Rules and semantic arrangements (for example owl:equivalentClass, owl:subClassOf, owl:sameAs) can be utilized to change and adjust the information to the pronounced ontologies. Contingent upon the expressiveness of the ontologies, thinking motors can additionally gather affiliations and connections into the information.

For information transmission and putting away in a Semantic Web setting, JSON-LD, a W3C suggestion from 2014, gives a advantageous approach to serialize RDF information. XML design is too accessible. Triplestores (for example Fuseki, StarDog) are utilized to store RDF significantly increases. The inquiry language for the Semantic Web is (SPARQL Protocol And RDF Query Language). It gives a helpful method to cross examine different triplestores over HTTP.

(2) Big Data and AI calculations.

Semantic Advances are a superb decision for IoT frameworks [19] for two reasons: (1) it permits sharing the information depiction through diagrams and ontologies and (2) it permits information encoding in ontologies by means of portrayal rationale develops. Anyway enormous amount of information joined with high expressive ontologies, can limit reasoners' exhibition in their inductions. IoT frameworks are commonly used to screen, analyze, foresee and suggest activities. In metaphysics based frameworks, the information is depicted from the earlier, this makes them less adjusted to frameworks where the goal is to foresee and dissect practices of various conditions and clients. Starting here of view, the coordination of AI calculations with very much portrayed information could give better esteem administrations and applications. An model utilizing both measurable learning and ontologies to separate private client action is introduced.

9.6.3 IoT Services and Applications

Ontologies and semantic comments can likewise upgrade the portrayal of the offered types of assistance. OWL-S gives semantic markup for web administrations. The Semantic Sensor Perception Service (SemSOS) and the utilization of ontologies for computerized organization are a few instances of semantic advances based applications and administrations in IoT conditions.

10 Ontologies for Smart Cities

KM4City, an Italian national task, demonstrated a cosmology intended for amassing static or dynamic brilliant city information. The creators reuse ontologies, for example, OWL-time, DC terms, FOAF, WGS84, GoodRelations, and cosmology transportation systems (OTNs). The undertaking is adaptable since they handle 81 million triples with a development of 4 million triples for each month. It gives a connected information chart, perception and investigation device what's more, administration map applications abusing the collected information.

Semantic traffic examination and thinking for CITY (STARCITY), an IBM venture, is conveyed in four keen urban communities, such as Dublin, Bologna, Miami, and Rio de Janeiro. The venture is centered around planning ontologies to analyze and anticipate street gridlocks. Information preparing misuses six heterogeneous sources.

(1) Road climate conditions.
(2) Weather data.
(3) Dublin transport stream.
(4) Social media takes care of.
(5) Road works and support.
(6) City occasions.

Semantic Web Rule Language (SWRL) rules have been intended to characterize rules, for example, substantial traffic stream. Celebration IoT is a H2020 European undertaking. The Celebration IoT cosmology is intended to bring together existing IoTrelated ontologies to structure information created by testbeds. The SmartSantander city or even keen structures are testbeds creating genuine information, which is semantically commented on as per the philosophy.

Imperative, a FP7 European undertaking, planned a cosmology to manage heterogeneous information streams produced by gadgets inside keen urban areas. The metaphysics models sensors and their, for IoT frameworks and administrations, and for keen city applications. Indispensable is creative since it gives a working framework to IoT to manage administration creation, coordination, and conventions. Fundamental gives the accompanying attributes: virtualization, measured quality, norms based (RDF and JSON-LD) and inexactly coupled, also, open-source.

CityPulse, a FP7 European undertaking, gives the SAO to bring together savvy city datasets. SAO has been structured to address ongoing viewpoints. Brilliant city philosophy (SCO) is a metaphysics distributed in 2015. It reuses a few ontologies, for example, SKOS, yet it doesn't reuse the SSN philosophy and absences of best practices. For example, the cosmology isn't partaken in a appropriate way.

Shrewd city SOFIA2 philosophy doesn't expand SSN cosmology be that as it may, reuses IoT.est philosophy. PRISMA venture planned a philosophy which reuses WGS81, NeoGeo, and assortments ontologies. Notwithstanding, it makes reference to neither the utilization of information created by gadgets nor the utilization of SSN cosmology. The cosmology is predominantly intended to bind together heterogeneous information.

(1) GeoData from the geographic data framework, information on lines, and stops of the open vehicle transport framework (REST Web administration in JSON group).
(2) Public lighting framework for the upkeep of the city (XML document).
(3) State of the streets, walkways, signs, and markings (Microsoft SQL Server database).
(4) Historical information on city squander assortment (Microsoft Exceed expectations document).
(5) Historical information on the urban flaw detailing administration (MySQL Server database).

The venture gives the LODView instrument to a HTML portrayal of RDF assets and the LODLive apparatus to peruse the RDF diagram. This paper doesn't concentrate on the depiction of the philosophy, however presents the need of this cosmology to give connected open information and executes Web administrations, SPARQL endpoints, browsable highlights, and representation on head of it. Brilliant city metaphysics (SCOnt) has been planned and utilized in a semantic-based system to control brilliant city information. In any case, the cosmology has not been shared online which ruins interoperability of shrewd city frameworks and the reuse of the philosophy. The cosmology reuses a populace philosophy, a geo-area metaphysics and the DBpedia metaphysics. Depictions with respect to the structure of the metaphysics and semantic planning are absent. The curiosity contrasted with existing brilliant city ventures isn't clearly clarified. SCOnt is utilized to control keen city information in an engineering including four layers.

(1) Data scratching layer assembles and refines information since duplication and inadequacy of meta-data and missing qualities issues are confronted.
(2) Data adjustment layer gives cosmology displaying and semantic planning.
(3) Data the board layer stores and records information inside a NoSQL database. Semantic Web administrations are referenced yet neither connection nor portrayals are given or on the other hand referenced.
(4) Applications layer gives dashboards and APIs.

Smart city ontologies are routinely upgraded which thwarts semantic interoperability. More ontologies identified with keen urban areas can be found on the LOV4IoT and OSC philosophy indexes.

10.1 Measures to Compare Smart Cities and IoT Ontologies

For examination of keen city ontologies in a lot of measures to think about smart city ontologies which can likewise be applied to IoT ontologies. Those standards are primarily centered around the reusability of the ontologies.

(1) Ontology objective ought to be obviously clarified. As a rule, the cosmology is intended for a task or an application.
(2) Ontology size shows the profundity of the cosmology. Little or lightweight ontologies would be simpler to reuse.
(3) Ontology documentation decreases the expectation to absorb information to comprehend and incorporate the metaphysics, and support its reusability. A well known practice is to give an on the web HTML documentation. A distribution, deliverable or any documentation is important to clarify in detail the metaphysics and its effect.
(4) Ontology accessibility is emphatically empowered. Metaphysics ought to be shared on the Web to energize semantic interoperability. Philosophy planners should make an exertion in coordinating past ontologies and staying alert of the metaphysics constraints.
(5) Ontology prominence shows the effect of the metaphysics and its genericity when the cosmology is utilized in different tasks.
(6) Ontology support should be accomplished. Ordinarily, at the point when the undertakings are done, the metaphysics isn't kept up. Be that as it may, cosmology creators may be responsive on the off chance that they keep on taking a shot at a similar exploration point.
(7) Ontology metadata is basically required for building programmed instruments.

11 Future Trends and Conclusion

Ontology and Vocabulary both are essential components for IoT. The cross linking and interoperability of data make it difficult to create a standard set difficult. It requires timely maintenance with up gradation in rules. Security on data is one of the most needed aspects but similar use of individuals can make it risky. The interoperability of data is still not possible due to different vocabulary present for IoT which have no interconnection among each other. Each platform uses its own set of Axioms and programming languages which make integration difficult. Scrutinize the evaluation pattern of ontology is still undefined and not considered.

References

1. Gyrard, A., Zimmermann, A., Sheth, A.: Building IoT-based applications for smart cities: how can ontology catalogs help? IEEE Internet Things J. **5**(5), 3978–3990 (2018)
2. Gyrard, A., Serrano, M.: A unified semantic engine for internet of things and smart cities: from sensor data to end-users applications. In: 2015 IEEE International Conference on Data Science and Data Intensive Systems, pp. 718–725. IEEE (2015)
3. Kamilaris, A., Gao, F., Prenafeta-Boldu, F. X., Ali, M.I.: Agri-IoT: A semantic framework for internet of things-enabled smart farming applications. In: 2016 IEEE 3rd World Forum on Internet of Things (WF-IoT), pp. 442–447. IEEE (2016)
4. Noura, M., Gyrard, A., Heil, S., Gaedke, M.: Automatic knowledge extraction to build semantic web of things applications. IEEE Internet Things J. **6**(5), 8447–8454 (2019)
5. Gyrard, A., Atemezing, G., Bonnet, C., Boudaoud, K., Serrano, M.: Reusing and unifying background knowledge for internet of things with LOV4IoT. In: 2016 IEEE 4th International Conference on Future Internet of Things and Cloud (FiCloud), pp. 262–269. IEEE (2016)
6. Ganzha, M., Paprzycki, M., Pawłowski, W., Szmeja, P., Wasielewska, K.: Semantic interoperability in the internet of things: an overview from the INTER-IoT perspective. J. Netw. Comput. Appl. **81**, 111–124 (2017)
7. Gyrard, A., Datta, S.K., Bonnet, C., Boudaoud, K.: Cross-domain internet of things application development: M3 framework and evaluation. In: 2015 3rd International Conference on Future Internet of Things and Cloud, pp. 9–16. IEEE (2015)
8. Szilagyi, I., Wira, P.: Ontologies and semantic web for the internet of things-a survey. In: IECON 2016-42nd Annual Conference of the IEEE Industrial Electronics Society, pp. 6949–6954. IEEE (2016)
9. Gyrard, A., Bonnet, C., Boudaoud, K.: Enrich machine-to-machine data with semantic web technologies for cross-domain applications. In: 2014 IEEE World Forum on Internet of Things (WF-IoT), pp. 559–564. IEEE (2014)
10. Bermudez-Edo, M., Elsaleh, T., Barnaghi, P., Taylor, K.: IoT-Lite: a lightweight semantic model for the internet of things and its use with dynamic semantics. Pers. Ubiquit. Comput. **21**(3), 475–487 (2017)
11. Gyrard, A., Bonnet, C., Boudaoud, K., Serrano, M.: LOV4IoT: a second life for ontology-based domain knowledge to build semantic web of things applications. In: 2016 IEEE 4th International Conference on Future Internet of Things and Cloud (FiCloud), pp. 254–261. IEEE (2016)
12. Gyrard, A., Serrano, M., Atemezing, G. A.: Semantic web methodologies, best practices and ontology engineering applied to Internet of Things. In: 2015 IEEE 2nd World Forum on Internet of Things (WF-IoT), pp. 412–417. IEEE (2015)

13. Gyrard, A., Datta, S. K., Bonnet, C., Boudaoud, K. Standardizing generic cross-domain applications in Internet of Things. In: 2014 IEEE Globecom Workshops (GC Wkshps), pp. 589–594. IEEE (2014)
14. Alaya, M.B., Medjiah, S., Monteil, T., Drira, K.: Toward semantic interoperability in oneM2M architecture. IEEE Commun. Mag. **53**(12), 35–41 (2015)
15. Compton, M., Barnaghi, P., Bermudez, L., GarcíA-Castro, R., Corcho, O., Cox, S., Graybeal, J., Hauswirth, M., Henson, C., Herzog, A., Huang, V.: The SSN ontology of the W3C semantic sensor network incubator group. J Web Semant. **17**, 25–32 (2012)
16. Daniele, L., den Hartog, F., Roes, J.: Created in close interaction with the industry: the smart appliances reference (SAREF) ontology. In: International Workshop Formal Ontologies Meet Industries, pp. 100–112. Springer, Cham (2015)
17. Soldatos, J., Kefalakis, N., Hauswirth, M., Serrano, M., Calbimonte, J. P., Riahi, M., Aberer, K., Jayaraman, P.P., Zaslavsky, A., Žarko, I.P., Skorin-Kapov, L.: Openiot: Open source internet-of-things in the cloud. In: Interoperability and open-source solutions for the internet of things, pp. 13–25. Springer, Cham (2015)
18. Bajaj, G., Agarwal, R., Singh, P., Georgantas, N., Issarny, V.: A study of existing Ontologies in the IoT-domain. arXiv preprint arXiv:1707.00112 (2017)
19. Lam, A.N., Haugen, Ø.: Applying semantics into service-oriented iot framework. In: 2019 IEEE 17th International Conference on Industrial Informatics (INDIN), vol. 1, pp. 206–213. IEEE (2019)
20. Mishra, S., Jain, S.: Ontologies as a semantic model in IoT. Int. J. Comput. Appl. **42**(3), 233–243 (2020)
21. Ganzha, M., Paprzycki, M., Pawłowski, W., Szmeja, P., Wasielewska, K.: Towards common vocabulary for IoT ecosystems—preliminary considerations. In: Asian Conference on Intelligent Information and Database Systems, pp. 35–45. Springer, Cham (2017)
22. Venceslau, A., Andrade, R., Vidal, V., Nogueira, T., Pequeno, V.: IoT Semantic interoperability: a systematic mapping study. In: International Conference on Enterprise Information Systems, vol. 1, pp. 535–544. SciTePress (2019)
23. Rhayem, A., Mhiri, M.B.A., Gargouri, F.: Semantic web technologies for the internet of things: systematic literature review. Internet Things 100206 (2020)
24. Zgheib, R., Conchon, E., Bastide, R.: Semantic middleware architectures for IoT healthcare applications. In: Enhanced Living Environments, pp. 263–294. Springer, Cham (2019)

PerfectO: An Online Toolkit for Improving Quality, Accessibility, and Classification of Domain-Based Ontologies

Amélie Gyrard, Ghislain Atemezing, and Martin Serrano

Abstract Sensor-based applications are increasingly present in our everyday life. Due to the enormous quantity of sensor data produced, interpreting data and building interoperable sensor-based applications is needed. There are several problems to address the heterogeneity of (1) data format, (2) languages to describe sensor metadata, (3) models for structuring sensor datasets, (4) reasoning mechanisms and rule languages to interpret sensor datasets, and (5) applications. Semantic Web technologies (a.k.a, knowledge graphs), are immersed in an increasing number of online activities we perform today (e.g., search engines for gathering information). There is a need to find better ways to share data and distribute more meaningful and more accurate information. Innovative methodologies are needed to link and associate the data from different domains to improve knowledge discovery. Semantic knowledge graphs, made of datasets and ontologies, are intended to describe and organize heterogeneous data explicitly. If an ontology is widely used to structure data of a particular domain, the accessibility and the efficiency in sharing and reusing that information will increase. For this reason, we focused on the *ontology quality* used when building sensor-based applications. We designed PerfectO, a **Knowledge Directory Services** tool, focusing on ontology best practices, which: (1) improves knowledge quality, (2) leverages usability, accessibility, and classification of the information, (3) enhances engineering experience, and (4) promotes engineering best practices. PerfectO implementation is applied to the Internet of Things (IoT)

Thanks to the Linked Open Vocabularies (LOV) team for sharing their expertise regarding the usage of validation tools.

A. Gyrard (✉)
Kno.e.sis, Wright State University, Dayton, USA
e-mail: amelie@knoesis.org

Trialog, Paris, France

G. Atemezing
Mondeca, 35, Boulevard Strasbourg, Paris 75010, France
e-mail: ghislain.atemezing@mondeca.com

M. Serrano
Insight Center for Data Analytics, National University of Galway, Galway, Ireland
e-mail: martin.serrano@insight-centre.org

R. Pandey et al. (eds.), *Semantic IoT: Theory and Applications*, Studies in Computational Intelligence 941, https://doi.org/10.1007/978-3-030-64619-6_7

domain because it covers more than 20 application domains (e.g., healthcare, smart building, smart farm) that use sensors. `PerfectO` enhances knowledge expertise quality implemented within any ontologies as demonstrated with the Linked Open Vocabularies for IoT (LOV4IoT) ontology catalog.

Keywords Knowledge directory · Knowledge directory service · Semantic data interoperability · Ontology quality · Methodology · Web of things · Internet of things · Semantic web of things · Semantic web technologies

1 Introduction

We have been witnessing a growing number of sensors embedded in smart objects (e.g., Fitbit watches). Sensor-based applications are increasingly present in our everyday life (e.g., Mother [61] reminds medication, Apple HealthKit [62] tracks fitness, nutrition, and sleep, and Foobot measures air quality). The hardware devices, protocols, and infrastructures are already deployed. However, these devices have not been engineered by the same company nor explicitly designed to be compatible with other devices. The devices produce data that is sent to the Web to build the 'Internet/Web of Things' (IoT/WoT) applications. According to Cisco's predictions [63], there will be more than 50 billion of devices connected to the Internet by 2020. Due to the enormous quantity of sensor data generated, interpreting data and building interoperable IoT applications is needed. There are several heterogeneity issues to address (1) data format, (2) languages to describe sensor metadata, (3) ontologies to structure sensor datasets, (4) reasoning mechanisms and rule languages to interpret sensor datasets, and (5) applications. The challenge today is finding better ways to reuse more meaningful and more accurate information, to get useful abstractions from sensor data. Innovative methodologies are required to link the data from different domains to improve knowledge discovery.

Semantic Web technologies are widely used in applications and in online activities that we perform every day. Google, reused the **Knowledge Graph** (KG) [64] term in 2012, which became popular, and demonstrated the impact of semantic web technologies. Everyday, we are using KG technologies without being aware of it. Indeed, when we are looking for information (e.g., a famous person, a restaurant) using the Google search engine, structured information appears on the right. According to Paulheim's KG survey [1], a KG (1) mainly describes real-world entities and their interrelations, organized in a graph, (2) defines possible classes and relations of entities in a schema (e.g., ontologies), (3) allows for potentially interrelating arbitrary entities with each other, and (4) covers various domains. KGs are based on Linked Data [2] mechanisms implemented by semantic web standards (i.e., XML, RDF, OWL, SKOS, SPARQL, etc.) to enable: (1) large-scale data integration, and (2) reasoning on information over the Web. Linked Data structures data according to ontologies. Ontologies [3] facilitate data exchange and interoperability within applications by modeling a specific domain. Many academic and industrial communities now understand the need for ontologies.

The design, integration, and re-usability of ontologies are widely encouraged by the semantic web community. However, in other communities (e.g., IoT, Sensor Web, Semantic Sensor Networks), reusing ontologies is still a significant challenge where most of the time, similar ontologies are re-designed regularly, and not aligned with existing ontologies. There is a need to reference and classify those ontologies. Ontology catalogs such as Linked Open Vocabularies (LOV) [4], LOV for the Internet of Things (LOV4IoT) [5–7], and BioPortal for medical ontologies encourage the reuse of ontologies. However, interoperability issues remain due to the lack of proper guidance to acquire knowledge for best practice and the lack of time.

Helping ontology designers in improving ontologies is essential for the following reasons: (1) Reusing domain knowledge in other projects/domains because most of the time the knowledge is already conceived and implemented as ontologies. (2) Reducing the exhaustive job of re-inventing ontologies. (3) Reducing development time. (4) Increasing semantic interoperability among systems when machines share the same ontology. (5) Providing cross-domain interoperability, the same ontology can be used in different domains. For instance, a smart home ontology is relevant in healthcare for Ambient Assisted Living (AAL) to help elderly people to stay independent at home.

Improving the quality of knowledge representation is gaining more attention, existing surveys are: (1) knowledge graphs in [1], (2) Linked Data in [8], and (3) ontology quality in [9–12]. Fernandez-Lopez et al. [13] study the LOV ontology catalog and demonstrate that: (1) 36% of the ontologies registered in LOV are not appropriately be loaded, (2) 78% of ontologies reuse W3C endorsed ontologies, and (3) 27% of the LOV ontologies do not have English-language identifiers such as labels and comments, or within URIs. Easy-to-use ontology improvement tools are still needed to ease ontology designers' life. We focus on the ontology quality by foster a subset of tools to encourage ontology reuse.

Contributions: Our proposed solution, called `PerfectO` (stands for Perfect Ontology), relies on organizing and helps improve the existing knowledge rather than generating more. `PerfectO`'s primary objective is to increase the ontology engineer/designer experience and facilitate the ontology quality improvement process by selecting a subset of useful validation tools providing web services or online GUIs. `PerfectO` enhances ontology experience, accessibility, ontology classification, ontology interlining, and the optimization of the available ontologies. In this chapter, we use `PerfectO` implementation in the Internet of Things domain enhancing ontologies from Linked Open Vocabularies for IoT (LOV4IoT) ontology catalog as a demonstration of how `PerfectO` can be used beyond the semantic web domain. `PerfectO` LOV4IoT use case is included, and it is deployed as the proof of concept and experimental validation in other fields.

Structure of the chapter: The rest of the chapter is structured as follows: Sect. 2 highlights the need to help ontology designers in improving ontology quality, accessibility, and knowledge classification. Section 3 investigates the literature to enhance ontologies by covering complementary research topics: quality, evaluation, methodologies, and metrics. Section 4 describes PerfectO, a web-based tool to improve ontology quality and its implementation. Section 5 provides evaluation, lessons learned, and discussion. Section 6 concludes the chapter and highlights future work.

2 Improving Ontology Quality, Accessibility and Knowledge Classification

Following the above explanations, we introduce the PerfectO methodology, which promotes ontology best practices to leverage quality and usability, semantic interoperability, and classification. PerfectO can be defined by the term of *Knowledge Directory Services* [14] that we enrich with the definition as follows: (1) encouraging the reuse of knowledge through ontology catalogs, (2) disseminating best practices to improve ontology quality, (3) referencing ontology validation tools to enhance ontology usability, and (4) facilitating semantic interoperability among ontologies.

2.1 Encouraging the Reuse of Knowledge Through Ontology Best Practices

The reuse of ontologies has always been the primary problem in ontology engineering. Tools and repositories where ontologies can be indexed are limited. Existing ontology catalogs need to be widely disseminated. We take inspiration from software engineering and programming languages that define best practices when developing such as starting a class with an uppercase, providing comments, documentation, etc. Technical aspects are essential (e.g., the Java programming language community provides certification to encourage programmers to follow best practices). There are also code repositories such as GitHub and BitBucket where developer community members are invited to reuse the code.

Similarly, the design of PerfectO takes inspiration from the software engineering world to disseminate best practices for *Semantic Interoperability* [15] where there are already ontology engineering research fields working on quality such as *ontology quality*, *ontology methodology*, *ontology evaluation*, etc. PerfectO enhances self-assessment for ontology improvement from technology, and also human perspective.

2.1.1 Improving the Quality of the LOV4IoT Ontology Catalog

LOV4IoT [7, 65] is an ontology catalog referencing 499 ontology-based IoT projects (in March 2019) integrating semantic web technologies. LOV4IoT covers more than 20 application domains: robotics, smart cities, Internet of Things (IoT), Web of Things (WoT), Ambient Assisted Living (AAL), wearable computing and Wireless Body Area Networks (WBANs), agriculture, smart energy, water management, waste management, environment, logistics, manufacturing, weather, home, tourism, healthcare, etc. The healthcare domain is the main exception being familiar with semantic web best practices, but it is not the case for other communities such as the smart home. LOV4IoT aims to encourage the reuse of existing IoT ontologies. LOV4IoT extends Linked Open Vocabularies (LOV) [4] because multiple ontologies that have been designed do not meet the best practices criteria preconized by the LOV community.

The experience acquired when referencing new ontologies on the LOV4IoT ontology catalog has been used to create the `PerfectO` ontology improvement methodology, explained in Sect. 4. Only a subset of 221 ontologies is available online that are integrated with `PerfectO`. Most of the ontologies after improvements will become more interoperable to be referenced in the LOV catalog. LOV4IoT is an ontology catalog incubator for those ontologies developed by multi-domain communities.

2.2 *Disseminating Tools for Ontology Quality*

There is a real need for the proper tools to improve ontology quality, guide, and educate ontology designers/engineers. A state of the art analysis (Sect. 3) for best practices, quality, and evaluation of ontologies has been done to fill this gap. Taking inspiration from the Linked Data blog [87], we innovatively disseminate state of the art: a website which encourages semantic interoperability by classifying and referencing reusable tools that can assist the improvement of ontologies.

2.2.1 Example for Dissemination and Ontology Quality

Federated Interoperable Semantic IoT Testbeds and Applications (FIESTA-IoT)[1] is an H2020 EU project focusing on IoT semantic interoperability. To integrate new testbeds (e.g., smart cities), a set of practices needs to be followed to semantically annotate and link the data, deduce new knowledge from data, and unify models, services, and applications. FIESTA-IoT ontology [16], aims to integrate a set of IoT-related ontologies that have been studied: 19 ontologies for IoT, 28 ontologies for Wireless Sensor Networks (WSNs), 8 ontologies for the Web of Things (WoT), and 16 ontologies for smart cities. Our goal was to design a *unified IoT ontology* by reusing well-designed and most common ontologies. Furthermore, the FIESTA-IoT dataset validation tool compliant with the FIESTA-IoT ontology has been implemented [17] to check interoperability among semantic datasets. When IoT-related ontologies have been found (e.g., reading scientific publications), we have suggested the ontology URLs to the LOV ontology catalog. We have learned from the LOV community (through email exchanges) that the required practices are not followed most of the

[1] http://fiesta-iot.eu/.

time and ontologies need to be improved with the set of tools presented in Table 3. From this experience, we realize the need to disseminate those practices and tools within the IoT community [18]. In the LOV journal publication [4], note the special acknowledgment for our dissemination work.[2] Further, there is a common co-author. Evaluating ontology quality helps decide which ontologies to reuse.

2.3 Development Time Optimization and Ontology Improvement

Re-using ontologies is highly encouraged by the ontology engineering research community [19]. What the communities urgently need, is an *ontology improvement methodology* that assists novice ontology designers in improving and publishing reusable ontologies. The key benefit is to enhance ontology methodology development [20], with a focus of best practices and help in the learning curve to reduce the development time. It can be done by integrating existing tools. The main novelty is to extend our analysis [18] on improving ontologies and automate the process of the usage of the tools.

2.4 Semantic Ontology Interoperability Methodology

IoT infrastructure requires information sharing for identifying, locating, organizing, and managing everyday "things", their services, and resources. The IoT data infrastructure relies on Big Data systems to have enough capacity to store and process the amount of information collected [21]. SEG 3.0 is a methodology defined to enable data interoperability across different data systems, software platforms, and applications [22]. SEG 3.0 implements data exchange using an ontological approach and allowing access to data. `PerfectO` is a platform designed to reduce the learning curve to support the SEG 3.0 semantic interoperability. `PerfectO` (explained in Sect. 4) focuses on ontological improvement and interoperability.

2.4.1 Semantic Interoperability According to Standards

IERC AC4 (European Research Cluster on the Internet of Things) [15, 23, 24] highlighted four levels of interoperability: (1) technical, (2) syntactic, (3) semantic, and (4) organizational. Technical and syntactical interoperability were the main concerns in research and development in recent years. IERC AC4 does not reference concrete tools encouraging—(i) the best practices, (ii) the use of methodologies to ensure interoperability among ontology-based IoT applications, and (iii) reuse of the

[2] "Thanks to Amelie Gyrard for the help on the project".

domain knowledge already designed within ontologies. For this reason, `PerfectO` provides a set of concrete tools to encourage semantic interoperability and reuse improved ontologies. Standardizations are demonstrating the need to help ontology developers, and ontology quality to achieve semantic interoperability. **AIOTI Working Group 3** is dedicated to IoT standardization and has confirmed that one of the most important topics are the semantic interoperability.[3] **OneM2M, an international standard for IoT and Machine to Machine (M2M)**, is looking for the best IoT semantic interoperability practices as well [25]. **Semantic Interoperability for the Web of Things White paper** highlights the main interoperability issues [26] and citing our research work as a baseline. Furthermore, in October 2019, two "Semantic Interoperability for IoT" White Papers [27, 28] have been released and disseminated by ETSI, W3C, AOITI, etc. to guide IoT and standard developers (also from OneM2M and ISO) to reuse and develop semantics-based IoT applications easily.

3 Related Work

We review the literature in this section which is summarized in Table 1. The literature review has been introduced earlier when the *Catalogs of Tools* depicted in Fig. 9 has been described in Sect. 6. Existing surveys are classified in Sect. 3.1, ontology methodology in Sect. 3.2, ontology evaluation in Sect. 3.3, ontology metrics in Sect. 3.4, and ontology quality in Sect. 3.5. Finally, the main limitations of the related work are highlighted in Sect. 3.6.

3.1 Existing Surveys

Existing surveys are covering complementary research topics for ontology quality, evaluation, ranking, metrics, usability, and methodologies that are reviewed within this section and summarized within Table 1.

McDaniel et al. [9] provide a set of metrics to evaluate ontologies, but it can be challenging to implement them. For instance, the recognition metric computes the number of times the ontology is downloaded which is not provided by catalogs such as LOV and LOV4IoT. A second example is the lack of explanations for the metric consistency implementation. Reasoners can explain if the consistency metric is satisfied but do not provide an explicit number. McDaniel et al. design the **Domain Ontology Ranking System (DoORS)** prototype [9] to query ontology catalogs with specific keywords and assess ontology quality. Metrics are implemented to automate the selection of the ontology within the prototype. The DoORS prototype provides the following quality assessment metric modules: (1) *Syntactic*: Quality,

---

[3] https://ec.europa.eu/digital-single-market/en/alliance-internet-things-innovation-aioti.

lawfulness, richness, structure, (2) *Semantic*: quality, consistency, interpretability, precision, (3) *Pragmatic*: quality, accuracy, adaptability, comprehensiveness, ease of use, relevance, (4) *Social*: quality, authority, history, recognition, and (5) *Overall*. We are expecting more explanations for why and how those metrics have been chosen, and the meaning of the evaluation numbers provided within the prototype.

However, this excellent survey paper, published in 2018, misses important references of pioneer work to define metrics (e.g., the authors in [29] outline metrics such as the ontology competency and completeness). Furthermore, compared to their work, we collect a set of tools to improve ontologies. We took into consideration others metrics such as documentation, visualization, dissemination on ontology catalogs. Since those tools have most of the time ontology validators integrated, it enables to evaluate ontology quality at the same time.

Ma et al. [30] propose the **Ontology Usability Scale (OUS)**, a ten-item Likert scale derived from statements prepared according to a semiotic framework and an online poll in the semantic web community to evaluate ontology usability. ISO 9241-11 defines usability as follows: "the extent to which a product can be used by specified users to achieve specific goals with effectiveness, efficiency, and satisfaction in a specific context of use." The authors estimate the costs of using ontologies when developing applications to create better ontologies. An ontology may be consistent (i.e., without any contradictory assertion, complete (i.e., without any missing definition) and concise (i.e., without any unnecessary definition) but still unusable or very cumbersome to use (e.g., due to lousy documentation). The authors classify ontologies into three categories: (1) pragmatics, (2) semantics, and (3) syntax.

Raad et al. [10] investigate what makes a good ontology by analyzing ontology evaluation methods and discuss their advantages. These methods are used to evaluate the quality of automatically constructed ontologies. A good ontology can contribute to the success of semantic services and various knowledge management applications. Four categories have been designed: (1) gold standard-based, (2) corpus-based, (3) task-based, and (4) criteria-based.

Hlomani et al. design competency questions for ontology evaluation with a focus on quality and correctness as follows [11]: (1) Does the model cover required context information? (2) Does the DL language provide the logical constructs required by the reasoner? (3) Are the adaptation purposes sufficient to describe a WoT app context?, and (4) Are the adaptation purposes redundant or overlapping?

Corcho et al. survey the existing ontology methodologies [31] and answer the following questions: (1) which methods and methodologies can I use for building ontologies?, (2) which tools give support to the ontology development process, and (3) which languages can I use to implement the ontology? This work can be used as a guideline for analyzing the existing methodologies and tools for ontology engineering but need to be updated since it has been published in 2003.

3.2 Ontology Methodologies and Ontology Design Patterns (ODPs)

Ontology methodologies encourage well-designed ontologies and their reuse. Reusing ontologies is challenging. Sometimes several ontologies are using the same concept that the ontology designers need. More guidance needs to be provided to assist in the selection of ontologies. We reviewed existing *ontology methodologies* (also referenced in Table 1). The NeON [19] and Noy et al.'s [20] methodologies are the most popular ones.

The Noy et al.'s **ontology development 101 methodology** encourages ontology designers to reuse existing domain knowledge (e.g., ontologies) [20]. The methodology consists of the following iterative steps: (1) determination of the domain and scope of the ontology, (2) reuse of existing ontologies, (3) enumeration of essential terms, (4) definition of the classes and the class hierarchy, and (5) definition of the properties and creation of instances.

The **Neon** project[4] recommends reusing available knowledge and proposes a set of methodologies [32]. The Neon project focuses on nine scenarios [32]: (1) from specification to implementation, (2) reusing and re-engineering non-ontological resources, (3) reusing ontological resources, (4) reusing and re-engineering ontological resources, (5) reusing and merging ontological resources: ontology matching tools enable ontology aligning or merging, (6) reusing merging, and re-engineering ontological resources, (7) reusing ontology design pattern (ODPs), (8) restructuring ontological resources, and (9) localizing ontological resources to translate of all the ontology terms into another natural language. We are mainly interested in scenario 3 to help IoT developers in reusing ontologies relevant for IoT. The other future steps are interesting for re-designing ontologies in an interoperable manner and not "reinventing the wheel at each ontology development" to speed up the ontology development process.

On-to-Knowledge is another methodology for designing ontologies comprised of four steps: (1) kick-off, (2) refinement, (3) evaluation, and (4) ontology maintenance [33].

Ontology methodologies such as NeOn [32] and METHONTOLOGY [34] have been taken into consideration when designing PerfectO since the methodologies highly encourage to document and reuse ontologies. For instance, METHONTOLOGY does not provide a set of tools. PerfectO enriches those methodologies with a set of tools (presented in Sect. 7.3) to improve ontologies. Our approach is to reuse as much as possible the existing tools by integrating them instead of redesigning a new tool from scratch.

Ontology Design Patterns is a research topic itself, an entire book is dedicated to this topic [35].

---

[4] http://www.neon-project.org/.

Table 1 Summary of ontology methodology, evaluation, metric, and quality works

Authors, framework and publication	Year of publication	Topic	Prototype URL
McDaniel et al. [9]	2018	Ontology **Quality** Survey	✓ URL [90]
Domain ontology		Ontology **Metrics** Survey	
Ranking system (DoORS)		Ontology **Ranking** Survey	
Ma et al. [30]	2018	Ontology **Usability** Survey	×
Ontology usability scale (OUS)			
Paulheim et al. [1]	2017	**Knowledge graph**	×
		Evaluation survey	
Hitzler et al. [35]	2016	**Ontology design patterns**	✓ ODPs wiki [91]
Zaveri et al. [8]	2015	Linked **Data Quality Survey**	×
Raad et al. [10]	2015	Ontology **Evaluation** Survey	×
Poveda et al. [36]	2014-2010	Ontology **Quality**	✓ URL [68] Integrated with PerfectO
Hlomani et al. [11]	2014	Ontology **Evaluation** Survey	×
Staab et al. [33] On-to-Knowledge	2013	Ontology **Methodology**	×
Duque-Ramos et al. [37, 38] OQuaRE	2013	Ontology **Quality**	✓ URL [86]
Suarez et al. (NeON) [32]	2012	Ontology **Methodology**	✓ URL [85]
Garcia et al. [12]	2010	Ontology **Metrics** Survey	×
Fernandez et al. [39]	2009	Ontology **Quality**	×
METHONTOLOGY [34]	1997	Ontology **Methodology**	
Tartir et al. [40, 41]	2007-2005	Ontology **Quality**	×
OntoQA		3 Ontology **Metrics**	
		9 Instance **Metrics**	
Brank et al. [42]	2005	Ontology **Evaluation** Survey	×
Burton et al. [43]	2005	10 ontology **Metrics**	×
Lozano et al. [44]	2004	160 ontology **Metrics**	×
OntoMetric			
Corcho et al. [31]	2003	Ontology **Methodology** Survey	× (Survey paper)
Noy et al. [20]	2001	Ontology **Methodology**	×
Ontology development 101			
Gruninger et al. [29]	1995	Ontology **Evaluation**	×

3.3 Ontology Evaluation

There is a need for tools supporting *ontology evaluation* research approaches, also referenced in Table 1.

Gruninger et al. [29] are the pioneer working on ontology evaluation in 1995 and define the competency of the ontology as a set of questions that an ontology answers. The competency of the ontology can be evaluated by proving completeness theorems for the competency questions.

Vrandecic et al. [45] design a conceptual framework for ontology evaluation to assess the quality of an ontology for the Web. Eight criteria have been defined which are accuracy, adaptability, clarity, completeness/competency, computational efficiency, conciseness, consistency/coherence and organizational fitness/commercial accessibility. Evaluating ontologies can be done to deal with six different aspects: vocabulary, syntax, structure, semantics, representation, and context. The implementation of the work has been done within the Semantic MediaWiki, an extension of the MediaWiki which provides collaborative creation and maintenance of ontologies. Another innovative idea is the introduction of unit tests for ontologies by the same authors [46].

3.4 Ontology Metrics

Ontology metrics have been defined in [9, 11]. However, there is a need to classify and prioritize metrics. Ontology metric works are also referenced in Table 1.

OntoMetric is a methodology to choose the appropriate ontology for a specific system [44]. Five dimensions are provided: (1) the content of the ontology and its organization, (2) the language used for the implementation, (3) the development methodology employed to build the ontology, (4) the software tools used to design the ontology, and (5) the costs to develop and maintain the ontology. The OntoMetric framework defines 160 characteristics according to those five dimensions.

Burton et al. [43] define ten metrics for ontology quality: lawfulness, richness, interpretability, consistency, clarity, comprehensiveness, accuracy, relevance, authority, and history.

3.5 Ontology Quality

Tools are emerging to evaluate the *ontology quality* (also referenced in Table 1), but not enough known outside from the semantic web community. There is a real necessity to disseminate those tools and test their usability.

Fernandez et al. [39] define the term of "good" and high-quality ontology which means the easy reuse of some parts of a given ontology instead of the entire ontology.

"Parts of ontologies" refer to Ontology Design Patterns (ODPs) [47] and modular ontologies [48] research approaches (not covered in this paper).

OQuaRE is a framework for evaluating the quality of ontologies [38] based on the SQuaRE standard for software quality evaluation. A quality model and quality metrics (structural, functional adequacy, reliability, operability and maintainability) have been defined. The framework has been evaluated with units of measurement ontologies. Future work of this paper highlights the needs of automated ontology evaluation.

OntoQA [40, 41] assists ontology developers and users to determine the quality of an ontology. It provides metrics to evaluate ontology design and instances. OntoQA provides three ontology metrics: (1) *Relationship richness*: an ontology that contains many relations other than class-subclass relations is more precious than a taxonomy with only class-subclass relationships, (2) *Attribute richness*: the number of attributes that are defined for each class can indicate both the quality of ontology design and the amount of information of instance data, and (3) *Inheritance richness*: a good indication of how well knowledge is grouped into different categories and subcategories in the ontology. OntoQA defines nine instance metrics: (1) *Class richness* for KB Metrics is related to how instances are distributed across classes, (2) *Average population* for KB Metrics (average distribution of instances across all classes) indicates the number of instances compared to the number of classes. It can be useful if the ontology developer is not sure if enough instances were extracted compared to the number of classes, (3) *Cohesion* for KB Metrics can be used to indicate what areas need more instances to connect instances more closely, (4) *Importance* for class metrics provides the percentage of instances that belong to classes at the subtree rooted at the current class for the total number of instances, (5) *Fullness* for class metrics is mainly used by an ontology developer interested in knowing how well the data extraction was with respect to the expected number of instances of each class, (6) Inheritance richness for class metrics indicates how well knowledge is grouped into different categories and subcategories under this class, (7) *Relationship richness* class metrics: measures how much of the properties in each class in the schema is being used at the instance level, (8) *Connectivity* for class metrics explains which classes play a more central role than other classes, and (9) *Readability* for class metrics indicates the existence of human-readable descriptions in the ontology, such as comments, labels or captions. From our point of view, is also really relevant for doing automation with user interfaces.

OntoQA provides three kinds of evaluations: (1) evaluation-based validation, (2) symbolic-based validation, and (3) attribute-based validation. OntoQA has been evaluated with three ontologies: SWETO, TAP and GlycO, and related datasets. As a future work, the need for a web-based tool to automatically measure the quality of the ontologies is explained. However, we did not find such tools available online that we can reuse and integrate with other tools.

3.6 Limitations of the Related Work

PerfectO and **Widoco** [49] (a tool not introduced yet) have the common objective of helping developers to create more reusable ontologies. Widoco is an open-source ontology documentation standalone Java application. Widoco is not integrated yet within our PerfectO approach since it does not provide a web service to ease a fast development. Widoco focuses on ontology documentation which refers to one step in our ontology improvement methodology explained in Sect. 4.2.

The current literature focused on methodologies or validation tools (used in Table 3) to design good ontologies. We did not find any work surveying the existing web-based tools and designing the methodology which integrates current web services to support better the design of reusable ontologies. Our PerfectO approach aims to assist the LOV project by providing a web site to provide more guidance to help ontology designers fixing ontology errors encountered with the validation tools.

4 PerfectO: Architecture and Implementation

The PerfectO architecture and its components are described in this section. PerfectO demonstrates its applicability and validated experience for enhancing ontologies on a particular use case for the Internet of Things.

4.1 PerfectO Architecture

The PerfectO architecture (depicted in Fig. 1) follows interoperability principles established with the intention to exchange data with common ontologies [22]. The architecture comprises GUIs, APIs, and core components. Ontology designers interact with our web-based software using Graphical User Interfaces (GUIs). Application Programming Interfaces (APIs) are either employed by the GUI or by the ontology designers and are developed according to RESTful principles [50]. The core components of the architecture are as follows:

1. **Ontology improvement methodology** encourages ontology designers to improve the ontology (explained in Sect. 4.2). Improvement methodologies for datasets, queries or reasoning mechanisms are left for future work.
2. **Catalogs of Tools** refers to state of the art (see Sect. 3). State of the art is also shared online through clickable links and interactive mind maps[5] (introduced in Sect. 6).
3. **Dr. PerfectO Availability of Validation Tools (DPAT)** checks the availability of tools (see Sect. 7.3).
4. **PerfectO Guidance** selects a subset of validation tools (introduced in Sect. 7.3).

[5] http://perfectsemanticweb.appspot.com/?p=ontology_sota.

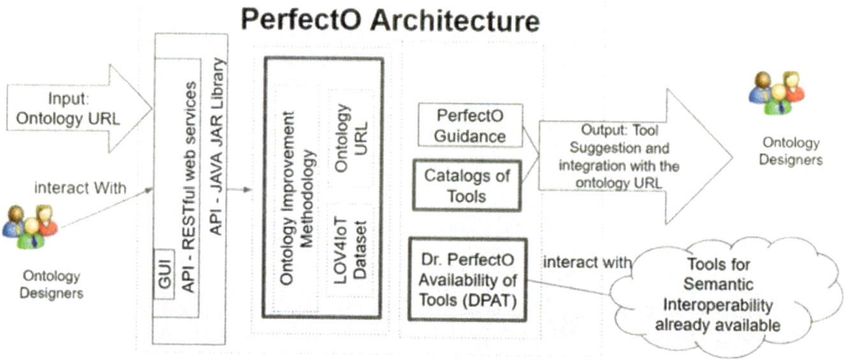

Fig. 1 PerfectO Architecture

Technologies used for the implementation: The RESTful web services have been implemented in Java using the Jersey web service library. For instance, the web service[6] returns the result status OK or NOT OK. The URL parameter is a web service pinging the tools mentioned in Sect. 7.3. The web services are queried using AJAX technology. The results returned are parsed in JavaScript and displayed in the HTML web pages. We highly encourage the readers to browse the PerfectO web site[7] to check out the different modules already implemented.

4.2 *Ontology Improvement Methodology*

To reduce the learning curve and time-consuming development task of designing reusable ontologies, we conceived the *ontology improvement methodology* (as depicted in Fig. 2). This methodology evaluates IoT and smart city ontologies (see section Evaluation [5]). The methodology comprises the following steps; some of them are interchangeable in their position:

1. **Syntactic validation** is necessary during the compilation, and the execution of ontologies with libraries to be processed by the *ontology quality methodology*. Tools such as OWL Manchester and Triple Checker can be used.
2. **Serialization** supports the OWL ontology format since it is a W3C recommendation.
3. **Interlinking** enhances interoperability, integration, and browsing among ontologies. Ontology matching tools such as LogMap[8] can be employed.
4. **Documentation** eases ontology understandability. Parrot and LODE have been chosen since web services for automatic documentation are provided. More and

[6] http://perfectsemanticweb.appspot.com/perfecto/statusTool/?url={url}.

[7] http://purl.org/perfecto.

[8] https://www.cs.ox.ac.uk/isg/projects/LogMap/.

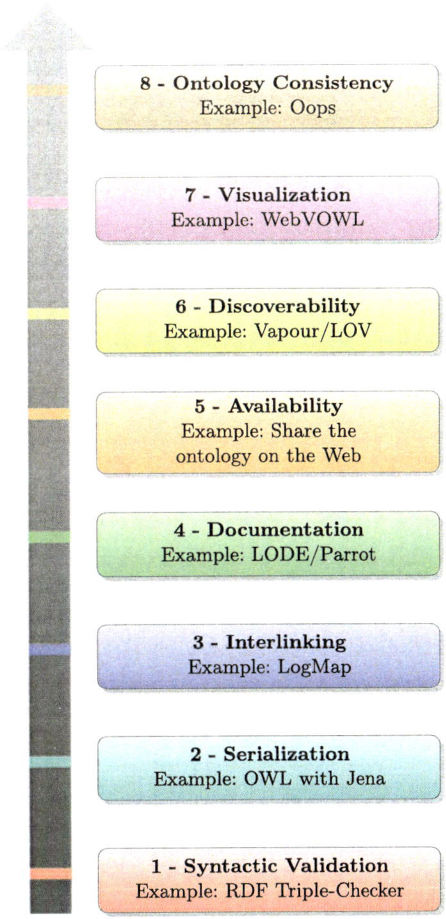

Fig. 2 PerfectO's ontology improvement check-up levels

more tools are being designed to provide such criteria to ease the task of developers (e.g., Widoco [49]). We designed the ontology documentation mindmap[9] to give an overview of existing ontology documentation tools.

5. **Availability** advocates sharing the resource on the web. Developers do not have time, resources, or administrative skills to manage the server. Ideally, an ontology catalog server hosts any ontologies and provide the right URL. Sharing the ontology code and documentation on the web encourage ontology reuse.
6. **Discoverability** improves the dissemination of ontologies. For instance, suggesting ontologies on ontology catalogs and semantic search engines support the dissemination. Several ontology catalogs are available which depends on the application domain. Ideally, each ontology provides dereferenceable URI and can be tested with the Vapour tool or the Curl command line.

[9] http://perfectsemanticweb.appspot.com/?p=ontology_sota#div_ontology_documentation_mindmap.

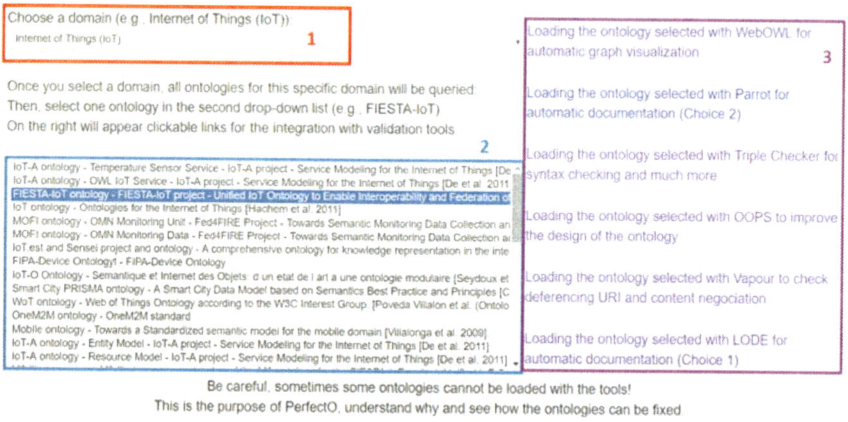

Fig. 3 The PerfectO ontology improvement methodology implemented and integrated with the LOV4IoT ontology catalog

7. **Visualization** eases the learning phase by providing a fast understanding of the ontology. WebVOWL tool is integrated to provide an automatic ontology graph visualization.
8. **Ontology Consistency** can be improved with the Oops! tool which detects numerous ontology pitfalls and suggests how to avoid/fix them.

PerfectO Ontology Improvement Tool Integrated with LOV4IoT. We have implemented the *ontology improvement* component[10] by integrating six tools (TripleCkecker, Oops!, Parrot, Vapour, LODE, and WebVOWL) with the ontologies referenced within the LOV4IoT RDF dataset.[11]

Figure 3 demonstrates the implementation of the PerfectO *Ontology Improvement* methodology (explained within the Architecture Sect. 4.1) integrated with the LOV4IoT ontology catalog. Figure 3 comprises three main parts:

1. A drop-down list is shown with all IoT applicative domains referenced within the LOV4IoT ontology catalog (more than 20 domains) introduced in Sect. 2.1.1. Once a domain is selected, the second drop-down list is filled in. In this example, the selected domain is IoT.
2. A second drop-down list provides all ontologies referenced within LOV4IoT for a specific domain. In this example, all ontologies for IoT have a URL which is shown within the tooltip.
3. Once a particular ontology is selected, tools mentioned above in Sect. 7.3 are automatically integrated. The integrated tools appear on the right part of the screen. Clickable links are provided to go to the web service tools with the selected ontology.

[10] http://perfectsemanticweb.appspot.com/?p=ontologyValidationLOV4IoT.

[11] http://purl.org/lov4iot-dataset.

Enter the URL of your ontology:

E.g. http://purl.org/iot/ontology/fiesta-iot# or https://w3id.org/seas/ or https://www.irit.fr/recherches/MELODI/ontologies/IoT-O

http://purl.org/iot/ontology/fiesta-iot#

Click on the button, to check your ontology with validation and improvement tools below

Click here to improve your ontology with all of those tools: WebVOWL, Parrot, TripleChecker, Oops, LODE, Vapour, etc.

Sometimes some ontologies cannot be loaded with those tools! This is the purpose of PerfectO, understand why and see how the ontologies can be fixed. For your information, the previous version of this tool is still available here

Name Tool and Logo	Tool Classification	Tool Description	Tool availability status	Integration with this tool	Status on PerfectO comments
WebVOWL	Visualization	WebOWL provides an automatic graph visualization	OK	Your ontology integrated with WebVOWL here!	Tool already here in V1
RDF Triple-Checker	Syntactic Validation	Triple Checker provides syntax checking and much more	OK	Your ontology integrated with TripleChecker here!	Tool already here in V1
	Documentation	LODE provides an automatic documentation (Choice 1)	OK	Your ontology integrated with LODE here!	Tool already here in V1
VAPOUR	Discoverability	Vapour checks deferencing URI and content negociation	OK	Your ontology integrated with VAPOUR here!	Tool already here in V1

Fig. 4 The PerfectO ontology improvement methodology implemented using an ontology URL—Part I

PerfectO Ontology Improvement Tool with an Ontology URL. A more sophisticated interface for ontology validation has been designed where the designers directly enter the ontology URL in case the ontology is not referenced yet within the catalog. Figures 4 and 5 show an extract of the online interface[12]

1. The user enters a specific ontology URL or can use the one by default.
2. The user clicks on the button to automatically integrate the ontology with the tools mentioned in Sect. 7.3. A table is created to integrate the tools with the given ontology automatically. The tools are accessible through the clickable links thanks to their web service. The generated table enables quick access to any tools to improve the ontology.

Figures 4 and 5 show a table that provides the name of the tool. The tool classification is provided in the second column, and the classification is introduced in our proposed PerfectO methodology explained in Sect. 4. The third column describes the usage of the tool. The fourth column checks the availability of the web service.

[12] http://perfectsemanticweb.appspot.com/?p=ontologyValidation.

oops!	Ontology Consistency	OOPS improves the design of the ontology	OK	Your ontology integrated with OOPS here!	Tool already here in V1
	Documentation	Parrot provides automatic documentation (Choice 2)	#NOT OKAY	Your ontology integrated with Parrot here!	Parrot hosted on Mondeca server but takes time Be patient!
	Discoverability	Linked Open Vocabularies (LOV) is an ontology catalog Suggesting your ontology on ontology catalogs will improve its dissemination	OK	Your ontology integrated with LOV SUGGEST here!	Tool added in V2 - April 2018
BioPortal	Discoverability	BioPortal is an ontology catalog for healthcare Suggesting your ontology on ontology catalogs will improve its dissemination	OK	Your ontology integrated with BIOPORTAL here!	Tool added in V2 - April 2018 Need login and password for automatic integration
LOV4IOT	Discoverability	Linked Open Vocabularies for Internet of Things (LOV4IoT) is an ontology catalog for Internet of Things (IoT) which includes more than 20 applicative domains (smart cities, smart home, robotics, environment, etc.) Suggesting your ontology on ontology catalogs will improve its dissemination	OK	Your ontology integrated with LOV4IoT here!	Tool added in V2 - April 2018
LogMAp	Interlinking	LogMap is an ontology matching tool	OK	Your ontology integrated with LOGMAP here!	Tool added in V2 - April 2018 Need name, institution and email for automatic integration

Fig. 5 The PerfectO ontology improvement methodology implemented using an ontology URL—Part II

The fifth column integrates the ontology to improve with the specific tool. The last column provides some comments regarding the integration of the tools (e.g., some maintenance issues are known, new tools recently added).

4.3 Use Cases

Semantic Interoperability for the Web of Things White paper highlights the main interoperability issues [26] and citing our research work as a baseline. Furthermore, in October 2019, two "Semantic Interoperability for IoT" White Papers [27, 28] have been released and disseminated by ETSI, W3C, AOITI, etc. to guide IoT and standard developers (also from OneM2M and ISO) to reuse and develop semantics-based IoT applications easily. The FIESTA-IoT ontology [16] has been employed within the FIESTA-IoT H2020 EU project involving 14 academic and industrial partners, it enables to process data generated by sensor within smart cities. The LOV4IoT dataset is used within experiments to rank the popularity of IoT ontologies in [51] and consider the following metrics: availability, believability, understandability, interlinking, PageRank, consistency, and richness. PerfectO is also refined and applied within the IEEE Autonomous robotics ontology community [52], and for well-being and affective science with the design of a recommender system for happiness which encourages the design and reuse of ontologies for emotion, food, obesity, fitness, sleep, stress, depression, acupuncture, etc. [53].

4.4 Limitations

The current demonstrators have some limitations. The current Graphical User Interfaces (GUI) can be more user-friendly. The demonstrators can be extended with additional tools such as the use of ontology matching tools. The overall PerfectO demonstrator is hosted on the Google Platform to avoid maintenance and security server issues. However, latency issues can appear to load the website and query web services. Sometimes, we notice that some tools are not maintained anymore. However, we regularly update our demonstrators. For instance, Parrot has been duplicated and hosted on another server by Mondeca Labs.

Ideally, in the future, PerfectO distinguish four types of resources:

- An *Ontology* is a set of concepts and relationships between concepts to describe a specific domain. We designed the *Ontology Improvement* methodology explained in Sect. 4.2 which is the main focus of this paper.
- A *Dataset* is structured according to ontologies.
- A *Reasoning* or inference engine executes a set of logical rules (e.g., IF THEN ELSE deductive rules) compliant with the ontologies to extract meaningful information from a dataset.
- A *Query* can be a SPARQL query compliant with the ontology and the dataset, used to retrieve a subset of the dataset required to design a specific application.

5 Evaluation, Lessons Learned and Discussions

The PerfectO approach is evaluated with a set of ontologies referenced by the LOV4IoT catalog in Sect. 5.1. Evaluation criteria defined by the Semantic Web community are reminded in Sect. 5.2. Discussions including lessons learned are provided in Sect. 5.3.

5.1 Ontology Quality Evaluation

We complete a detailed evaluation with **26 IoT or smart cities ontologies** referenced within the LOV4IoT ontology catalog (summarized in Table 4 and accessible online[13]). However, the PerfectO approach is generic enough to be applied to any ontologies (e.g., from the LOV ontology catalog). Those ontologies are tested with **six ontology improvement tools** (Parrot, WebVOWL, Oops, TripleChecker, LODE, and Vapour) mentioned in Table 3. Numerous ontologies cannot be successfully loaded with all of the tools and show that multiple errors are encountered. The LODE tool is preferred because more ontologies can be automatically documented, as compared to the Parrot tool. Furthermore, in March 2019, the Parrot automatic documentation web service does not seem maintained anymore.

As an example, a set of errors encountered with those tools:

- **[Parrot]** Unable to read input document: invalid mimeType "application/octet-stream" (returned by URI) for parrot.
- **[Parrot]** No error encountered when loaded the ontology, but nothing is displayed.
- **[Parrot]** I/O Error: Server returned HTTP response code: 403 for URL.
- **[Parrot]** Unable to read input document: applicationrdfxml parse error: Content is not allowed in prolog.
- **[TripleChecker]** Misuse of terms from Dublin Core namespace and date format with TripleChecker.
- **[TripleChecker]** ERROR: VERY close match to "license."
- **[TripleChecker]** Error loading—No parser available.
- **[WebVOWL]** ERROR "There is nothing to visualize."
- **[OOPS]** OOPS Pitfall P36: URI contains file extension.
- **[OOPS]** OOPS Pitfall P37: Ontology not available on the Web.
- **[LODE]** Reason: An empty sequence is not allowed as the value of variable $rdf.
- **[LODE]** Reason: A sequence of more than one item is not allowed as the @select attribute of xsl:sort.
- **[LODE]** Reason: org.xml.sax.SAXParseException; lineNumber: 1; columnNumber: 1; Content is not allowed in prolog.
- **[LODE]** The source can't be downloaded in any permitted format. Connection reset Connection reset Connection reset Connection reset Connection reset Connection reset.
- **[LODE]** The source can't be downloaded in any permitted format. Received fatal alert: protocol_version.
- **[LOV]** Bad IRI: http://example.com Code: 57/REQUIRED_COMPONENT_MISSING in HOST: A component that is required by the schema is missing.

Some errors are more complicated to fix such as the server configuration. Also, when the syntax is checked with tools such as OWL Manchester validators, errors still arise when loading the ontologies.

[13] http://perfectsemanticweb.appspot.com/?p=evaluation_lov4iot_perfecto.

5.2 Semantic Web Community Evaluation Criteria

To leverage and identify value for our PerfectO methodology, we discuss the evaluation criteria as outlined by the International Semantic Web Conference (ISWC) conference and their call for resources (e.g., ontologies) track papers [54].

Impact Criteria. PerfectO has a significant impact outside of the semantic web community. It has been mainly applied to the IoT community but is generic enough to be applied in other domains. As demonstrated in Sect. 4.2, PerfectO has been used with the LOV4IoT ontology catalog which covers more than 20 applicative domains (introduced in Sect. 2.1.1). It is the first innovative platform for semantic interoperability and disseminates best practices and existing tools as a set of catalogs following "Knowledge Directory Services" approaches. Our first step is to focus on ontology interoperability and improvement. Instead of saying that our tool is better than existing ones, we exploit their potential when they already provide web services. Our literature review shared in a structured way on the web through FAQs and mindmaps explained in Sect. 6 is also useful for teaching, tutorials, etc. We have in mind to have a similar approach for datasets, reasoning, and querying, as explained in Sect. 4.1. We are aware that researchers can consider our methodology as not innovative since we reuse and integrate tools, but we are following software engineering research methodologies [14].

Availability Criteria. The PerfectO software is under the GNU GPLv3 license. A GitHub repository has been initiated https://github.com/perfectkb/perfecto which stands for perfect knowledge bases. The LOV4IoT catalog has been designed to extend the LOV ontology catalog since LOV inserts only new ontologies when best practices are followed. Due to this experience, we realized that the learning curve to improve ontologies and use the tools needs to be reduced and disseminated within other communities (e.g., IoT community). This work is aligned with our current activities at improving ontology quality in any projects designed at the Knoesis research center until 2020. PerfectO is hosted on Google Application Engine to avoid any server and DNS maintenance related issues.

Reusability Criteria. We provide a "Semantic Web Best Practices for Dummies" documentation [88] to improve ontologies. More than 200 ontology URLs have been referenced with the LOV4IoT dataset which has been integrated with PerfectO. PerfectO can be applied to any ontology URL. It provides online tools; no set up is required which reduces the development time. It can be extended with more tools to improve ontologies. For instance, we can integrate ontology matching tools such as LogMap, and ontology catalogs such as BioPortal and LOV to automatically suggest a new ontology. We are aware of the current limitations of PerfectO: some tools cannot load some ontologies. We try to understand why to help beginners solve the issues and report the solutions within the documentation.

Design and Technical Quality Criteria. We classify, reference, synthesize, vulgarize and disseminate best practices learned from the semantic web community to other communities (e.g., IoT). We reuse existing tools for automatic visualization, documentation and ontology quality as explained in Sect. 7.3.

5.3 Discussions and Lessons Learned

Lessons Learned. We shared the lessons learned to fix the issues to improve the ontologies with the tools suggested by PerfectO (see Table 3) in the "Semantic Web Best Practices for Dummies" documentation mentioned above. A summary is presented as a set of rules in Table 2, where we synthesize 16 rules to disseminate best practices. For each rule, we provide examples of bad practices and best practices to help beginners in their learning journey in a set of slides entitled "Step-by-step tutorial to improve the ontology quality, dissemination, reuse" [89]. This work is an enhancement of our previous work [18]. We are highly encouraging validation tools to provide REST APIs. However, REST APIs need to be maintained to ease PerfectO development. PerfectO's main limitation is the usage of external web services which can be offline.

Table 2 Ontology best practices: check list summary

Rule number	Description	Difficulty
Rule 1	Finding a good ontology name	*
Rule 2	Finding a good ontology namespace	**
Rule 3	Sharing your ontology online	**
Rule 4	Adding ontology metadata	**
Rule 5	Adding rdfs:label, rdfs:comment, dc:description for each concept and property	*
Rule 6	All classes start with an uppercase and properties with a lowercase.	*
Rule 7	Submitting your ontology to ontology catalogs	**
Rule 8	Reusing and linking ontologies	***
Rule 9	Deferenceable URI copy paste the namespace URL of your ontology in a web browser to get the code	**
Rule 10	Checking syntax validator	*
Rule 11	Adding ontology documentation	*
Rule 12	Adding ontology visualization	*
Rule 13	Improving Ontology Design	***
Rule 14	Improving dereferencing URI and content negotiation	***
Rule 15	Ontology can be loaded with ontology editors (e.g., Protege)	**
Rule 16	Registering your ontology on prefix catalogs	*

Discussions. Ontologies designed within projects (e.g., European projects) are more impactful, since they are used by different partners and for real use cases. For instance, the FIESTA-IoT ontology has been employed within the FIESTA-IoT H2020 EU project involving 14 academic and industrial partners. However, the major issue of European projects is the lack of ontology maintenance once the project is finished.

PerfectO improves ontology quality. For instance, in July 2019, best practices have been disseminated to the robotics community: IEEE P1872.2 Standard for Autonomous Robotics (AuR) ontology.

6 Conclusion and Future Work

Designing good quality ontologies is often neglected due to the lack of knowledge regarding the best practices. Getting familiar with ontology quality tools and best practices requires a steep learning curve and a lot of effort (e.g., for the IoT community). `PerfectO` assists ontology designers in this process towards improving ontologies to be reused in other projects. `PerfectO` selects and classifies a sub-set of tools providing an online interface or a web service simple to use. Those tools help to enhance ontologies and synthesize a set of practices. It has a significant impact since it has been designed to be applied to the LOV4IoT ontology catalog which covers more than 20 application domains.

As future work, we plan to use the lessons learned from the evaluated ontologies to fix the ontology automatically and provide a ranking system to suggest the ontology fitting the ontology developer's needs.

Acknowledgements This work has partially received funding from the European Union's Horizon 2020 research and innovation programme under grant agreement No 857237 (Interconnect), Science Foundation Ireland (SFI) under Grant Number SFI/12/RC/2289, Insight Centre for Data Analytics and H2020 FIESTA-IoT-CNECT-ICT-643943. The opinions expressed are those of the authors and do not reflect those of the sponsors.

7 Appendix

7.1 *Catalogs of Tools*

The PerfectO web site provides the *Catalogs of tools* menu (depicted in Fig. 8) to easily give access to software URLs and publications. The *Catalogs of tools* references tools for (1) ontology improvement which is the focus of this paper, (2) dataset quality, (3) querying, and (4) reasoning. Figure 9 demonstrates an extensive literature survey within the table of contents classified in two different ways:

- A bullet list referencing tool's URLs and scientific publications are displayed simply with clickable links (see Fig. 6 as an example).

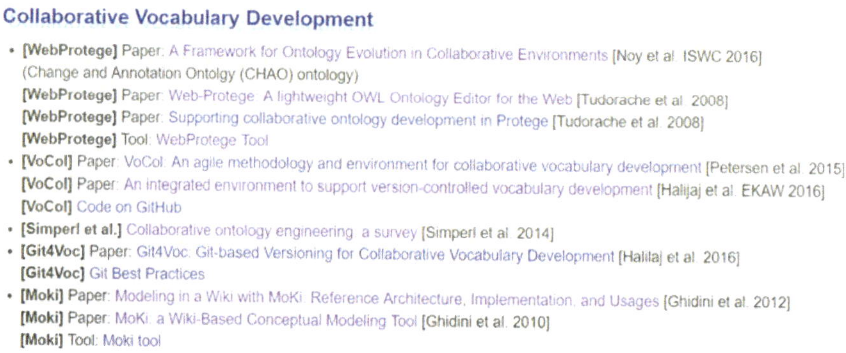
Collaborative Vocabulary Development

- [WebProtege] Paper: A Framework for Ontology Evolution in Collaborative Environments [Noy et al. ISWC 2016] (Change and Annotation Ontolgy (CHAO) ontology)
  [WebProtege] Paper: Web-Protege: A lightweight OWL Ontology Editor for the Web [Tudorache et al. 2008]
  [WebProtege] Paper: Supporting collaborative ontology development in Protege [Tudorache et al. 2008]
  [WebProtege] Tool: WebProtege Tool
- [VoCol] Paper: VoCol: An agile methodology and environment for collaborative vocabulary development [Petersen et al. 2015]
  [VoCol] Paper: An integrated environment to support version-controlled vocabulary development [Halilaj et al. EKAW 2016]
  [VoCol] Code on GitHub
- [Simperl et al.] Collaborative ontology engineering: a survey [Simperl et al. 2014]
- [Git4Voc] Paper: Git4Voc: Git-based Versioning for Collaborative Vocabulary Development [Halilaj et al. 2016]
  [Git4Voc] Git Best Practices
- [Moki] Paper: Modeling in a Wiki with MoKi: Reference Architecture, Implementation, and Usages [Ghidini et al. 2012]
  [Moki] Paper: MoKi: a Wiki-Based Conceptual Modeling Tool [Ghidini et al. 2010]
  [Moki] Tool: Moki tool

Fig. 6 Literature survey for collaborative vocabulary development with clickable links

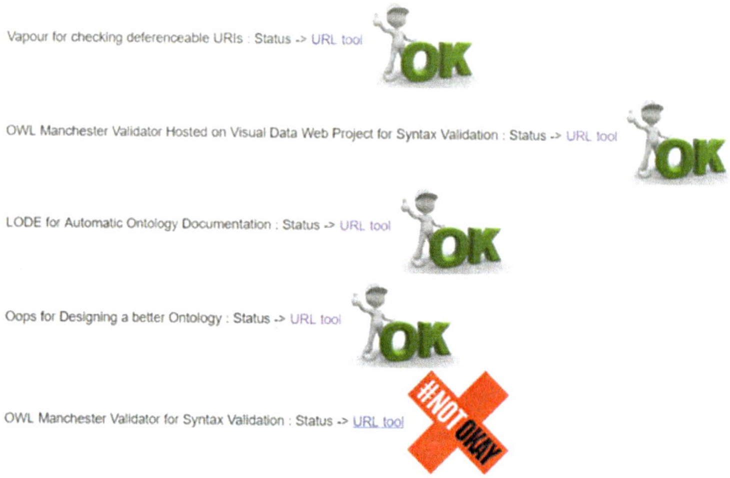

Fig. 7 Dr. PerfectO availability of tools (DPAT)

- Mind maps [55] are recognized as a useful methodology and a powerful graphics technique used to translate what's in the mind into a visual picture. Since mind mapping works like the brain, it allows us to organize and understand information faster and better. We designed mind maps to answer Frequently Asked Questions (FAQs). Figure 11 shows the Mind maps for ontology catalogs. Figure 12 illustrates the Mind maps for ontology methodologies. Figure 13 illustrates the Mind maps for ontology validation.

The catalogs of tools are available online[14]; it covers numerous research domains: (1) ontology documentation, (2) ontology catalogs and semantic search engines, (3) ontology methodologies, (4) ontology validators, (5) ontology validators for IoT, (6) ontology visualization, (7) collaborative vocabulary development, (8) ontology

[14] http://perfectsemanticweb.appspot.com/?p=ontology_sota.

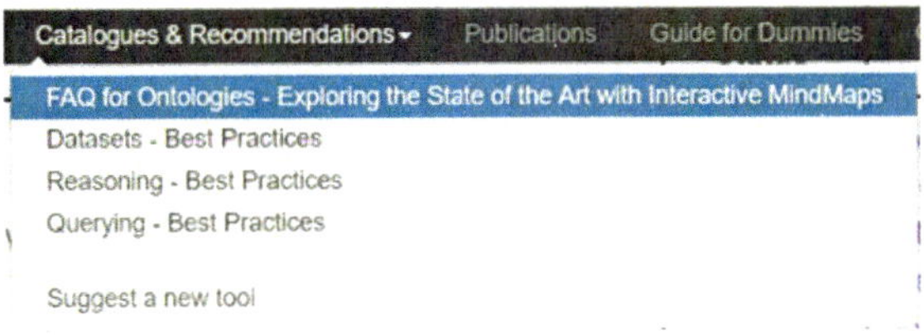

Fig. 8 Catalogs of tools

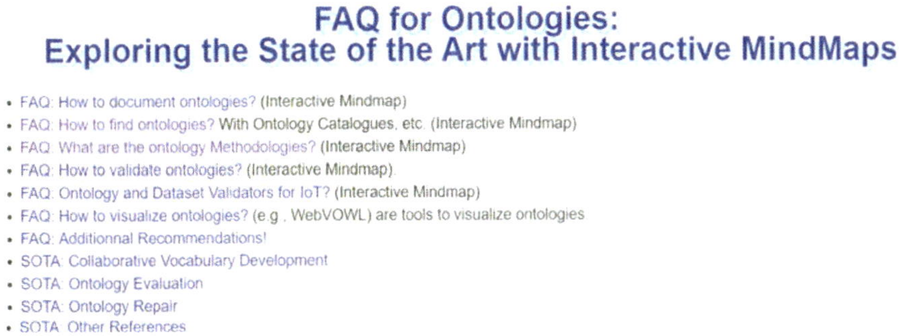

Fig. 9 State of the art classifying ontology improvement tools

evaluation, and (9) ontology repair. Extensive work has been done in ontology documentation (not covered in the Related Work Section) and structured within a mind map, but referenced by the *Catalogs of Tools* as displayed in Fig. 10. We guide ontology designers with LODE and Parrot tool since web services are provided, as shown in Sect. 7.3 and Table 3. We cover some of those topics (ontology methodology, ontology evaluation, ontology metrics, quality, and relevant tools) in Sect. 3. Ontology designers can contribute to enrich the set of *Catalogue of Tools* by using a Google Form interface.[15]

7.2 *Dr. PerfectO Availability of Tools (DPAT)*

DPAT tool (depicted in Fig. 7 and introduced in Sect. 4.1), online http://perfectsemanticweb.appspot.com/?p=availability_tools, checks the availability of reusable tools in case the server is down. Each row provides: (1) the name of the tool, (2) its usage, (3) the clickable tool's URL, and (4) the tool availability (displayed as images: OK or NOT OKAY). For instance, the LODE ontology documentation web service runs well when DPAT has been deployed.

[15] http://perfectsemanticweb.appspot.com/?p=updateCatalogueForm.

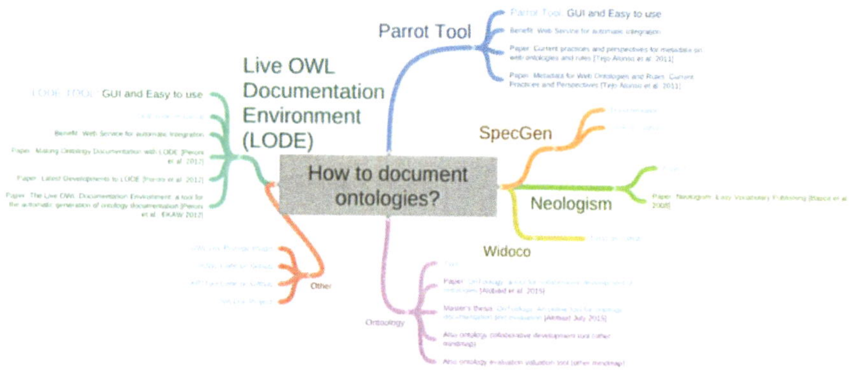

Fig. 10 Mind map classifying ontology documentation tools

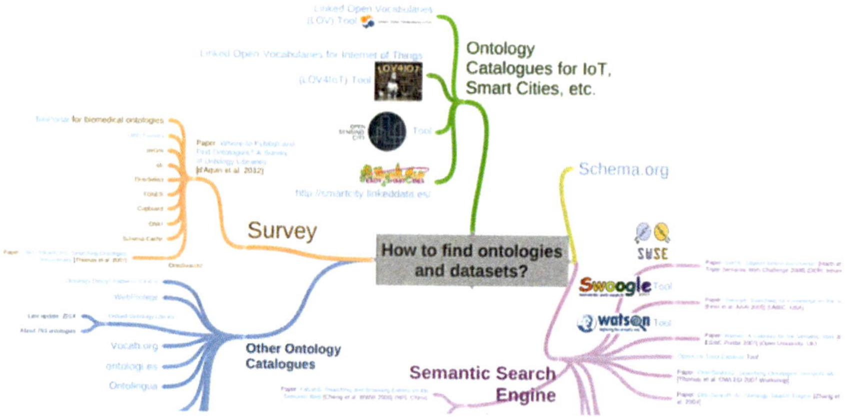

Fig. 11 Mind map classifying ontology catalog tools

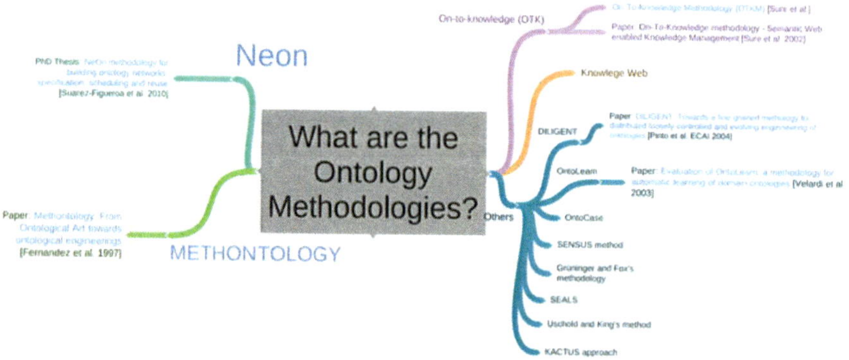

Fig. 12 Mind map classifying ontology methodologies

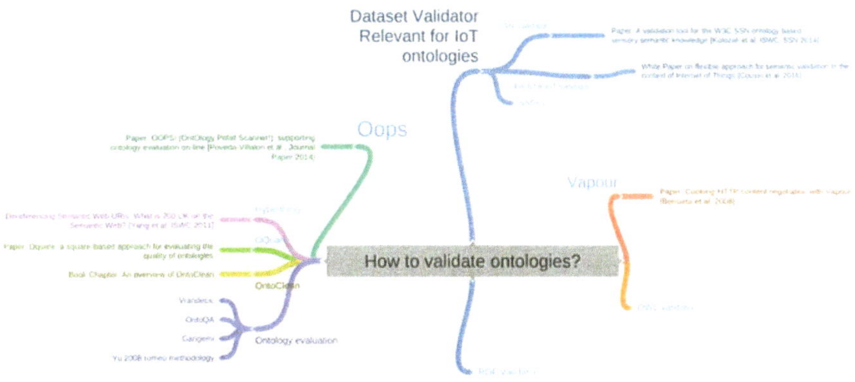

Fig. 13 Mind map classifying ontology validation tools

Table 3 Reusable tools for the ontology improvement

Tool name and publication	Validation requirement	GUI	Web service	Code availability	Maintained
Jena [56]	Serialization Syntactic	[66]	✓ API	✓ GitHub Java code [67]	✓ High
Oops [36]	Ontology Consistency	[68]	✓ [69]	×	✓ High
Triple Checker	Syntactic	[70]	✓	✓ GitHub PHP code [71]	✓ High
LOV [4]	Discoverability	[72]	✓	✓ Github Back end—Java [73] GUI—Javascript [74]	✓ High
Parrot [57]	Documentation	[75]	✓	✓ Bitbucket Java [76]	✓ Hosted by Mondeca now
LODE [58]	Documentation	[77]	✓	✓ Java on GitHub [78]	✓
WebVOWL [59]	Visualization	[79]	✓	✓ GitHub JavaScript [80]	✓ High
Vapour [60]	Discoverability	[81]	✓ Java JavaScript API [83]	✓ [82]	✓ Medium
OWL Manchester	Syntactic	[84]	×	×	✓ Medium

7.3 PerfectO Guidance: The Most Accessible Tools for Ontology Engineering

The learning curve for software engineering can be extremely high when developers are not familiar with the same programming languages and libraries used to build the tools. A set of tools that can be considered <easy-to-use> if: (1) The tools provide GUIs and web services, (2) the documentation is available and well-explained, (3) the ontologies can be evaluated with tools offering diverse functionalities, and (4) software setup configuration is not required.

The classification of tools that we have selected is available within Table 3 to support the *Ontology Improvement* methodology (explained in Sect. 4.2). The table is a way to organize the multiple tested technologies and if there is an available source for documenting it. For instance, WebVOWL tool can be used to provide automatic ontology graph visualization, Parrot for automatic documentation, etc. In the table, the first column is dedicated to the tool name, and scientific publication is available. The second column explains the requirement satisfied. The third column provides the GUI interface URL. The fourth column indicates the web service or API if available. The fifth column contains the code URL if accessible. The sixth column explains the maintainability of the tools. The web services are more convenient to

Table 4 Evaluation: IoT ontologies with tools for ontologies

Ontology	Domain	Last Tested	TripleChecker (Syntax)	Vapour (URIs)	Oops (Design)	WebVOWL (Visualization)	Parrot (Documentation)	LODE (Documentation)
SEAS ontology root	Energy	5/30/2017	OK Ontology loaded	Success 6/14 passed	Error OOPS, something went wrong	Err - mail sent 5/30/2017	The ontology is loaded but no documentation provided. Email 15 May Answered 15 May	Ontology loaded but not really well documented
SSN V1	IoT/SSN	5/15/2017	Ontology loaded, one minor error (lice…	Success: 10/14 tests passed	Critical Pitfalls P10 …	Error: There is nothing to visualize	It works. The ontology is loaded and the documentation is provided	OK
SSN V2	IoT/SSN	5/15/2017	Ontology loaded, no error	Success: 4/9 tests passed	Important pitfalls	Error: There is nothing to visualize	It works. The ontology is loaded and the docu…	LODE error Reason: org.xml.sax.SAXParseException; lineNumber: 1; columnNumber: 1; Content is not allowed in prolog.
SSN V2 SOSA	IoT/SSN	5/15/2017	Ontology loaded, no error	Success: 4/9 tests passed	Important pitfalls	Error: There is nothing to visualize	It works. The ontology is loaded and the docu…	LODE error Reason: org.xml.sax.SAXParseException; lineNumber: 1; columnNumber: 1; Content is not allowed in prolog.
IoT-lite	IoT/SSN	5/15/2017	Ontology loaded, one minor error (license) to fix	Success: 6/9 tests passed	Critical pitfalls	OK	It works. The ontology is loaded and the documentation is provided	OK
IoT-o Seydoux	IoT/SSN	5/12/2017	Ontology loaded, minor errors (license + literals) to fix	Success: 12/19 tests passed	Critical Pitfalls	Ontology cannot be loaded. Error: "There is nothing to visualize"	The ontology is loaded but no documentation provided. Email 15 May Answered 15 May	OK
SAN Seydoux	IoT/SSN	5/12/2017	Ontology loaded, one minor error (license) to fix	Success: 8/9 tests passed	Critical Pitfalls	Ontology cannot be loaded. Error: "There is nothing to visualize"	It works. The ontology is loaded and the documentation is provided	OK
M3 ontology	IoT/SSN	5/30/2017	NO ontology loaded - ERROR loading no parser available for""	Failed: 5/9 passed	Minor pitfalls	Error: There is nothing to visualize - mail sent 5/23/2017	Unable to read input document; invalid mimeType "application/octet-stream" [returned by URI …] for parrot Email 15 May Answered 15 May	LODE error Reason: A sequence of more than one item is not allowed as the @select attribute of …
M3-lite	IoT/SSN	5/30/2017	Ontology loaded - Minor errors	Success 6/14 passed	Critical Pitfalls	OK	OK	OK
FIESTA-IoT	IoT/SSN	5/30/2017	Ontology loaded - Minor errors	Success 11/14 passed	Critical Pitfalls	OK	OK	OK
WoT	IoT/SSN	5/30/2017	OK Ontology loaded	Failed 5/9	Important Pitfalls	OK	OK works with owl extension	Ontology loaded but not really well documented
WoT	IoT/SSN	5/30/2017	OK Ontology loaded	Success 10/11	Important Pitfalls	OK	Otherwise Error: Unable to read input document: application/n-triples parse error: Content is not allowed in prolog.	Ontology loaded but not really well documented
OpenIoT	IoT/SSN	5/30/2017	OK Ontology loaded	Failed 5/9	Minor pitfalls	OK	OK Parrot seems better than LODE	Ontology loaded but not really well documented
VITAL	IoT/SSN	5/30/2017	Ontology loaded but numerous errors	Failed 5/9	Oops Critical Pitfalls	OK	The ontology is loaded but no documentation provided.	LODE error Reason: An empty sequence is not allowed as the value of variable …
OneM2M	IoT/SSN	5/31/2017	OK Ontology loaded	Success: 2/9 passed	Minor Pitfalls P36 file extension	Error nothing to visualize	I/O Error: Server returned HTTP response code: 400 for URI; Email 15 May Answered 15 May	OK
SAREF	IoT/SSN	5/31/2017	Ontology loaded - Minor errors	Success: 9/19 passed!	Critical Pitfalls	OK	The ontology is loaded but no documentation provided.	OK LODE better than Parrot
IoT Hachem	IoT/SSN	5/12/2017	Ontology loaded - Minor errors	Failed: 5/9 tests passed	Minor Pitfal. P36. URI …	Ontology cannot be loaded. Error: "There is nothing to visualize"	OK	OK
spitfire	IoT/SSN	5/15/2017	Ontology loaded, some errors to fix	Failed: 5/9 tests passed	Critical pitfalls	OK	Error: Unable to read input document: application/rdf+xml parse error: Content is not allowed in prolog. Email 15 May Answered 15 May	LODE error Reason: org.xml.sax.SAXParseException; lineNumber: 1; columnNumber: 1; Content is not allowed in prolog.
SAO	IoT/SSN	5/18/2017	OK Ontology loaded	Success 6/14 passed	Critical Pitfalls	OK	The ontology is loaded but no documentation provided.	OK
kotis	IoT/SSN	5/16/2017	OK Ontology loaded	Success 11/14 passed	Critical Pitfalls P38 …	Ontology cannot be loaded. Error: "There is …	The ontology is loaded but no documentation …	Nothing happened?
[illegible]	IoT/SSN	5/16/2017	OK Ontology loaded	Success 12/14 passed	Critical pitfalls	It works	It works	OK
Hypercat	IoT/SSN	5/17/2017	OK Ontology loaded	Failed: 5/9 tests passed	Critical Pitfall (P3?) Ontology not available on the Web	It works	Error: Unable to read input document: I/O Err…	Reason: The ontology can't be download…
OpenSensingCity Weather		5/18/2017	Ontology loaded - Minor errors to fix …	Success: 8/9 tests passed	Critical Pitfalls (P40: Nam…	It works	It works	Reason: org.xml.sax.SAXParseExcep…
OpenSensingCity Bike		5/18/2017	Ontology loaded - Minor errors to fix …	Success: 8/9 tests passed	Critical Pitfalls (P40: Nam…	It works	It works	Reason: org.xml.sax.SAXParseExcep…
OpenSensingCity Parking		5/18/2017	Ontology loaded - Minor error to fix …	Success: 4/9 tests passed	Critical Pitfalls (P40: Nam…	It works	It works	Reason: org.xml.sax.SAXParseExcep…
OpenSensingCity Pollution		5/18/2017	Ontology loaded - Minor errors to fix …	Success: 8/9 tests passed	Ontology loaded but not …	It works	The ontology is loaded but no documentation	Reason: org.xml.sax.SAXParseExcep…

Also accessible online see footnote 13

integrate when developing the methodology, but the implementation depends on web reliability and the maintenance of web services. Sometimes the servers hosting the web services are down, or when new versions are released, it has an impact on the implementation. When the tools are open source, such dependencies are avoided, but it is more time-consuming for developers to get into the code based on various languages and technologies. It is another reason demonstrating the needs to help ontology designers. In Table 3, within the maintained column: *High* means that the community behind the tools is reactive when issues arise such as server down, fixing bugs, answering questions or adding new functionalities. *Medium* means that the tools is frequently down, due to server issues.

More tools will be integrated later since we are facing the issues of the availability of tools as well. For this reason, a parallel work was to develop the *Dr. PerfectO Availability of Tools (DPAT)* component demonstrator (introduced in Section 7.2).

References

1. Paulheim, H.: Knowledge graph refinement: a survey of approaches and evaluation methods. Semant. Web **8**(3), 489–508 (2017)
2. Bizer, C., Heath, T., Berners-Lee, T.: Linked data-the story so far. Int. J. Semant. Web Inf. Syst. (2009)
3. Gruber, T.R.: Toward principles for the design of ontologies used for knowledge sharing? Int. J. Hum.-Comput. Stud. (1995)
4. Vandenbussche, P.Y., Atemezing, G.A., Poveda-Villalón, M., Vatant, B.: Linked Open Vocabularies (LOV): a gateway to reusable semantic vocabularies on the Web. Semant. Web J. (2016)
5. Gyrard, A., Zimmermann, A., Sheth, A.: Building IoT based applications for Smart Cities: how can ontology catalogs help? IEEE Internet Things J. (2018)
6. Gyrard, A., Bonnet, C., Boudaoud, K., Serrano, M.: LOV4IoT: a second life for ontology-based domain knowledge to build Semantic Web of Things applications. In: IEEE International Conference on Future Internet of Things and Cloud (2016)
7. Gyrard, A., Atemezing, G., Bonnet, C., Boudaoud, K., Serrano, M.: Reusing and unifying background knowledge for internet of things with LOV4IoT. In: IEEE International Conference on Future Internet of Things and Cloud (2016)
8. Zaveri, A., Rula, A., Maurino, A., Pietrobon, R., Lehmann, J., Auer, S.: Quality assessment for linked data: a survey. Semant. Web J. **7**(1):63–93 (2015)
9. McDaniel, M., Storey, V.C., Sugumaran, V.: Assessing the quality of domain ontologies: metrics and an automated ranking system. Data Knowl. Eng. **115**, 32–47 (2018)
10. Raad, J., Cruz, C.: A survey on ontology evaluation methods. In: KEOD (2015)
11. Hlomani, H., Stacey, D.: Approaches, methods, metrics, measures, and subjectivity in ontology evaluation: a survey. Semant. Web J. (2014)
12. García, J., Jose'García-Peñalvo, F., Therón, R.: A survey on ontology metrics. In: World Summit on Knowledge Society. Springer (2010)
13. Fernández-López, M., Poveda-Villalón, M., Suárez-Figueroa, M.C., Gómez-Pérez, A.: Why are ontologies not reused across the same domain? J. Web Semant. (2018)
14. Rus, I., Lindvall, M.: Knowledge management in software engineering. IEEE Softw. J. **19**, 26–38 (2002)
15. Serrano, M., Barnaghi, P., Carrez, F., Cousin, P., Vermesan, O., Friess, P.: Internet of Things IoT Semantic Interoperability: Research Challenges, Best Practices, Recommendations and Next Steps. Technical report, IERC AC4 (2015)

16. Agarwal, R., Fernandez, D.G., Elsaleh, T., Gyrard, A., Lanza, J., Sanchez, L., Georgantas, N., Issarny, V.: Unified IoT ontology to enable interoperability and federation of testbeds. In: IEEE World Forum on Internet of Things (2016)
17. FIESTA IoT Consortium, E.: FIESTA-IoT project Deliverable 6.1 Design of Global Market Confidence Programme on IoT interoperability (2016)
18. Gyrard, A., Serrano, M., Atemezing, G.: Semantic web methodologies, best practices and ontology engineering applied to internet of things. In: IEEE World Forum on Internet of Things (2015)
19. Suárez-Figueroa, M.C.: NeOn Methodology for Building Ontology Networks: Specification, Scheduling and Reuse. PhD thesis, Universidad Politecnica de Madrid, Facultad de Informatica, Departamento de Inteligencia Artificial (2010)
20. Noy, N.F., McGuinness, D.L.: Ontology Development 101: A Guide to Creating your First Ontology (2001)
21. Zaslavsky, A., Perera, C., Georgakopoulos, D.: Sensing as a Service and Big Data. arXiv preprint arXiv:1301.0159 (2013)
22. Gyrard, A., Serrano, M.: Connected Smart Cities: interoperability with SEG 3.0 for the Internet of Things. In: 30th IEEE International Conference on Advanced Information Networking and Applications Workshops (2016)
23. Rezaei, R., Chiew, T.K., Lee, S.P., Aliee, Z.S.: Interoperability evaluation models: a systematic review. Comput. Ind. (2014)
24. Serrano, M., Barnaghi, P., Cousin, P.: Semantic Interoperability: Research Challenges, Best Practices, Solutions and Next Steps, IERC AC4 Manifesto. Technical report, European Research Cluster on the Internet of Things, AC4 (2014)
25. Gyrard, A., Bonnet, C.: Semantic Web best practices: Semantic Web Guidelines for domain knowledge interoperability to build the Semantic Web of Things. OneM2M International Standard, Management, Abstraction and Semantics (MAS) Working Group 5, April 2014, Eurecom (2014)
26. Murdock, P., Bassbouss, L., Bauer, M., Alaya, M.B., Bhowmik, R., Brett, P., Chakraborty, R.N., Dadas, M., Davies, J., Diab, W., et al.: Semantic Interoperability for the Web of Things (2016)
27. Bauer, M., Baqa, H., Bilbao, S., Corchero, A., Daniele, L., Esnaola, I., Fernandez, I., Franberg, O., Garcia-Castro, R., Girod-Genet, M., Guillemin, P., Gyrard, A., Kaed, C.E., Kung, A., Lee, J., Lefrançois, M., Li, W., Raggett, D., Wetterwald, M.: Semantic IoT Solutions—A Developer Perspective (Semantic Interoperability White Paper Part I) (2019)
28. Bauer, M., Baqa, H., Bilbao, S., Corchero, A., Daniele, L., Esnaola, I., Fernandez, I., Franberg, O., Garcia-Castro, R., Girod-Genet, M., Guillemin, P., Gyrard, A., Kaed, C.E., Kung, A., Lee, J., Lefrançois, M., Li, W., Raggett, D., Wetterwald, M.: Towards semantic interoperability standards based on ontologies (Semantic Interoperability White Paper Part II) (2019)
29. Grüninger, M., Fox, M.S.: Methodology for the Design and Evaluation of Ontologies (1995)
30. Ma, X., Fu, L., West, P., Fox, P.: Ontology usability scale: context-aware metrics for the effectiveness, efficiency and satisfaction of ontology uses. Data Sci. J. (2018)
31. Corcho, O., Fernández-López, M., Gómez-Pérez, A.: Methodologies, tools and languages for building ontologies. Where is their meeting point? Data Knowl. Eng. J. **46**, 41–64 (2003)
32. Suarez-Figueroa, M.C., Gomez-Perez, A., Fernandez-Lopez, M.: The NeOn methodology for ontology engineering. In: Ontology Engineering in a Networked World. Springer (2012)
33. Staab, S., Studer, R.: Handbook on Ontologies. Springer, Heidelberg (2013)
34. Fernández-López, M., Gómez-Pérez, A., Juristo, N.: Methontology: From Ontological Art Towards Ontological Engineering (1997)
35. Hitzler, P., Gangemi, A., Janowicz, K.: Ontology Engineering with Ontology Design Patterns: Foundations and Applications. IOS Press (2016)
36. Poveda-Villalón, M., Gómez-Pérez, A., Suárez-Figueroa, M.C.: OOPS!(Ontology Pitfall Scanner!): an on-line tool for ontology evaluation. Int. J. Semant. Web Inf. Syst. (2014)
37. Duque-Ramos, A., Fernández-Breis, J.T., Iniesta, M., Dumontier, M., Aranguren, M.E., Schulz, S., Aussenac-Gilles, N., Stevens, R.: Evaluation of the oquare framework for ontology quality. Expert Syst. Appl. (2013)

38. Duque-Ramos, A., Fernández-Breis, J.T., Stevens, R., Aussenac-Gilles, N.: OQuaRE: a SQuaRE-based approach for evaluating the quality of ontologies. J. Res. Pract. Inf. Technol. (H Index=21) (2011)
39. Fernández, M., Overbeeke, C., Sabou, M., Motta, E.: What makes a good ontology? A case-study in fine-grained knowledge reuse. In: Asian Conference on The Semantic Web. Springer (2009)
40. Tartir, S., Arpinar, I.B.: Ontology evaluation and ranking using OntoQA. In: Semantic Computing, 2007. ICSC 2007. International Conference on. IEEE (2007)
41. Tartir, S., Arpinar, I.B., Moore, M., Sheth, A.P., Aleman-Meza, B.: OntoQA: Metric-Based Ontology Quality Analysis (2005)
42. Brank, J., Grobelnik, M., Mladenić, D.: A Survey of Ontology Evaluation Techniques (2005)
43. Burton-Jones, A., Storey, V.C., Sugumaran, V., Ahluwalia, P.: A semiotic metrics suite for assessing the quality of ontologies. Data Knowl. Eng. (2005)
44. Lozano-Tello, A., Gómez-Pérez, A.: OntoMetric: a method to choose the appropriate ontology. J. Database Manag.(2004)
45. Vrandečić, D.: Ontology evaluation. In: Handbook on Ontologies. Springer (2009)
46. Vrandečić, D., Gangemi, A.: Unit tests for ontologies. In: On the Move to Meaningful Internet Systems OTM Workshops. Springer (2006)
47. Gangemi, A., Presutti, V.: Ontology design patterns. In: Handbook on Ontologies. Springer (2009)
48. Bezerra, C., Freitas, F., Euzenat, J., Zimmermann, A.: ModOnto: a tool for modularizing ontologies. In: Proceedings of 3rd Workshop on ontologies and Their Applications (Wonto) (2008)
49. Garijo, D.: WIDOCO: a Wizard for Documenting Ontologies. In: International Semantic Web Conference (ISWC, A-rank Conference). Springer (2017)
50. Fielding, R.T., Taylor, R.N.: Principled design of the modern web architecture. ACM Trans. Internet Technol. (TOIT) (2002)
51. Kolbe, N., Kubler, S., Le Traon, Y.: Popularity-driven ontology ranking using qualitative features. In: International Semantic Web Conference. Springer (2019)
52. Olivares-Alarcos, A., Beßler, D., Khamis, A., Goncalves, P., Habib, M.K., Bermejo, J., Barreto, M., Diab, M., Rosell, J., Quintas, J., Olszewska, J., Nakawala, H., Pignaton, E., Gyrard, A., Borgo, S., Alenya, G., Beetz, M., Li, H.: A Review and Comparison of Ontology-Based Approaches to Robot Autonomy (2019)
53. Gyrard, A., Sheth, A.: IAMHAPPY: Towards An IoT Knowledge-Based Cross-Domain Well-Being Recommendation System for Everyday Happiness (2019)
54. Lecue, F., Tamma, V.: ISWC 2017 Resources Track: Author and Reviewer Instructions (2017)
55. Buzan, T., Buzan, B.: The Mind Map Book: How to Use Radiant Thinking to Maximize Your Brain's Untapped Potential (1996)
56. McBride, B.: Jena: a semantic web toolkit. Internet Comput. **6**, 55–59 (2002)
57. Tejo-Alonso, C., Berrueta, D., Polo, L., Fernández, S.: Metadata for web ontologies and rules: current practices and perspectives. In: Metadata and Semantic Research. Springer (2011)
58. Peroni, S., Shotton, D., Vitali, F.: Tools for the automatic generation of ontology documentation: a task-based evaluation. In: Computational Linguistics: Concepts, Methodologies, Tools, and Applications. IGI Global (2014)
59. Lohmann, S., Link, V., Marbach, E., Negru, S.: WebVOWL: Web-based visualization of ontologies. In: Knowledge Engineering and Knowledge Management. Springer (2014)
60. Berrueta, D., Fernández, S., Frade, I.: Cooking http content negotiation with vapour. In: 4th Workshop on Scripting for the Semantic Web (SFSW), Citeseer (2008)

Web References

61. Mother IoT device: https://sen.se/store/mother/
62. Apple HealthKit: http://bit.ly/2xBFo8x
63. IoT Cisco's predictions:http://bit.ly/2JqJLdj
64. Google Knowledge Graph: https://www.youtube.com/watch?v=mmQl6VGvX-c
65. LOV4IoT: http://lov4iot.appspot.com/
66. Jena Framework Documentation: https://jena.apache.org/
67. Jena on GitHub: https://github.com/apache/jena
68. Oops GUI: http://oops.linkeddata.es/
69. Oops Web Service: http://oops-ws.oeg-upm.net/
70. Triple Checker GUI: http://graphite.ecs.soton.ac.uk/checker/
71. Triple Checker on GitHub: https://github.com/cgutteridge/TripleChecker
72. LOV GUI: http://lov.okfn.org/dataset/lov/
73. LOV Back End Java code on Github:https://github.com/pyvandenbussche/lovScripts
74. LOV JavaScript code for the GUI on GitHub:https://github.com/pyvandenbussche/lov
75. Parrot GUI: http://ontorule-project.eu/parrot/parrot
76. Parrot Java code on Bitbucket: https://bitbucket.org/fundacionctic/parrot/wiki/Home
77. LODE GUI: http://www.essepuntato.it/lode
78. LODE Java code on GitHub: https://github.com/essepuntato/LODE
79. WebVOWL GUI: http://vowl.visualdataweb.org/webvowl.html
80. WebVOWL JavaScript code on GitHub: https://github.com/VisualDataWeb/WebVOWL
81. Vapour GUI: http://linkeddata.uriburner.com:8000/vapour
82. Vapour code on Bitbucket: https://bitbucket.org/fundacionctic/vapour/wiki/Home
83. Vapour JavaScript API: http://vapour.sourceforge.net/api/
84. OWL Manchester GUI: http://visualdataweb.de/validator/
85. NeON ontology methodology: http://neon-toolkit.org/
86. OQuaRE ontology quality tool: http://miuras.inf.um.es:9080/oqmodelsliteclient/
87. Linked Data blog: http://linkeddata.org/home
88. Semantic Web Best Practices for Dummies Documentation: http://bit.ly/2XB9jsa
89. Slides step-by-step tutorial to improve the ontology quality, dissemination, reuse, etc. Semantic Web Best Practices: https://goo.gl/Rg4cGr
90. Domain Ontology Ranking System (DoORS) prototype: https://owlparser.herokuapp.com/
91. Ontology Design Patterns (ODPs) wiki: http://ontologydesignpatterns.org/

Discovering Critical Factors Affecting RDF Stores Success

Gianfranco E. Modoni and Marco Sacco

Abstract Technologies for the effective and efficient handling of RDF data are one of the main success factors for a larger scale take-up of Semantic Web Technologies in real scenarios. In this regard, several software components (RDF Stores) devoted to the semantic data persistence and retrieval are available in literature. However, each of them may be appropriate and usable for some kinds of tasks and not for others, and a one-size-fits-all killer application for this type of solutions is still not (and probably will never be) available. The large number of available solutions and the lack of widely accepted benchmarks for their rigorous evaluation do not help the selection and the adoption of an appropriate RDF store compliant with the identified needs of a specific case study. In order to contribute to fill this gap, a methodological approach to evaluate and rank the relevant features of the RDF stores is presented in this paper. Such an approach can help on one hand other researchers to discover the factors affecting the success of the RDF stores and the other hand software architects to select which RDF stores best fits the requirements of a certain application scenario.

1 Introduction

Semantic Web Technologies (SWT) [2] are increasingly being adopted to model data (and knowledge) in a variety of fields such as, for example, manufacturing, biology, medicine and healthcare, and Public Administration. In the growing landscape of "Polyglot Persistence" where enterprises exploit multiple technologies for data management [31], SWT can play a key role to aggregate and integrate heterogeneous data distributed across many sources. This is due to their aptitude to enhance

G. E. Modoni (✉)
Institute of Intelligent Industrial Technologies and Systems for Advanced Manufacturing, National Research Council of Italy, via Lembo 38F, Bari, Italy
e-mail: gianfranco.modoni@stiima.cnr.it

M. Sacco
Institute of Intelligent Industrial Technologies and Systems for Advanced Manufacturing, National Research Council of Italy, via Previati 1/E, Lecco, Italy
e-mail: marco.sacco@stiima.cnr.it

R. Pandey et al. (eds.), *Semantic IoT: Theory and Applications*, Studies in Computational Intelligence 941, https://doi.org/10.1007/978-3-030-64619-6_8

the semantic interoperability of a technological system, i.e. the latter's capability to exchange information with other systems [17]. In this regard, the SWT provide a valid support to the Linked Open Data (LOD) [3], i.e. the paradigm more and more used for connecting data to be published both on and off the Web. Another key advantage of SWT adoption consists in their capability to support (some form of) reasoning, which allow to entail and infer new meaningful knowledge about the already defined concepts and their linking relationships. Through a concept map, main aspects related with RDF stores are reported in Fig. 1.

The Resource Description Framework (RDF) is the W3C standard model for data interchange on the basis of SWT [19]. It offers a flexible and domain-agnostic syntax to abstract the information as lists of statements which are in turn represented under the form of triples consisting of a subject, a predicate, and an object. In this regard, one of the RDF strengths is that it allows to express virtually any type of information, without the need to previously adapt its data reference structure as in the relational schema-based approach. As the flip side, the expressivity and flexibility poses the need to review, for information represented in RDF syntax, traditional data management issues such as efficient persistence, query processing and optimization. In particular, the fine grain of the RDF data models (triples instead of whole records) increases the number of the joins included in the queries, making not trivial the formulation of complex queries and also introducing several issues of scalability. The databases for the persistence and management of any type pf data specified in RDF syntax are termed RDF stores (also known as Triple stores). Their main components are the repository and the Application Programming Interface (API), which communicates with the underlying repository to programmatically expose the main database services. The first solutions of RDF stores have been implemented leveraging existing databases which are in turn based on the more traditional and widely proven relational model. However, despite this approach allowed to build working solutions with little effort, the well-defined and rigid structure of the relational model on which this approach is based does not fit well with a flexible model such RDF [25]. For this reason, the current efforts of researchers and technicians are addressed towards the implementation of so called native RDF stores, i.e. purpose-built databases which, not depending on rigid schemas, fits more properly to the flexible structure of RDF data.

Native RDF stores are usually considered representatives of the NoSql databases, whose market is large and not homogeneous. Their most well-known classification groups existing implementations in terms of the following categories: (a) Column Databases: e.g. Cassandra, HBase, etc.; (b) Document Databases: e.g. MarkLogic, MongoDB, etc.; (c) Key-value Databases: e.g. Project Voldemort, Dynamo, etc.; (d) Graph Databases: e.g. Neo4J, AllegroGraph, etc. [8, 11]. This last category comprises the native RDF stores, since RDF data can be thought in terms of a directed labeled graph, where each graph's node can represent the subject or object of a triple, while the arc is the predicate that links subject and object. Moreover, it should be noted that, in addition to the native triple stores, other solutions (e.g. "native" graph databases or belonging to other NoSql category) are available to handle RDF data, even they are not designed mainly for this purpose [7].

Each of these solutions may be suitable and usable for some kinds of tasks and not for others, while a one-size-fits-all killer application for this type of solutions is not available. Under these conditions, the large number of available solutions to handle RDF data and also the lack of valid benchmarks for their rigorous evaluation make not trivial the task of selection of a valid RDF store during the design of a Semantic infrastructure. In order to choose the most suitable database to be used within a use case scenario, it is important to know the offered features, and related advantages and drawbacks. The main objective of this work is to identify criteria that help organizations in their evaluation, selection, and adoption of an RDF store.

To achieve this objective, it is introduced in this paper an empirical approach which allows to gain an understanding of the main quality characteristics of RDF stores and thus to discover, in different application domains, their critical success factors in the role of backbone of a semantic-based architecture. In particular, according to the defined requirements and needs of three real different self-conducted case studies, these factors are elicited and their relative criticality is evaluated. Afterwards, each of them is analyzed in terms of an up-to-date state of the art based on the literature review.

The remainder of this paper is structured as follows. Section 2 reviews the literature related to this research study, whereas Sect. 3 introduces the methodological approach used to rank the RDF stores features. In addition, Sect. 3 illustrates the application of the proposed methodology. Finally, Sect. 4 draws the conclusions, summarizing the main outcomes.

2 Related Works

The evaluation of RDF stores has been recently studied in various research works. In particular, a core topic has been their comparison in terms of performance, on the basis of specific metrics such as query duration. In this regard, several benchmarks based on the evaluation of these metrics (such as LUBM [4, 13], and Linked Data Benchmark [5], etc.) have been defined and formalized. However, a framework for the definitive and rigorous comparison of RDF store is still missing, since the current proposed solutions only partially meet all the needed expectations for a valid benchmark (i.e. verifiable, fair, repeatable, relevant, and economical) [5].

Aside the performances, various other attributes influence the success of an RDF store acting as backbone of a complex semantic-based architecture. In fact, like for other Information Systems, organizations might have different requirements and expectations of an RDF store, depending on the specific use case scenario. This is the reason why it is essential to pair the quantitative analysis of the RDF store performance with a qualitative analysis of its features. In this regard, Haslhofer et al. combined in [15] a qualitative and quantitative evaluation of a set of prominent triple stores leveraging various selected criteria. Modoni et al. [25] presented a qualitative analysis and comparison of six different triple store solutions in terms of their capability to support streaming and security. Another interesting evaluation has been

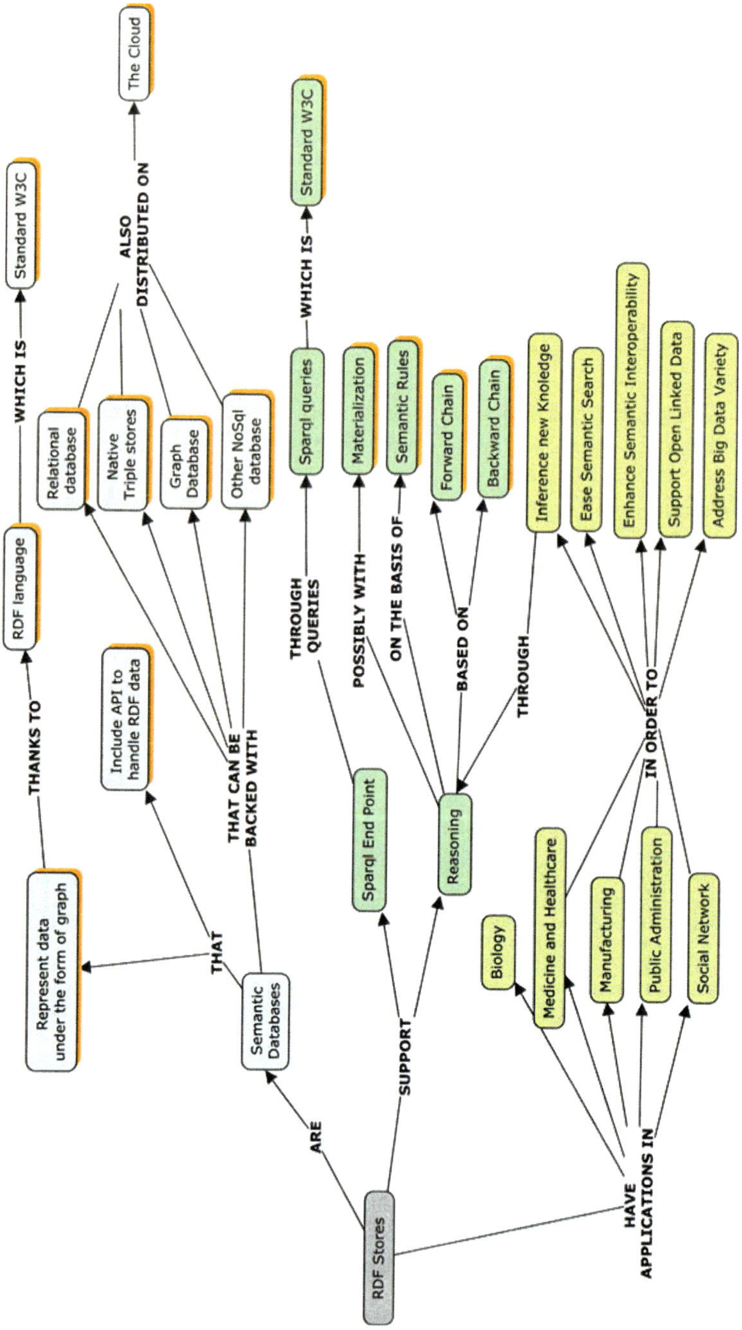

Fig. 1 A concept map which illustrates main aspects related with RDF stores

conducted by Cudré-Mauroux et al. in [7]; they studied the applicability of NoSQL databases as backend of RDf store, stating that they represent a valid choice when the workloads are limited. Other studies also surveyed RDF stores in terms of specific quality attributes [8, 11]. Moreover, the evaluation of RDF store fits nicely into the study of Delone et al. [9], which evaluates a generic Information System (IS) on the basis of six success dimensions. According to the model provided by this study, an IS can be analyzed and evaluated taking into account various characteristics that affect in turn the subsequent use of the IS and the perceived user satisfaction (such as the service quality). Even this view is rather abstract, the herein presented work concur with it, focusing mainly to find various measures for RDF stores that can be classified under the dimensions of system and information quality defined by Delone et al.

As the evaluation of an RDF store in terms of its feature can be treated as a multi-criteria decision problem (MCDM) [36], it can be analyzed in a systematic way through MCDM, which is an approach widely used to find the best option choice selected from several and also eventually conflicting alternatives. The MCDM methods allow to clearly and systematically analyze a specific problem, enabling the decision makers to analyze a problem with a methodological approach and scale it according to the evolving requirements of the problem. In this regard, one of the best-known and most widely proven MCDM techniques is the Analytic Hierarchy Process (AHP) [29], which allow to bring back a complex decision to a set of pair-wise comparisons. From the answers to such comparisons, the AHP method allows to calculate the weight for each factor in the overall decision problem. Even if the MCDM approach has been widely used in literature, none of the available research works, to the best of authors knowledge, have applied it to support systematic analysis within the RDF stores evaluation.

3 The Methodological Approach

The main objective of this work is to explore the factors affecting the success of an RDF store. The factors are selected through the analysis of three real self-conducted case-studies in different application domains, while their criticality is evaluated through a literature review. Specifically, the proposed approach consists of two main steps (Fig. 2):

- **Analysis of three different self-conducted case studies in the areas of Ambient Assisted Living, Manufacturing and eHealth**, in order to elicit the critical success factors of the RDF stores, selected from functional and non-functional features.
- **Analysis of each of the identified factors** in terms of an up-to-date state of the art based on the literature review.

Each of the steps are described in details in the following sections.

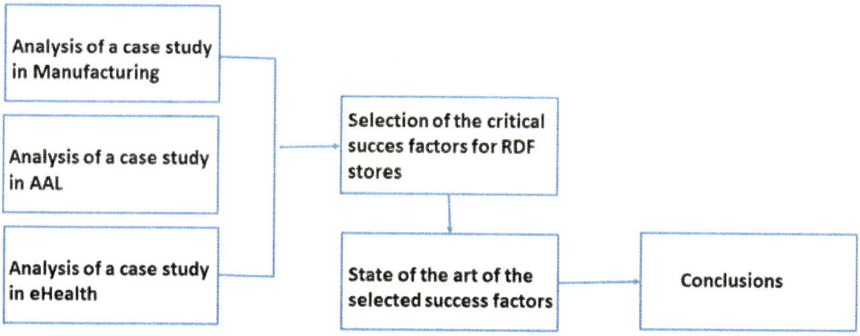

Fig. 2 The research approach

3.1 *Analysis of the Case Studies*

The analysis of a case study can help to create a bridge between theory and design, thus allowing to in understand the various dynamics present within specific settings [10], supporting explorative researches which allows to show which of the generic factors could be critical for use in a particular context. For this reason, the herein approach leverages the analysis of three real self-conducted case studies conducted respectively in the areas of Manufacturing (Apps4aME project), Ambient Assisted Living (D4All project), and eHealth (Pegaso project), in order the elicit the critical success factors of the RDF stores. All the three case studies are characterized by the fact they leverage the SWT, which are however applied (in the three cases) for different scopes and purposes. Specifically, we outline a set of technical requirements for each case study, and present open research issues, challenges, and future research directions concerning these requirements. The case studies are illustrated in details in the following subsections.

3.1.1 Apps4aME Case Study

Apps4aME is a European research project which aims at enabling the integration of product design, process development, factory production planning and factory operation through Engineering Apps (eApps) [35]. In order to realize this integration, the Apps4aME approach proposes the development of a solution based on a common reference model that provides a detailed overview of all relevant domain specific and inter-domain interdependencies. This reference model is formalized and expressed exploiting the SWT. In this regard, in order to properly manage the data adhering to the adopted reference model, it is needed to find an effective RDF store acting as backbone of an enterprise architecture comprising different software components. One of the main needs of such an architecture is its capability to handle streams of data in motion. Indeed, this capability allows the specific company to gather huge amount of information concerning its products, customers, productive resources. These information can be then (also in near real time) processed to optimize the production resources usage or to increase customer satisfaction. Nonetheless, the implementation of this capability is only the first needed item towards the strengthening of the approach proposed by Apps4ME. In addition to it, the scenarios analyzed in the project [35] pose the need of specific key-features, such as the capability to guarantee different security users profiles (corresponding to different access levels to the resources) and the capability to manage various historical versions of data (as a version management system) [23]. Finally, since the enterprise typically owns different type of information, it must be capable to handle multimodal information (e.g. text, structured information, audios, images, other binary files, etc.).

3.1.2 PEGASO Case Study

PEGASO is an European research project which aims the improvement of motivation and self-awareness of the teenagers towards a healthful lifestyle [24]. In particular, the project aims at reducing their obesity-related risks as behavioural habits of the teenagers can significantly affect their health status when become adults.

One of the challenges of the project is the handling of large amounts of heterogeneous data coming from various sources (e.g. sensors, apps, etc.). In order to face this issue, a solution has been conceived and adopted based on a common shared data model, which aims at representing a conceptualization of the overall PEGASO knowledge. On the basis of the SWT, it is defined a meta-model expressed under the form of application ontology that allows to capture the obesity-related features of teenagers, thus enabling formally organizing, searching and sharing their behavioural information. In this regards, the application ontology provides a representation of the main concepts and the relationship among these concepts in the specific analyzed domain. To manage this ontology, a solution of cloud-based RDF store is needed. For this reason, a significant stage of this project has been focused on the identification and adoption of an efficient RDF store capable to manage semantic data, also under the form of big data. These data that must be managed include: (a) an application ontology (the meta-mode), which represent the knowledge about teenagers behaviours; (b) the ontological individuals, which adhere to the domain ontology; (c) the inference entailment needed to derive new knowledge. The major required features for an ad hoc RDF store comprises the capability to manage both data in motion and data at rest, to guarantee the security of these data (most of the handled data is sensitive). In addition, it is important that RDF store provides the capability of reasoning in order to allow to infer new information, leveraging the data acquired from sensors and apps which monitor the behaviours of the teenagers. Finally, spatio-temporal data, generated to monitor teenagers, must be properly managed.

3.1.3 Design for All Case Study

The Italian research project Design for All aims at conceiving a framework which enables a design of an Ambient Assisted Living (AAL) focused on the real human being, considering each person in his peculiarities and particularities [30]. On the one hand, this framework manages the information concerning the domestic environment, also through the modeling of different scenarios describing the most prominent states of the home dwellers. On the other hand, the project enhances the interoperability of various digital tools supporting both the home's design and working, which in this way can exchange information in an efficient way. In order to achieve these objectives, Design for All investigates the potential of SWT, which provide a method for representing and integrating the knowledge concerning the home environment. In such a scenario, it is essential to persist and handle this knowledge through a proper semantic repository [26]. In particular, the latter must be capable to entail new information from continuous and heterogeneous streams of data produced by the various devices scattered in the home environment. In addition, scalability, real-time, and security requirements must be taken into account in order to extend the approach in real environment. Finally, in consideration of the high number of smart objects connected to the network in the home environment, it is also essential to identify valid strategies for distributing intelligence and data among the different hardware and software components distributed within the infrastructure (sensors, microcontrollers, services and databases on the cloud, etc.), relying in particular on the paradigms of fog and edge Computing. In order to identify a scalable architecture model based on these distributed paradigms, a federated repository model was used that best supports a distributed approach such as that offered by fog and edge Computing.

3.2 Selection of the Critical Success Factors

On the basis of the analysis of the three case studies reported in the previous section, a set of functional and non-functional features for RDF store solutions and that affect the latter's success. These features can then be used as criteria to conduct a methodological comparison of a group of RDF stores, which have before been identified and selected. The complete list of the adopted criteria is reported in Table 1, which also reports the case study from which the criteria has been elicited. In addition, the reported criteria are categorized in table depending on whether the criteria is a functional or non-functional feature (third column).

Table 1 List of evaluated criteria and corresponding case studies from which they have been elicited

ID	Criteria	Case studies	Is funct.
C1	Handling data in motion	D, P	N
C2	Security of the graph	All	N
C3	Versioning of the graph	A	Y
C4	Handling multimodal data in a knowledge repository	A, P	Y
C5	Zero impedance mismatch	D	Y
C6	(Distributed) reasoning	P	Y
C7	Federation and replication to support data distribution	D, P	N
C8	Spatio temporal database	All	N

D stands for D4All, A for Apps4ME, P for PEGASO

3.3 Analysis of the Critical Success Factors

3.3.1 Handling Data in Motion

In IoT based scenarios it is continuously growing the amount of live data in motion (also known as streaming data), i.e. data continuously generated by different sources (e.g. sensors, etc.) to be then lively transmitted, collected and analysed on a remote point. These data in motion are typically produced at high frequencies, e.g. once per millisecond or even higher in specific contexts. The processing of these data provides both short term, closed-loop and live decision making capabilities, and scalable long term off-line analysis capabilities. The advantage of using these data is that the analysis is done on fresh data, which is temporal close to the event (e.g., cyber-security attack) or the condition (e.g., machinery in degraded status) that must be detected. The disadvantage is that specific computing capabilities are needed to process the data without delay [37]. In addition, the data variety (i.e., heterogeneity) and their velocity (i.e., frequency of change over time) drive respectively to the problem of the interpretation of these data and the management of the digital tools performance. In particular, regarding the first problem, the lack of integration among the data produced by various sources typically separate crucial streams of data and increase the issue of too much data but not enough knowledge [32]. In order to contribute to overcome this gap, it is relevant to understand if SWT (and in particular RDF stores) are ready and mature to support data in motion.

3.3.2 Security of the Graph

An increasing number of threats and vulnerabilities can jeopardise cybersecurity of information systems such RDF stores. In particular, data confidentiality, integrity, and availability (so called CIA) are at risk. Thus, it is essential to identify and apply new security measures to mitigate the risks of attacks against the systems to an accept-able level, thus protecting the information processed by those systems. One crucial mechanism to handle security of data stored within an RDF store consists in applying the access controls at graph level or even at lower triple level through a more fine-grained security policy, thus allowing to grant users a granular data access [25]. In this regard, RDF stores must provide proper tools to manage these privileges to allow users to perform specific operations.

3.3.3 Versioning of the Graph

In any collaborative scenario, many different involved stakeholders can interact concurrently by applying continuous changes to the information concerning the ongoing activities. Under these conditions, it is essential to have a mechanism capable to version these information, thus allowing to ensure the consistency among different versions of these persisted information. Specifically, in the context of the SWT, the versioning capability allows to keep track of different versions of RDF graphs, which are also defined in terms of complex hierarchies of other RDF graphs [25]. The latter can be in turn versioned to handle all their evolutions. Moreover, since different versions of graphs can be potentially linked, the versioning should allow to specify how these different versions are linked [18]. Finally, the versioning system should support users to apply and manage changes, by allowing to compare different versions of graphs, highlight the differences, and also merge the different users involved in the design process [28].

3.3.4 Handling Multimodal Data in a Knowledge Repository

In various fields, information concerning the same phenomenon can be collected through different modalities of acquisition by using specific equipment [20]. In fact, due to the rich set of properties characterizing the real phenomena, it is difficult that a single modality offers a holistic knowledge about the interesting phenomenons. The growing availability of these different modalities referring to the same phenomenon poses the need to properly manage and combine these multimodal data in the repositories where they are persisted [33]. Among the technical challenges to realize the management of these multimodal data, there is its organization, link, and fusion of the interdisciplinary information as within a knowledge repository [48]. Under these conditions, as the information can be represented in this system under the form of different data models, it is needed the reconciliation and mediation of these models. In this regard, one potential solution can be based on the alignment of the different data models following the approach provided by the INTER-IoT project [12, 22], which defines mapping rules to perform semantic translation from these models to another model which plays the role of central hub to enable shared understanding.

3.3.5 Zero Impedance Mismatch

Even the SWT are enabling technologies for information integration, they are affected by the issue of the Impedance Mismatch, i.e. the discrepancy both in language syntax and semantics between the models of the programming languages and the RDF semantic models [6]. In traditional relational-database applications, various solutions have been implemented in order to bypass the Impedance Mismatch. In particular, the object-relational mapping (ORM) software solutions (such as Hibernate) offer programmatic access towards relational data sources through the mapping of the data objects into programmatic objects, thus allowing developers to programmatically handle data exploiting their habitual application programming interface [27]. However, this approaches is not applicable to Semantic Web based applications since the conceptual model represented through the SWT languages differ substantially from the relational model and from the object-oriented model which characterizes the ORM. For this reason, it is essential to identify new strategies in order to provide an immediate programmatic access to Semantic Web data, thus limiting or eliminating the Impedance Mismatch [27].

3.3.6 (Distributed) Reasoning

It is the capability of a specific component (the reasoner) to derive new knowledge directly from assertions contained in the store, applying an appropriate set of logical rules [16]. The deductions that follow can be used with the dual purpose of extending the basic knowledge available for queries and calculations, enlarging the network of concepts, and validating the knowledge base itself. Often the inference capability is integrated directly into the RDF store and is transparent to the users. In other applications, inference is implemented as an external tool that is manually initiated and the entailments that follow are manually added to the store. The latter approach can be used as a way of mitigating the computational cost of performing inference if it negatively impacts the overall performance of the store. In fact, the scalability is one of the main barrier of currently existing reasoning systems. Thus, one of the aim of researchers is to overcome this obstacle, exploiting, for instance, the benefits of parallel computation techniques applied to reasoning algorithms. In addition, an approach based on distributing reasoning can improve performance reasoning with large data sets. Many of the existing RDF stores perform inference using rules-based reasoning engines. When evaluating reasoning engines, it should be checked key-features such as the compliance with standard languages for reasoning (e.g. OWL 2 RL, OWL 2 QL, and OWL 2 EL), the inferencing strategies (forward chaining or backward chaining) and the life cycle of the inferred tripled (they are materialized or not). In the recent years, the efforts of researchers are also addressed towards stream reasoning, which aims to process dynamic (data in motion) data, unlike traditional reasoning which processes static data [21].

3.3.7 Databases Federation and Replication to Support Data Distribution

Distributed approaches based on paradigms such as fog and edge Computing offer valid mechanisms to distribute data and intelligence among the various software and hardware components of an internet of things based architecture (microcontrollers, sensors, services, and databases on the cloud, etc.), thus offering more suitable methods for respond to the uninterrupted rise of the number of devices which more and more are connected to the network. Leveraging these distributed approaches, it is possible to transfer part of the computing power and storage close to the sources of data, thus improving the bandwidth consumption and increasing the data availability

In order to identify a scalable architecture model based on these distributed paradigms, one significant contribution can come from the implementation of a federation of different RDF stores (also called virtual integration) which allow to perform complex and structured queries against a federated set of data sources [14]. In fact, the federated approach goes into the direction of a distributed approach such the one that offered by fog and edge computing. It should be underlined that the federation query processing is enabled by the nature of linked data inherent in the semantic models managed by the RDF stores. The federation of the databases can be also combined with their (partial) synchronization replication, eventually limited to some specific data models [1], thus organizing a peer-to-peer network where different autonomous actors can collaboratively update and enrich the knowledge base stored in the semantic datasets.

3.3.8 Handling Spatio-temporal Data

A growing number of software applications generate spatiotemporal data which track the position of moving objects (e.g. cars, etc.) or moving people. To integrate spatiotemporal data with other relevant information coming from other sources, it can be useful to represent them under the form of RDF and stored in the RDF stores [34]. The spatiotemporal data are characterized by the following two significant features: (a) the values of these data can continuously change, also with a high velocity; (b) the volume of these data is high. Despite spatiotemporal data processing is a consolidated feature, it is not easy to integrate it with RDF data management and for this reason efforts of researchers must mainly addressed towards scalability issues to manage large scale and dynamic spatiotemporal RDF data [34].

4 Conclusions

As an outcome of this research, it can be stated that various features are relevant for the success of an RDF store depending on the application context where SWT are applied. Their lack can be a prominent gap that must be investigated by researchers and technicians in the close future. However, further features of an RDF store must be investigated, also analyzing other relevant domains in addition to the three already analyzed. In particular, one of these features consists in the capability of an RDF store to be linked with legacy systems. For this reason, first of all, it will be essential to extend and update this study by adding new criteria. In addition, this study will be complemented with the rating of the selected criteria using an AHP method, on the basis of the individual preferences of a selected sample of end-users. The goal is to face this challenge also in future works.

Acknowledgements The present paper has been developed within CasAware project, approved by Lombardy region (id 147152) within the Call "Bando Linea RS per aggregazioni".

References

1. Aslan, K., Molli, P., Skaf-Molli, H., Weiss, S.: C-set: A Commutative Replicated Data Type for Semantic Stores (2011)
2. Berners-Lee, T., Hendler, J., Lassila, O.: The semantic web. Sci. Am. **284**(5), 34–43 (2001)
3. Bizer, C., Heath, T., Berners-Lee, T.: Linked data: the story so far. In: Semantic Services, Interoperability and Web Applications: Emerging Concepts, pp. 205–227. IGI Global (2011)
4. Bizer, C., Schultz, A.: The berlin SPARQL benchmark. Int. J. Semant. Web Inf. Syst. (IJSWIS) **5**(2), 1–24 (2009)
5. Boncz, P., Fundulaki, I., Gubichev, A., Larriba-Pey, J., Neumann, T.: The linked data benchmark council project. Datenbank-Spektrum **13**(2), 121–129 (2013)
6. Calvanese, D., De Giacomo, G., Lembo, D., Lenzerini, M., Poggi, A., Rosati, R., Ruzzi, M.: Data integration through DL-LiteA ontologies. In: International Workshop on Semantics in Data and Knowledge Bases, pp. 26–47. Springer (2008)
7. Cudré-Mauroux, P., Enchev, I., Fundatureanu, S., Groth, P., Haque, A., Harth, A., Keppmann, F.L., Miranker, D., Sequeda, J.F., Wylot, M.: NOSQL databases for RDF: an empirical evaluation. In: International Semantic Web Conference, pp. 310–325. Springer (2013)
8. Curé, O., Blin, G.: RDF Database Systems: Triples Storage and SPARQL Query Processing. Morgan Kaufmann, Boston (2014)
9. Delone, W.H., McLean, E.R.: The DeLone and McLean model of information systems success: a ten-year update. J. Manag. Inf. Syst. **19**(4), 9–30 (2003)
10. Eisenhardt, K.M.: Building theories from case study research. Acad. Manag. Rev. **14**(4), 532–550 (1989)
11. Faye, D.C., Cure, O., Blin, G.: A Survey of RDF Storage Approaches (2012)
12. Ganzha, M., Paprzycki, M., Pawłowski, W., Szmeja, P., Wasielewska, K.: Semantic interoperability in the internet of things: an overview from the inter-IoT perspective. J. Netw. Comput. Appl. **81**, 111–124 (2017)
13. Guo, Y., Pan, Z., Heflin, J.: LUBM: a benchmark for OWL knowledge base systems. J. Web Semant. **3**(2–3), 158–182 (2005)
14. Haase, P., Mathäß, T., Ziller, M.: An evaluation of approaches to federated query processing over linked data. In: Proceedings of the 6th International Conference on Semantic Systems, pp. 1–9 (2010)

15. Haslhofer, B., Momeni Roochi, E., Schandl, B., Zander, S.: Europeana RDF store report. Tech. rep., University of Vienna (2011)
16. Horrocks, I., Patel-Schneider, P.F., Boley, H., Tabet, S., Grosof, B., Dean, M., et al.: SWRL: a semantic web rule language combining OWL and RuleML. W3C Member Submission **21**(79), 1–31 (2004)
17. IEEE: A Compilation of IEEE Standard Computer Glossaries. IEEE Standard Computer Dictionary (1990)
18. Klein, M., Fensel, D., Kiryakov, A., Ognyanov, D.: Ontology versioning and change detection on the web. In: International Conference on Knowledge Engineering and Knowledge Management, pp. 197–212. Springer (2002)
19. Klyne, G., Carroll, J.J., McBride, B.: Resource Description Framework (RDF): Concepts and Abstract Syntax, 2004. http://www.w3.org/TR/2004/REC-rdf-concepts-20040210 (2009)
20. Lahat, D., Adali, T., Jutten, C.: Multimodal data fusion: an overview of methods, challenges, and prospects. Proc. IEEE **103**(9), 1449–1477 (2015)
21. Margara, A., Urbani, J., Van Harmelen, F., Bal, H.: Streaming the web: reasoning over dynamic data. J. Web Semant. **25**, 24–44 (2014)
22. Modoni, G.E., Caldarola, E.G., Sacco, M., Wasielewska, K., Ganzha, M., Paprzycki, M., Szmeja, P., Pawlowski, W., Palau, C.E., Solarz-Niesłuchowski, B.: Integrating the AAL CasAware platform within an IoT ecosystem, leveraging the inter-IoT approach. In: Proceedings of First International Conference on Computing, Communications, and Cyber-Security (IC4S 2019). Springer (2020)
23. Modoni, G.E., Doukas, M., Terkaj, W., Sacco, M., Mourtzis, D.: Enhancing factory data integration through the development of an ontology: from the reference models reuse to the semantic conversion of the legacy models. Int. J. Comput. Integr. Manuf. **30**(10), 1043–1059 (2017)
24. Modoni, G.E., Sacco, M., Candea, G., Orte, S., Velickovski, F.: A semantic approach to recognize behaviours in teenagers. In: SEMANTICS Posters & Demos (2017)
25. Modoni, G.E., Sacco, M., Terkaj, W.: A survey of RDF store solutions. In: 2014 International Conference on Engineering, Technology and Innovation (ICE), pp. 1–7. IEEE (2014)
26. Modoni, G.E., Veniero, M., Sacco, M.: Semantic knowledge management and integration services for AAL. In: Italian Forum of Ambient Assisted Living, pp. 287–299. Springer (2016)
27. Oren, E., Heitmann, B., Decker, S.: ActiveRDF: Embedding semantic web data into object-oriented languages. J. Web Semant. **6**(3), 191–202 (2008)
28. Papavassiliou, V., Flouris, G., Fundulaki, I., Kotzinos, D., Christophides, V.: On detecting high-level changes in RDF/S KBs. In: International Semantic Web Conference, pp. 473–488. Springer (2009)
29. Saaty, T.L.: The Analytic Hierarchy Process, Planning, Priority Setting. Resource Allocation. McGraw-Hill, London (1980)
30. Sacco, M., Caldarola, E.G., Modoni, G., Terkaj, W.: Supporting the design of AAL through a SW integration framework: the D4All project. In: International Conference on Universal Access in Human–Computer Interaction, pp. 75–84. Springer (2014)
31. Sadalage, P.J., Fowler, M.: NoSQL Distilled: A Brief Guide to the Emerging World of Polyglot persistence. Pearson Education, Upper Saddle River, NJ (2013)
32. Sheth, A., Henson, C., Sahoo, S.S.: Semantic sensor web. IEEE Internet Comput. **12**(4), 78–83 (2008)
33. Soleymani, M., Lichtenauer, J., Pun, T., Pantic, M.: A multimodal database for affect recognition and implicit tagging. IEEE Trans. Affect. Comput. **3**(1), 42–55 (2011)
34. Vlachou, A., Doulkeridis, C., Glenis, A., Santipantakis, G.M., Vouros, G.A.: Efficient spatio-temporal RDF query processing in large dynamic knowledge bases. In: Proceedings of the 34th ACM/SIGAPP Symposium on Applied Computing, pp. 439–447 (2019)
35. Volkmann, J.W., Landherr, M., Lucke, D., Sacco, M., Lickefett, M., Westkämper, E.: Engineering apps for advanced industrial engineering. Procedia CIRP **41**, 632–637 (2016)
36. Weistroffer, H.R., Smith, C.H., Narula, S.C.: Multiple criteria decision support software. In: Multiple Criteria Decision Analysis: State of the Art Surveys, pp. 989–1009. Springer (2005)
37. Zhang, Y., Duc, P.M., Corcho, O., Calbimonte, J.P.: SRBench: a streaming RDF/SPARQL benchmark. In: International Semantic Web Conference, pp. 641–657. Springer (2012)

Creation of Ontological Knowledge Bases in the Semantic Web by Analyzing Table Structures

Vitalina Babenko, Igor Shostak, Mariia Danova, and Olena Feoktystova

Abstract The active development of the Semantic Web initiative to create expressive models for representing knowledge distributed on the Web in the form of ontologies raises a number of problems associated with the development of information structures of ontological knowledge bases for automatic processing of data and knowledge. The subject of the research is the ontological knowledge base and methods of their formation within the framework of the Semantic Web. Moreover, various tabular structures are considered as sources of knowledge. The problem arises as a result of the contradiction between the wide variety of tabular structures used to organize the content of knowledge sources in a hypermedia environment and the insufficient efficiency of classical methods for analyzing sources of this type. In the course of research, this problem was decomposed into a number of tasks:

- analysis of existing approaches to the formation of ontological knowledge bases based on the sources of tabular structures;
- development of a formal model of ontological knowledge bases;
- development of a method for the formation of databases of ontological knowledge based on targeted enumeration and its mathematical support;
- development of a formal model of the sources of knowledge of table structures;

V. Babenko (✉)
International E-Commerce and Hotel and Restaurant Business Department, V.N. Karazin Kharkiv National University, 4 Svobody Sq, Kharkiv 61022, Ukraine
e-mail: vitalinababenko@karain.ua

I. Shostak · M. Danova
Department of Software Engineering, National Aerospace University "KhAI", 17 Chkalova Street, Kharkiv 61070, Ukraine
e-mail: iv.shostak@gmail.com

M. Danova
e-mail: danovamariya@gmail.com

O. Feoktystova
Department of Management, National Aerospace University "KhAI", 17 Chkalova Street, Kharkiv 61070, Ukraine
e-mail: e.i.shostak@gmail.com

R. Pandey et al. (eds.), *Semantic IoT: Theory and Applications*, Studies in Computational Intelligence 941, https://doi.org/10.1007/978-3-030-64619-6_9

- development of a method for analyzing the sources of knowledge of tabular structures based on targeted enumeration and its mathematical support;
- development of a method for generating instances of objects of subject areas based on knowledge sources of tabular structures and its mathematical support;
- application of the developed methods for the implementation of a set of software tools for the formation of ontological knowledge bases.

In the course of solving the first of the stated tasks, it was found that historically the first was the approach to the formation of ontological knowledge bases based on the methods of structural analysis. The effectiveness of methods of this kind is limited by the small number of tabular structures analyzed and the inconsistency of interpretation of the structural components of the knowledge sources of tabular structures and their visual representation. The need to solve the problem of creating ontological knowledge bases based on the sources of tabular structures, characterized by a high level of complexity of organizing the contents of these sources, has led to the emergence of a new generation of intelligent methods for forming ontological knowledge bases based on top-level ontologies. The approach to the formation of ontological knowledge bases on the basis of upper-level ontologies involves the formation of such bases in accordance with the terminology defined in the upper-level ontology. Thus, the boundaries of the presentation of the components of domain objects, in contrast to structural analysis, are found as a possible combination of terms defined in the ontology by calculating measures of semantic similarity. This approach allows you to build procedures for obtaining new knowledge, abstracting from the method and format of storing the contents of structured sources of knowledge. The methodological basis of research includes the ideas and principles of artificial intelligence, elements of the hypertext technologies of the Semantic Web, tools for knowledge engineering, in particular ontological engineering. Experimental studies were carried out on test examples and on real sources of knowledge of tabular structures in the form of documents that are widely used in the Semantic Web environment for organizing workflows. The implementation of the theoretical results of the study in the form of algorithmic, mathematical support, as well as experimental studies conducted to determine the upper bound and the nature of the growth of complexity of the method of forming the ontological knowledge bases based on targeted enumeration, confirm the validity of the hypothesis adopted at the beginning.

Keywords Semantic web · Ontological knowledge bases · Tabular structures · Organizing workflows · Hypermedia environment

1 Goal and Objectives of the Research

Important tasks that arise in the implementation of modern artificial intelligence systems include the development of methods and means of forming the knowledge bases of these systems. The need for this is due, first of all, to the active development of the Semantic Web initiative [1] to create expressive models for representing knowledge distributed over the Web in the form of ontologies, which are sets of explicit specifications for object-domain-related (ODR) relationships for organizing automatic processing of data and knowledge. Within the framework of the Semantic Web, ontologies, together with the instances of their constituent objects, form the ontological knowledge base (OKB) [2–5].

As one of the main sources of knowledge for the Semantic Web, various types of data sets are organized using tabular structures, the analysis of which can add additional knowledge about the ODR,—sources of knowledge of tabular structures (SKTS) widely used in Intranet/Internet environments. However, the wide variety of tabular structures used to organize the content of sources constitutes a serious obstacle to the effective application of classical methods for analyzing sources of this type to create ODR objects, their instances, and the formation of OKB on their basis. In addition, the orientation of existing projects and means of forming the OKB towards the manual input of knowledge about ODR in some cases is economically and organizationally not always advisable in practical applications.

At the same time, it should be noted that studies of this problem do not yet allow us to propose a set of effective means of forming OKB different purposes. Thus, relevant and appropriate is the development of effective methods of formation OKB, focused on the automatic input of knowledge through analysis SKTS.

The purpose of this article is to present an approach to improving the efficiency of electronic document management in the Semantic Web by developing methods and means of forming ontological knowledge bases based on sources of tabular structures.

To achieve the stated goal during the study, the following tasks were implemented:

- analysis of existing approaches to the formation of ontological knowledge bases based on the sources of tabular structures;
- development of a formal model of ontological knowledge bases;
- development of a method for the formation of ontological knowledge bases based on targeted enumeration and its mathematical support;
- development of a formal model of the sources of knowledge of table structures;
- development of a method for analyzing the sources of knowledge of tabular structures based on targeted enumeration and its mathematical support;
- development of a method for generating instances of objects of subject areas based on knowledge sources of tabular structures and its mathematical support;
- application of the developed methods for the implementation of a set of software tools for the formation of ontological knowledge bases.

To solve the problems formulated at the beginning of the study, the ideas and principles of artificial intelligence, Semantic Web hypertext technologies, the theory of knowledge engineering, ontological engineering, and computational complexity were used.

The methods and means of forming OKB on the basis of SKTS developed in the course of research allow us to solve a wide class of problems related to the organization of workflow on the basis of OKB with the possibility of obtaining new knowledge in the form of objects related by relations ODR by analyzing SKTS different formats, as well as organizing the automatic processing of distributed knowledge on the semantic web.

2 Comparative Analysis of Modern Approaches to the Formation of Ontological Knowledge Bases in Semantic Web Systems

The current stage of development of theoretical and practical research in the field of artificial intelligence is characterized by increased attention to the development of industrial technologies of knowledge bases [6, 7]. The most promising area in this area, according to the W3C consortium and other research groups, is the development of ontological knowledge bases, which can become the basis for the creation of the next generation Web, namely, the basis of the Semantic Web. In this connection, there is an active development of development tools for web-oriented knowledge bases. Existing projects to create such tools focus mainly on manual input of knowledge about ODR. Moreover, the expert acts as a source of knowledge [8, 9]. At the same time, in [10] it is noted that any object of the material world, the analysis of which can add additional knowledge about ODR to OKB, can potentially be sources of knowledge for intelligent systems running on the Semantic Web.

According to studies [11, 12], the content of about 52% of electronic documents on the Internet is organized using tabular structures. The sources of knowledge of table structures include databases and data warehouses, various types of printed and electronic table documents [13].

Historically, the first approach to the formation of OKB was based on the methods of structural analysis of SKTS, which is carried out by processing the structure of these SKTS and identifying their structural components [14, 15]. The effectiveness of such methods is limited by the small number of tabular structures analyzed and the inconsistency of interpretation of SKTS structural components and their visual representation [16].

The need to solve the problem of creating ontological knowledge bases based on the sources of tabular structures, characterized by a high level of complexity of organizing the contents of these sources, has led to the emergence of a new generation of intelligent methods for forming ontological knowledge bases based on top-level ontologies [17, 18]. Currently, this direction is at an intensive stage of development, as evidenced by the flow of publications within the framework of IEEE Intelligent Systems. Below we will consider approaches to the formation of ontological knowledge bases on the basis of various structured sources, implemented by means of the modern theory of artificial intelligence.

By the formation of ontological knowledge bases is meant the implementation of a complex of works on converting the contents of knowledge sources into an object-oriented or property-centric representation of ODR objects and their instances—components of ontological knowledge bases.

An analysis of articles [19–21] shows that many approaches to the formation of OKB are based on structural analysis methods, some of which are based on the use of top-level ontologies.

So, an approach to the formation of OKB from existing data warehouses is described by analyzing the ER-diagrams of these data warehouses and constructing a mapping of the relational schema to an object based on top-level ontologies and workflow technologies. A feature of this approach is the need to prepare a user description of the organization scheme for the content of complexly structured sources, which complicates the practical implementation of OKB generation tools built on the basis of this approach.

The study of the types of SKTS received by municipal services from legal entities and the methods for presenting these SKTS in modern information-analytical systems made it possible to develop a unified SKTS layout for the presentation of reporting documents and their storage in the information storage. The structures of database tables, data being analyzed, and ways of interconnecting with system metadata are also defined. The necessity of creating a base of used terms and structures for organizing the analysis of SKTS unified layouts is noted.

An overview of the main problems that arise when organizing the extraction of data from spreadsheet documents in MS Excel format for the intellectual analysis and formation of ODR objects is given in [19]. The main problem, according to the authors of this article, is that the structure of user files in Excel format is difficult to formalize, which leads to the need for reconciliation, cleaning and quality control of the extracted data with the help of a human specialist. It is also noted that the resource consumption of the processes of forming ODR objects and their subsequent storage in the data warehouse is directly proportional to the complexity of the structure of the SSK (structured source of knowledge) and exponentially depends on their number. As a solution to the problem, we propose fixing the used source structures in a centralized meta-database of control information.

Problems arising in the structuring of data extracted from SSK based on the analysis of ODR ontological models for organizing further software processing were considered in [20]. In addition, the problem of establishing the boundaries of the terminological representation of ODR objects during SSK analysis is noted, which impedes the creation of tools for automatic formation of OKB.

An approach to the semantic coordination of SSK data in knowledge management systems, which significantly determine the quality of the conducted mining using Data Mining methods, is given in [5]. It is noted that it is impossible to coordinate the components of the structure and meta-information of the source with the ontological models of subject areas used in knowledge management systems without human intervention (automatically).

To date, issues related to the automatic analysis of complexly organized table structures as sources of knowledge for the formation of the Semantic Web have not been sufficiently developed.

A technology for automatically extracting ontological specifications of ODR objects from HTML tables of unknown structure based on top-level ontologies to facilitate the processes of semantic linking of document contents with models of subject areas of information systems was proposed in [2]. The possibility of generating new knowledge based on the analysis of ontological descriptions of documents, which means the evolution of information systems in the direction of ontological knowledge bases, is noted.

The approach to the formation of knowledge bases based on the use of structural analysis methods involves the identification of objects based only on the analysis of the structural characteristics of SSK, which allows you to build specialized procedures for the formation of knowledge bases focused on specific presentation structures and SSK storage formats.

The limitations of structural analysis methods are manifested in the wide variety of structures necessary for organizing data and attracting a person to "clean" the extracted data.

The approach to the formation of OKB based on top-level ontologies involves the formation of OKB in accordance with the terminology defined in the top-level ontology. Thus, the boundaries of the representation of the components of ODR objects, in contrast to structural analysis, are found as a possible combination of terms defined in the ontology by calculating measures of semantic similarity. To establish a correspondence between the terminological representations of source objects and ontologies, the methods of Lee, Wu, and Palmer or Reznik are often used. This approach allows you to build procedures for obtaining new knowledge, abstracting from the method and format of storing the contents of structured sources of knowledge. The process of forming an OKB based on this approach is that for each SSK a structural-logical data scheme is constructed, which, like an ontology, is represented in the form of a graph of objects and relationships, objects of a structural-logical scheme are characterized by properties. Each property corresponds to a domain, which, in addition to the set of acceptable values, determines the recording format adopted in a particular source of knowledge.

After constructing a structural logical diagram, an expert can modify it by deleting part of objects or narrowing domains. This makes it possible to simplify the work with SSK in the case when it is necessary to use only a part of the data contained in it, while the structural-logical scheme of SSK is compared with the top ontology by searching for matching terms of sources and ontology, as mentioned above. Considering the fact that it is not always possible to establish a direct correspondence between the terms of the ontology and the structural-logical scheme of SSK, additional restrictions may be imposed on the scheme, in particular, restrictions on the values of properties. Next, a comparison is made of the properties associated with domains by constructing a formula for transforming source values. This formula plays the role of a translator between ontology and source formats and eliminates differences in the forms of writing values, languages, units, etc. Next, the SSK data is converted to the format used by OKB.

The approach under consideration is characterized by high quality of the formation of knowledge bases on predefined templates of tabular structures and terminology used to represent concepts. At the same time, the complexity of organizing the process of forming knowledge bases and the need to first create a basic specification used in terminology sources in the form of a top-level ontology creates a problem for the widespread introduction of this approach. This study is devoted to eliminating the shortcomings of the approach presented in this subsection.

3 The Methodological Basis for the Presentation of Table Structures as Sources of Knowledge in Semantic Web Systems

The extensive development of the Semantic Web initiative led to the emergence of new structures for the representation of knowledge—ODR objects, descriptions of which are distributed on the Web; value properties; properties-objects that provide a high level of semantics-level interoperability between software applications. It should be noted that the ODR objects in the Semantic Web are a descriptive specification of objects, which does not allow them to be processed in machine processing as something active, containing methods for processing the data of these objects, as is customary in the object paradigm. The methods for processing ODR objects in the Semantic Web are external to these ODR objects. In addition, the lack of tools for specifying and accounting for changes in the state of ODR objects causes certain difficulties during the organization of the formation of OKB based on SKTS.

OKB is based on the concept of a class-subclass object hierarchy, where a class is the formal structure of some ODR entity, a subclass is a class that inherits the structure of one or more superclasses, a superclass is a class whose structure is inherited by other classes, type is a range of valid values that can take an instance of a class, an instance of a class is a specific ODR entity.

Given the above, OKB will be formally represented as a set of explicit specifications of knowledge about ODR in the form of structural logic diagrams of the relationships (R) of ODR objects (C) of ontology (Ont) and instances of the ODR objects that make it up (I^{U}):

$$\mathrm{BOZ} = < \mathrm{Ont}, \mathrm{I}^{\mathrm{U}} > . \quad (1)$$

The OKB ontology structure is defined as a combination of three interconnected components:

$$\mathrm{Ont} = < \mathrm{C}^{\mathrm{U}}, \mathrm{D}^{\mathrm{U}}, \mathrm{F}^{\mathrm{U}} >, \quad (2)$$

where: C^U—hierarchy of interconnected ODR objects by relations R; D^U—universal set of value properties; F^U—a universal set of properties-objects, and each property-object F^U is the following logical structure $< f, t : c'' \in C^U >$, where f—property, the value type (t) of which the object is a class $c'' \in C^U$.

Let us consider in more detail the ontology component C^U of the ontological knowledge base.

This component is the following hierarchical object-oriented structure:

$$C^U = < C, R >, \tag{3}$$

where C—the set of ODR objects entering R.

To specify the mathematical implementation of the ontology of the ontological knowledge base, we define the set of relations R as follows:

$$R = \{sup, sub\}, \tag{4}$$

where: sup—the relation "to be an object-superclass"; sub—the relation "to be an object-subclass".

We modify the set of R ontologies of the ontological knowledge base by adding the relationship "be an instance" between objects to organize communication between class objects and specific instances of these class objects, as follows:

$$R = \{sup, sub, inst\} \tag{5}$$

where inst—the relation "to be an instance of a class".

Then the ODR object hierarchy is formally defined as follows:

$$C^U = < C, sup, sub, inst >, \tag{6}$$

where C—the set of ODR objects entering into relations sup, sub, inst.

Each ODR object represents a formal structure of an ODR entity, defined as follows:

$$\forall c \in C : c = < S, A, B, D, F, E, \Omega >, , \tag{7}$$

where: c—the ODR object; C—the set of ODR objects; S—the symbolic name of the ODR object; $A = \{c' \in C : c' \, sup c\}$—the set of superclass objects of the object c; $B = \{c' \in C : c' \, sub \, c\}$—the set of subclasses of c; $D = \{< d_0, v_0 >, \ldots, < d_m, v_m >\}$—the set of properties-values of the object c, where $D \subseteq D^U$, where D^U—the finite set of all properties-values of the ontology; d_0, d_m—symbolic names of property values; ν_0, ν_m—values of property-values; $F = \{< f_0, t_0 : c'' \in C, v_0 : c^{inst} \in E^{c''} >, \ldots, < f_q, t_q : c'' \in C, v_q : c^{inst} \in E^{c''} >\}$—the set of object properties of the object c, where $F \subseteq F^U$, where F^U—the finite set of all OKB object properties; $f_0, f_q, t_0, t_q, v_0, v_q$—symbolic names, types and values of properties; c^{inst}—instance object; c''—class object; $E = \{c' \in C : c' \text{instc}\}$—the set of instances of the object c, and $E \subseteq I^U$, where I^U—the finite set of instance objects OKB; $\Omega = \{\omega_0, \ldots, \omega_z\}$—attached procedures of the ODR object.

The set D and F of these objects can be represented in a simplified form by the formulas:

$$D = \{< d_o, \emptyset >, \ldots, < d_m, \emptyset >\} \tag{8}$$

$$F = \{< f_0, t_0 : c''_0 \in C, \emptyset >, \ldots, < f_q, t_q : c''_q \in C, \emptyset >\}. \tag{9}$$

We use a simplified record of properties, then the property-values of the object-class will be defined as $D = \{d_0, \ldots, d_m\}$, and the property-objects.

$$F = \{< f_0, t_0 : c''_0 \in C >, \ldots, < f_q, t_q : c''_q \in C >\} \tag{10}$$

Thus, the formal OKB model is adequately represented by formulas (1–10). The synthesized model is necessary to develop a method for generating OKB based on targeted enumeration.

4 Method for the Formation of Ontological Knowledge Bases in Semantic Web Systems Based on Targeted Enumeration

The need to develop a method for building ontological knowledge bases based on targeted enumeration is due to the fact that ODR objects obtained as a result of SKTS analysis and their instances obtained as a result of analysis of SKTS datasets from which OKB is formed should be hierarchically organized according to the principle of "inheritance", according to which the root object of the hierarchy (the base object) contains the most common properties, and already within the framework of subclass objects there is a more accurate specification of ODR entities.

Given the above and the features of the property-centric representation of knowledge in OKB, the author proposes the formation of OKB by iteratively adding ODR objects to OKB based on requests to enter ODR objects in OKB containing structural logical diagrams of ODR objects, as represented by the expression below. Thus, the formation of an OKB based on SKTS requires solving sub-tasks related to the analysis of incoming requests for entering ODR objects in OKB, searching for the most relevant or fully matching ODR objects in OKB, and adding new related objects and their instances.

The search for the most relevant ODR object query in OKB, if any, will be carried out on the basis of targeted enumeration of ODR objects in OKB by finding the most relevant branch of the object hierarchy and identifying the ODR object in it that fully matches the query, or the most relevant query. At the same time, we will analyze ODR objects by processing the structural and logical circuits of request objects and OKB.

In order to avoid a complete enumeration of ODR objects in OKB when forming an OKB, we will search for the most relevant or exactly matching object by querying through targeted ODR objects in OKB as follows: a request to enter ODR objects in OKB is sent to the base ODR object; Further, the set of subclass objects of the base object is analyzed to determine the branch of OKB objects that is most relevant for processing in order to reduce the search space. In this case, the search for the most relevant branch of the object hierarchy is carried out at each level of the hierarchy of objects until the most relevant or fully relevant ODR object is found in OKB.

The result of targeted enumeration is the ODR object in OKB that is most relevant or matches the request exactly, processing the structural-logical scheme, which will allow adding new knowledge to OKB in the form of objects connected with relations and/or their new instances. If the most relevant object is found in the OKB, i.e. there is no ODR object that fully matches the query, then, based on the structural logical diagram of the query, a new ODR object is created and entered into the OKB by a subclass or superclass of the object most relevant to the query. Otherwise, if the number of logical characteristics of the entity, i.e. the power of property-objects and property-values is greater than the number of powers of property-objects and property-values of the most relevant request for an object, then the new object becomes an object-subclass of the most relevant request for the object, otherwise, the new object becomes an object-superclass. Thus, as a result of the search for the most relevant ODR object query in the OKB, there is always an ODR object, existing or most relevant, which can be further processed.

The method of generating an OKB based on targeted enumeration consists of the following set of actions aimed at finding the ODR object that most satisfies the request for OKB, and adding, depending on the structure of the source data, new subclass objects, superclass objects, and their instance objects:

1. A preliminary analysis of the request to enter ODR objects in OKB by checking it for inconsistency in order to build a structural-logical diagram of ODR objects, the input of which we will carry out.
2. A purposeful search for the most relevant branch of the object hierarchy and the most relevant ODR object request in the OKB by establishing a correspondence between the structural logical diagrams of the request objects and the OKB to determine the object whose subclass or superclass will be the request object.
3. Adding ODR objects and their instances to OKB to replenish new knowledge about ODR.

The implementation of the described method involves access to the OKB through the base object. We will apply for this object to enter ODR objects in OKB. We define the functionality of each object by the attached procedures Ω.

To solve the problem of forming an OKB, using the principle of inheriting object hierarchies, we define for all ODR objects the following attached procedures necessary to find the most relevant object request and add ODR objects to the OKB:

1. The procedure for checking the inconsistency of the request ω_{ap};
2. The procedure for forming OKB ω_{form};
3. The procedure for the targeted search for the most relevant ODR object query in OKB ω_{fop};
4. Procedure for verifying the existence of a more relevant ODR object request ω_{def};
5. The procedure for processing the most relevant ODR object request $\omega_{compare}$;
6. The procedure for the formation of sets of matching properties of objects ω_M;
7. The procedure for the formation of sets of matching properties-values ω_R;
8. The procedure for determining the powers of sets of matching properties ω_L;
9. The procedure for determining the serial number of the most relevant object request ωmax;
10. The procedure for adding an object—a superclass ω_A^+;
11. The procedure for adding an object—a subclass ω_B^+;
12. The procedure for adding new properties-objects ω_f^+;
13. The procedure for adding new value properties ω_d^+.

Thus, the functionality of each OKB object will be represented by a variety of attached procedures Ω of the following form:

$$\Omega = \{\omega_{ap}, \omega_{form}, \omega_{fop}, \omega_{def}, \omega_{compare}, \omega_M, \omega_R, \omega_L, \omega max, \omega_A^+, \omega_B^+, \omega_f^+, \omega_d^+, \}. \tag{11}$$

When adding new ODR objects to OKB, each ODR object is automatically endowed with the functionality presented above by expression (11), which allows you to create an active OKB formation mechanism that allows you to establish a correspondence between the structural and logical circuits of request objects and OKB with the possibility of additively organizing the hierarchy of OKB objects in accordance with relations R represented by the expression (5).

Let be Q—structural-logical diagram of a request to enter ODR objects in OKB, on the basis of which an ontological knowledge base is formed.

The request to enter ODR objects in OKB is formally presented in the form of consistent sets of property-values, object-properties and a possible symbolic name of the ODR object corresponding to some SKTS:

$$Q =< S, D, F > . \tag{12}$$

where: S—the symbolic name of the ODR object; D—many properties-values of the ODR object; F—a lot of properties-objects of the ODR object, where $F = \{< f_1, a_1'' >, \ldots, < f_\lambda, a_\lambda'' >\}_j$, where λ—the number of properties-objects of the

ODR object, and $a''_1, \ldots, a''_\lambda$—ODR objects formally built on the basis $q_1, \ldots, q_\lambda$—structural-logical diagrams of other ODR objects, instances of which are part of the ODR request object.

5 Formal Model of Table Structure Knowledge Sources

Modern office applications such as MS Office, Open Office, Lotus SmartSuite provide effective software tools for creating and maintaining documents, the contents of which can be organized using tabular structures with the ability to convert these documents into various formats. But unfortunately, the format of the documents supported by these packages is different: MS Office—doc, Open Office—sxw, Lotus SmartSuite—lwp. In addition, due to the dominance of the Windows desktop market with the MS Office suite, developers of alternative office suites are trying to add support for the doc format. MS Office. But today, support for the doc format. is limited. At the same time, all office suites provide excellent support for the HTML format, both for processing documents and for saving processing results, which allows expressing the contents of documents of various formats in HTML markup while maintaining the structure and without loss of formatting.

Using HTML to represent the contents of documents provides a syntax-compatible format for exchanging documents between software systems.

Recently, HTML has become a subset of the XML language, which has a wider set of tools for ensuring compatibility at the level of document syntax, but to date, XML support is not provided by all electronic document processing packages. Thus, it seems appropriate to use tabular models of the HTML language to develop a formal model of the sources of knowledge of the table structure. Further, taking into account the provisions of the integrated tabular model (ITM) of HTML, a formal model of the sources of knowledge of table structures will be developed by adding the necessary functionality to analyze the sources of knowledge of table structures in an integrated table model of the HTML hypertext language.

Imagine SKTS as follows:

$$TC = < TZ, TB^g >, \tag{13}$$

where: TC—SKTS; TZ—SKTS header; TB^g—SKTS dataset.

According to the provisions of KTM HTML, more than one data set can be defined within one SKTS, then TB^g mathematically it can be represented by the formula:

$$TB^g = \{TB_1, \ldots, TB_\beta\}, \tag{14}$$

where β—the TB number in SKTS.

Given that the number of TB greater than 1 is used solely to provide portioned downloading of SKTS content via the Internet to a client computer, which is irrelevant in the context of the ongoing research, in the future, if this does not lead to contradictions, many sets of TB, constituting TB^g, we will consider it as a whole, as presented below:

$$TB^g = \bigcup_{\beta=1}^{\gamma} TB_\beta, \quad (15)$$

where: β—the TB number in SKTS; γ—the total number of TB in SKTS.

The SKTS header and datasets are many rows, each row of which consists of cells, as defined below by formulas (16–18).

The formal SKTS header expression is presented below:

$$TZ = \{TR_1, \ldots, TR_\kappa\}, \quad (16)$$

where: $TR_1, \ldots, TR_\kappa$—the SKTS header lines; κ—the number of lines in the SKTS header.

The formal expression of the SKTS dataset can be represented as follows:

$$TB = \{TR_1, \ldots, TR_\varpi\} \quad (17)$$

where: $TR_1, \ldots, TR_\varpi$—the rows of the SKTS dataset; ϖ—the number of rows in the SKTS dataset.

According to KTM HTML, each SKTS line consists of cells whose structural-logical diagram contains information about the name of the cell, its contents, the number of columns and rows covered by the cell. In addition, it defines the concepts of a unique identifier and a header cell identifier describing the contents of a cell. Considering the need to use the concepts of "object", "property", "instance", etc., in the analysis of SKTS, we modify the structural-logical diagram of the cell by introducing the object characteristics of the description of the contents of the cells, namely, the type of the content of the header cell and the attribute of belonging to the ODR object. Mathematically define a cell as follows:

$$\forall TR \in TZ \wedge \forall TR \in TB : CL = < ac, bc, cr, dc, ec, fc, lp, sc, isa > \quad (18)$$

where: ac—the name of the CL cell; bc—cell contents; cr—the number of columns covered by the cell; dc—the number of rows covered by the cell; ec—unique identifier of the cell; fc—identifier of the header cell, which provides the current header information; lp—the content type of the header cell can take the following values "property", "object"; sc—an attribute defining a set of data for which the current cell provides header information; is a—an attribute that determines which part the cell is part of.

We will distinguish between two types of cells:

1. A data cell is a cell CL, its contents correspond to the value of a property whose symbolic name is determined by the header cell, the identifier of which is determined by the attribute f. In addition, we take into account that the data cell does

not provide header information for other cells, i.e. the attributes lp, sc of the structural logic diagram of this cell will be declared equal ∅. Formally, we define a data cell as a cell of the following structure:

$$CL^{TD} = < ac, bc, cr, dc, ec, fc, \emptyset, \emptyset, isa > \quad (19)$$

2. A header cell is a cell whose contents are header information describing the values of other cells. The header cells contain the symbolic names of the properties or ODR objects. Formally, the header cell is defined as a cell of the following structure:

$$CL^{TH} = < ac, bc, cr, dc, ec, \emptyset, lp, sc, isa > \quad (20)$$

Moreover, the sc attribute defining a set of data cells for which the header cell provides header information must take one of the following values:

row—the cell provides header information for the rest of the row of the table containing it.
col—the cell provides header information for the rest of the column containing it.

In view of the foregoing, the formal model of SKTS, focused on the presentation of arbitrary tabular structures, has the means of describing the object characteristics of the contents of SKTS cells and can be defined by formulas (13–20).

Based on the model presented above, we will carry out the development of formal criteria for belonging to the types of table structures to specify the classes of table structures analyzed in the work and take into account the specific features of data organization specific to these classes.

6 The Method of Analysis of the Sources of Knowledge of Table Structures Based on Targeted Enumeration

In the context of solving the problem of forming an OKB, the SKTS analysis takes on the following meaning—the structure of the SKTS is examined with the aim of creating a "processing-oriented" view—the structural logical diagram of this source and a request for entering ODR objects in the OKB, instances of which make up the contents of this SKTS. Within the framework of such an approach to the analysis of the sources of knowledge of a table structure, it is understood that a set of actions is taken to create structural logical schemes for the header and SKTS data set; mapping between header and dataset cells; as well as the formation of a structural-logical diagram of a request to enter ODR objects in OKB.

To reduce the algorithmic complexity of the proposed method, we will conduct SKTS analysis based on targeted enumeration of SKTS cells by setting user-defined criteria for the analysis of SKTS structural logic circuits.

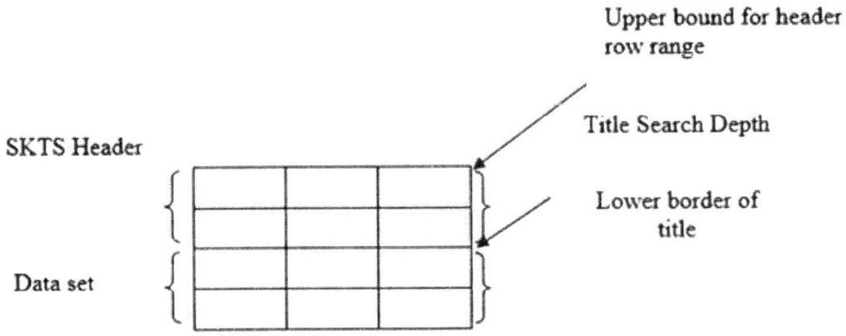

Fig. 1 SKTS header borders

As criteria for the analysis of SKTS structural logic circuits, we will use the following:

1. ε—depth of viewing the data set (in rows). This criterion sets the boundary for viewing the rows of the SKTS dataset, sufficient to make a decision in machine analysis of the structure of the dataset.
2. g_{table} —SKTS header search depth (in lines). This criterion sets the possible range of SKTS strings, as shown in Fig. 1, the cells of which contain header information, i.e., in essence, sets the possible bounds of the header. The beginning of the header is the first top line of the SKTS. Lower bound i.e. the number of the last line of the header is set by the user depending on the number of levels of the header structure.
   Thus, the SKTS header is a set of SKTS lines, starting from the first line and ending with the SKTS line with a number g_{table} , and the data set is from $g_{table} + 1$ and to the end of SKTS.
3. The criterion for the interpretation of the types of header cells according to their structure ω_{int}

Let $< ac, bc, cr, dc, ec, fc, lp, sc, isa >^{CL}$—the structural-logical diagram of a cell, then if this cell spans more than one column, i.e. when $cr^{CL} > 1$, then its contents will be interpreted as the symbolic name of the object. If the cell spans more than one row, i.e. when $dc^{CL} > 1$, then the contents of this cell will be interpreted as the symbolic name of the property. Otherwise, i.e. when there is one cell at the intersection of a column and a row, we will also interpret the contents of the cell as the symbolic name of the property. Mathematically, this criterion can be represented using the following function, the result of which is 1 if the cell content matches the symbolic name of the object, and 0 in other cases:

$$\omega_{int}(CL) = \left\{ \begin{array}{l} < ac, bc, cr, dc, ec, fc, lp, sc, isa >^{CL} |cr > 1 : 1, \\ < ac, bc, cr, dc, ec, fc, lp, sc, isa >^{CL} |dc > 1 : 0, \\ else : 0, \end{array} \right\}. \quad (21)$$

where $< ac, bc, cr, dc, ec, fc, lp, sc, isa >^{CL}$—the structural-logical diagram of the cell CL.

Thus, the analysis of SKTS taking into account the above criteria for the analysis of SKTS structural logic circuits will be carried out by setting the criteria for interpreting SKTS structural logic circuits and performing a number of actions aimed at determining the type of tabular structure, forming SKTS structural logic circuits and requesting input of objects ODR in OKB, which will allow for targeted cell traversal on specified types of table structures and generate requests for entering ODR objects in OKB.

Thus, the sequence of the SKTS analysis method based on targeted enumeration, taking into account the above criteria for the analysis of SKTS structural logic schemes, will be as follows:

1. Preliminary analysis of SKTS, aimed at determining the type of table structure used to organize the contents of SKTS and checking the possibility of machine analysis of SKTS in order to generate a request for entering ODR objects in OKB.
2. The formation of the structural logic diagram of the SKTS header by analyzing the structural features of SKTS based on the analysis criteria of the SKTS structural logic diagrams.
3. The formation of the structural logic diagram of the SKTS dataset in accordance with the method of graphical structuring of tabular data used in SKTS.
4. Establishing correspondence between header cells and data cells to create a connection between data and information describing them by analyzing their structural logic diagrams.
5. Determination of the symbolic name of the ODR object by analyzing the SKTS structural logic diagram to generate a request for entering objects in the OKB.
6. Determination of the structural-logical schemes of properties by analyzing the structural-logical scheme of the SKTS header based on the criteria for interpreting the types of header cells according to their structure according to expression (21).
7. Determining the ownership of properties of ODR objects by correlating the contents of header cells containing the properties of the ODR object with the contents of header cells containing the symbolic names of ODR objects in accordance with the method of graphical structuring of tabular data used in SKTS.
8. Formation of a request to enter ODR objects in OKB by composing the structural-logical diagrams of property-values, property-objects and the symbolic name of the ODR object into a single structure.

Steps 1–4 are intended for the formation of the SKTS structural logic diagram, further processing of which will allow generating a request for entering ODR objects in OKB, as well as instances of ODR objects that make up the contents of SKTS.

Steps 5–8 are intended for processing the SKTS structural logic diagram obtained in the previous steps to generate a request to OKB to enter ODR objects from which OKB will be generated.

7 Experimental Studies of the Effectiveness of the Method of Targeted Enumeration in the Formation of OKB

The upper estimate of the computational complexity (UECC) of the algorithmic support of the OKB formation method based on targeted enumeration from ODR objects obtained on the basis of the SKTS analysis by the proposed method and their instances obtained on the basis of the analysis of the SKTS data set will be calculated on the basis of the time cost criterion.

Let be n^q—the number of requests for input of ODR objects from which OKB will be formed; d^{mn}—the maximum number of value properties in ODR objects; f^{mn}—the maximum number of object properties in ODR objects; t^{mn}—the maximum number of nesting of property-objects in the structural-logical circuits of ODR objects; u^{mn}—the maximum number of subclass objects in ODR class objects; v^{mn}—the maximum number of levels of the object hierarchy in relation to "be a subclass"; i^{mn}—the maximum number of instance objects in ODR class objects, then the total number of class objects added to the generated OKB can be defined as $r = n^q \cdot t^{mn} \cdot f^{mn}$.

To form the OKB, we will use the following elementary operations: comparing the symbolic names of ODR objects; comparing property values; comparing properties of objects; adding an object; Add value properties adding object properties; adding instance objects.

For each of these operations, we determine the value of their UECC.

So, the UECC operation of symbolic comparison of ODR (ct_s^{cp}) object names, used when moving through the tree hierarchy of ODR objects when searching for the most relevant OKB object query, will be equal to the time spent on comparing two symbolic names (zt_s^{cp}):

$$ct_s^{cp} = zt_s^{cp} \tag{22}$$

The UECC of the property-value comparison (ct_d^{cp}) operation used to compare the value-property of the request objects and OKB will be equal to the time (zt_d^{cp}) for comparing the value-property of the request objects and OKB:

$$ct_d^{cp} = d^{mn} \cdot zt_d^{cp} \tag{23}$$

The UECC operation of comparing the properties of objects-objects (ct_f^{cp}), used to compare the properties of objects of request objects and OKB, will be equal to the time spent (zt_f^{cp}) on comparing the properties-values of request objects and OKB:

$$ct_f^{cp} = f^{mn} \times zt_f^{cp}. \tag{24}$$

UECC operations of adding an object (ct_o^{+}) is used after finding the most relevant OKB object request for entering the object in OKB and equals the time spent on performing this operation (zt_o^{+}):

$$ct_o^+ = zt_o^+. \quad (25)$$

The UECC operation of adding property-values (ct_d^+), intended to enter new property-values in OKB, is equal to the time (zt_d^+) on checking for uniqueness and entering property-values in OKB:

$$ct_d^+ = d^{mn} \times zt_d^+. \quad (26)$$

The UECC operation of adding property-objects (ct_f^+), intended to enter new property-objects in OKB, is equal to time (zt_f^+) on checking for uniqueness and entering property-objects in OKB:

$$ct_f^+ = f^{mn} \cdot zt_f^+. \quad (27)$$

The UECC of the operation of entering instance objects (ct_i^+), intended to enter instances of the request object in OKB, is equal to the product of the time spent on adding one instance object and the total number of instance objects added to OKB (i^{mn}):

$$ct_i^+ = i^{mn} \times zt_i^+. \quad (28)$$

Considering the above, we present UECC of forming OKB (T^{max}) as the time spent on processing r ODR objects from which OKB is formed. Moreover, during the processing of each of the r ODR objects, it is necessary to carry out a targeted search of the ODR objects already entered into the OKB by selecting at each level of the object hierarchy from u^{mn}—subclass objects the most relevant object and search, if possible and necessary, more relevant or exactly matching the request OKB object in ($\nu^{mn} - \tau$)—levels of the object hierarchy of OKB objects, where is the number of the current level of the object hierarchy.

Mathematically T^{max} determined by the following expression:

$$T^{max}(r) = \sum_{m=1}^{r} u^{mn}\nu^{mn}(ct_s^{cp} + ct_d^{cp} + ct_f^{cp}) + ct_o^+ + ct_d^+ + ct_f^+ + ct_i^+. \quad (29)$$

where $ct_s^{cp}, ct_d^{cp}, ct_f^{cp}, ct_s^+, ct_p^+, ct_d^+, ct_f^+, ct_i^+$—UECC operations for comparing symbolic names of ODR objects, comparing property values, comparing object properties, adding a superclass object, adding a subclass object, adding a value property, adding an object property, adding objects- copies respectively.

Thus, the formation time of an OKB T from r ODR objects is limited from above by the value of the function T^{max}:

$$T(r) \le T^{max}(r). \quad (30)$$

An analysis of expression (29) taking into account expressions (22–28) shows that the parameter has the greatest impact on the complexity of the proposed algorithmic support by the criterion r of time costs, and the increase in the complexity of OKB formation with an increase in the number of objects has a polynomial character.

For experimental verification of the effectiveness of the algorithmic support proposed by the author in comparison with other implementations, we consider the implementation of the algorithmic support for the formation of OKB based on the widely used StringSearch object search method described in WebProtege. The StringSearch method searches for objects by the symbolic names of ODR objects without taking into account their structural and logical characteristics. Thus, UECC of forming OKB (T_{wb}^{max}) based on StringSearch, by analogy with the algorithmic software developed by the author, is represented as the time spent on processing r—ODR objects from which OKB is formed. A feature of the implementation of this algorithmic support is that when adding the k-th ODR object to OKB, it is necessary to loop through at most k-1 OKB objects and add this (k-th) ODR object to it with a superclass object or subclass object. The mathematical expression of the UECC of this algorithmic support is presented below:

$$T_{wb}^{max}(r) = \sum_{m=1}^{r} (m-1)ct_s^{cp} + ct_o^{+} + ct_d^{+} + ct_f^{+} + ct_i^{+}. \quad (31)$$

Let us compare experimentally the above implementations of the algorithmic support for the formation of OKB.

The experimental values of the UECC OKB formation depend on the software implementation and configuration of computer technology. Table 1 presents the average values of time spent on operations.

The maximum values of the time costs of operations (22–28) are presented below:

$$ct_s^{cp} = 118.619(mcs)$$

Table 1 Time spent on operations (22–28), MCS

ct_s^{cp}	ct_d^{cp}	ct_f^{cp}	ct_o^{+}	ct_d^{+}	ct_f^{+}	ct_i^{+}
106.426	561.46	354.474	1234.37	638.56	667.89	75,713.3
110.346	582.41	331.848	1248.62	663.696	693.03	66,369.6
106.256	563.56	333.365	1264.43	638.556	687.12	63,101.4
118.619	582.44	316.764	1238.69	643.597	673.43	63,855.6
106.726	561.46	332.344	1280.46	648.559	657.39	64,903.1
110.686	583.47	317.654	1249.12	669.611	697.86	64,484.1
106.425	561.46	324.798	1235.41	632.665	687.59	64,903.1
110.116	582.41	306.708	1252.32	683.611	663.07	64,884.8
106.436	561.46	331.12	1272.92	631.536	657.49	64,400.3
110.623	582.69	316.764	1276.27	661.196	643.08	63,384.1

Table 2 UECC experimental values forming OKB

r	T^{max}, mcs	T^{max}_{wb}, mcs
15	1,617,387	1,343,246
21	2,264,342	2,142,100
27	2,911,297	3,205,711
33	3,558,252	4,610,945
39	4,205,207	6,434,667
45	4,852,161	8,753,741
51	5,499,116	11,645,034
57	6,146,071	15,185,410
63	6,793,026	19,451,734
69	7,439,981	24,520,871

$$
\begin{aligned}
ct_d^{cp} &= 583.47(\text{mcs}) \\
ct_f^{cp} &= 354.474(\text{mcs}) \\
ct_o^{+} &= 1280.46(\text{mcs}) \\
ct_d^{+} &= 683.611(\text{mcs}) \\
ct_f^{+} &= 697.86(\text{mcs}) \\
ct_i^{+} &= 75,713.3(\text{mcs})
\end{aligned}
$$

The experimental values T^{max} and T^{max}_{wb} obtained on the basis of the maximum values of the time costs of operations (22–28) depending on the parameter r with $t^{mn} = 1$, $u^{mn} = 4$ and $\nu^{mn} = 6$ are presented in Table 2.

Graphs of the dependence of the complexity of OKB formation on the number of ODR objects from which OKB is formed are presented in Fig. 2.

Thus, a comparison of the time spent on OKB formation (Table 1) and the graphs of the dependence of the complexity of OKB formation on the number of ODR objects from which OKB is formed (Fig. 2) show that with an increase in the number of objects, ODR is characterized by exponential growth, and T^{max} linear (polynomial 1-th order), which confirms the effectiveness of the developed approach.

8 Conclusion

The article presents the results of a study of ways to solve the problem of improving the efficiency of electronic document management in the Semantic Web by developing methods and means of forming ontological knowledge bases based on sources of tabular structures. The results obtained are of great scientific and practical importance both for the development of modern industrial technology of the Semantic Web initiative knowledge bases, and for solving specific problems of organizing electronic document management of enterprises based on ontological knowledge bases with the

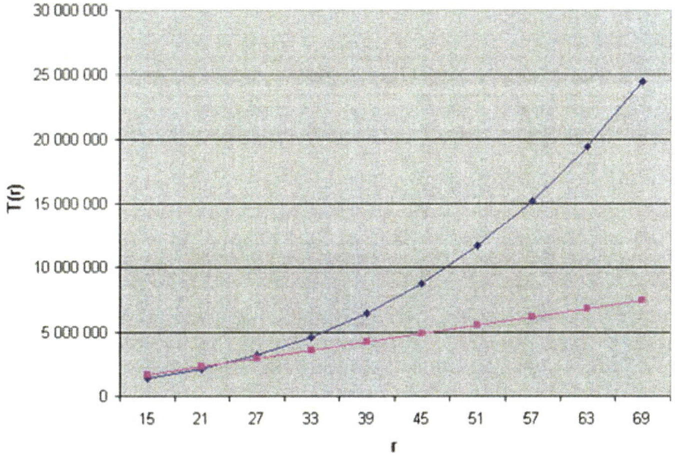

Fig. 2 Graphs of the dependence of the complexity of OKB formation on the number of ODR objects from which OKB is formed: —◆—UECC of forming OKB based on StringSearch (T_{wb}^{max}); —■—BOBS formation of OKB based on targeted enumeration (T^{max})

possibility of obtaining new knowledge in the form of ODR related objects from table knowledge sources structures of various formats.

References

1. Berners-Lee, T., Handler, J., Lassila, O.: The Semantic Web. Sci. Am. **284**(5), 34–43 (2001)
2. Shostak, I., Volobuyeva, L., Danova, M.: Ontology based approach for green software ecosystem formalization. In: Abstracts of the DEpendable Systems, SERvices and Technologies—DESSERT'2018, IEEE Ukraine Section, Kyiv 24–27 May 2018
3. Shostak, I., et al.: Ontological approach to the construction of multi-agent systems for the maintenance supporting processes of production equipment. In: Abstracts of the IEEE International Scientific and Practical Conference «Problems of Infocommunications. Science and Technology» (PICS&T-2018), Kharkiv, 9–12 Oct 2018, pp 209–214 (2018)
4. Pavlenko, V., et al.: Information support for business processes on the virtual enterprises with the use of multi-agent technologies. In: Abstracts of the DEpendable Systems, SERvices and Technologies—DESSERT'2018, IEEE Ukraine Section, Kyiv 24–27 May 2018
5. Noy, N.F., et al.: The knowledge model of protégé-2000: combining interoperability and flexibility. In: 2th International Abstracts of the Conference Knowledge Engineering and Knowledge Management, Springer, Juan-les-Pins, pp. 17–32 (2000)
6. Intelligent, A.H.: E-Buisness: from technology to value. IEEE Intell. Syst. **16**(4), 8–10 (2001)
7. Bast, R.: Learning the business of business. IEEE Intell. Syst. **16**(4), 4–7 (2001)
8. Hendler, J., Berners-Lee, T., et al.: Integrating applications on the semantic web. J. Inst. Electr. Eng. Jap **122**(10), 676–680 (2002)
9. Noy, N., Sintek, M., Decker, S., et al.: Creating semantic web contents with protege-2000. IEEE Intell. Syst. **2**(16), 60–71 (2001)
10. Rector. A.L.: Modularization of domain ontologies implemented in description logics and related formalisms including OWL. In: Abstracts of the 2nd International Conference on Knowledge Capture, Sanibel Island (USA): ACM Press, pp. 51–59 (2003)

11. Embley, D.W., Campbell, D.M., Jiang, Y.S., et al.: Conceptual-model-based data extraction from multiple-record web data. Data Knowl. Eng. **31**(3), 227–251 (1999)
12. Lopresti, D., Nagy, G.A.: Tabular survey of automated table processing. In: Proceedings of the Third IAPR Workshop on Graphics Recognition, Jaipur (India), pp 93–120. Springer, Berlin/Heidelberg (2000)
13. Hammer, J., Garcia-Molina, H., Cho, J., et al.: Extracting semistructured information from the Web. In: Proceeding of the Workshop on Management of Semistructured Data, Tucson (USA), p 50. AIII Press/MIT Press (1997)
14. Yourdon, E.: Modern Structured Analysis. Yourdon Press/Prentice Hall, N.J. (1989)
15. Liddle, S., Embley, D.W., Yau, D.S.: Extracting data behind web forms. In: Proceeding of the Joint Workshop on Conceptual Modelling Approaches for E-business: A Web Service Perspective, Tampere (Finland), pp 38–49. Springer, Berlin (2002)
16. Cowie, J., Lehnert, W.: Information extraction. Commun. ACM **39**(1), 80–91 (1996)
17. Biscup, J., Embley, D.W.: Extraction information from heterogeneous information sources using ontologically specified target views. Inf. Syst. **28**(3), 169–212 (2003)
18. Gordijn, J., Akkermans, H.: Designing and evaluating e-business models. IEEE Intell. Syst. **16**(4), 11–18 (2001)
19. Hu. J., Kashi, R., Lopresti, D., et al: Why table ground-truthing is hard. In: Proceeding of the 6th International Conference on Document Analysis and Recognition, Washington (USA), pp. 129–133. IEEE Computer Society (2001)
20. Kuznetsov, A., et al.: Performance of hash algorithms on GPUs for use in blockchain. In: 2019 IEEE International Conference on Advanced Trends in Information Theory, ATIT 2019 – Proceedings Kyiv, Ukraine, pp. 166–170 (2019). https://doi.org/10.1109/atit49449.2019.9030442
21. Green, E., Krishnamoorthy, M.: Model-Based analysis of printed table. In: Proceeding IAPR International Conference on Document Analysis & Recognition III. Montreal (Canada), pp. 80–91. Springer, Berlin (1995)

Semantic Techniques to Support IoT Interoperability

Beniamino Di Martino and Antonio Esposito

Abstract Smart devices and sensors have reached a very high level of pervasiveness: we are practically surrounded by intelligent items, which continuously communicate with each other and collect information. One of the most challenging issues regarding the use of such sensors regards the possibility to seamlessly make them interoperate to reach a specific goal. This objective could be difficult to achieve, due to the lack of a universally accepted standard for sensor communications. In this paper, we present a prototype tool for the analysis of sensors' API that, through a semantic graph representation, tries to overcome the possible interoperability issues that may arise in a sensor network, and provides instrument to support sensors' orchestration and management.

1 Introduction

In modern society IT technology has reached an unprecedent level of pervasiveness, as sensors, smart devices and intelligent systems have become of extremely common use. Just think bout smartphones: these devices come with a large number of embedded sensors, which are exploited by applications to deliver a variety of functionalities and services. The term "Internet of Things" (IoT) has surely become a "buzz word", as it ultimately represents the ubiquity of modern technologies and their fusion with every-day life. IoT comes with many opportunities, as the application scenarios of such technology are almost limitless: traffic control and automotive, domotics, health monitoring and patient caring, environmental monitoring are just a few of the possible cases in which IoT can be usefully applied. However, as the number of sensors' and smart devices' manufactures grows, together with the com-

B. Di Martino (✉) · A. Esposito
Department of Engineering, University of Campania "Luigi Vanvitelli", Aversa (CE), Italy
e-mail: beniamino.dimartino@unina.it

A. Esposito
e-mail: antonio.esposito@unicampania.it

B. Di Martino
Asia University, Taichung, Taiwan

R. Pandey et al. (eds.), *Semantic IoT: Theory and Applications*, Studies in Computational Intelligence 941, https://doi.org/10.1007/978-3-030-64619-6_10

plexity of the networks which can be built upon such technology, issues are bound to arise. One of the main problems which needs to be addressed regards the ability of sensors and devices, coming from different manufacturers, to immediately interoperate with each other. Whereas the adoption of open standards for devices' interconnection such as OPC-UA [12] or OMG Data-Distribution Service for Real-Time Systems (DDS) [17] could solve, or at least ease, the interoperability problem, often manufacturers prefer to resolve to proprietary solutions, or they simply introduce variations to differentiate their products. Another aspect to be considered is the possibility to automatically build and manage complex sensor topologies, given a set of specific requirements and a description of the sensors' functionalities and interfaces. Semantic technologies have proved in the past to be particularly suitable for the representation of heterogeneous data, and their consequent homogenization, and have also been adopted in the representation of complex systems and services [8, 15]. The application of semantic technologies to the representation of sensors' API can lead to interesting results and lead to the resolution of possible interoperability issues which may arise when trying to compose complex sensor networks. Also, semantic matchmaking can represent the base for advanced sensor composition and the management and orchestration of services based on smart devices. This paper aims at describing a semantic based framework which, starting from the description of Rest-full interfaces used to communicate with smart sensors, is able to derive a semantic representation of the sensors' functionalities and input/output parameters. Such a representation is then used to match different sensors and identify possible combinations thereof, having a specific goal in mind, and to match their input/output data. The presented tool is an improvement of the framework presented in [9]. The remainder of this paper is organized as follows: Sect. 2 reports a State of the Art in the field of sensors' interoperability; Sect. 3 introduces the semantic technologies used to implement the tool; Sect. 4 provides a description of the methodology applied to build the tool; Sect. 5 focuses on the creation of the Graph behind the semantic representation of the sensors' API; Sect. 6 provides information on the WSD and OWL-S generation; Sect. 7 closes the paper with final considerations.

2 State of the Art

Interoperability among services and sensors provided by different vendors is considered a hot topic, and related problems have been addressed at various levels, generally involving the device, network and data levels. To tackle such issues, one of the most desirable solutions to possibly enable complete interoperability is to adopt shareable and open (when available) standards. This could also represent a good opportunity for consortiums and organizations of difference sizes to engage in the development of new technologies, imposing de-facto standards in an ever changing and chaotic market.

The OPC Foundation developed the **OPC Unified Architecture (OPC-UA)** [14], a communication technology completely based on TCP/IP, belonging to the family of Machine to Machine (M2M) communication protocols, that allows sensors and devices to exchange information seamlessly. Despite being first employed in the field of industrial automation, OPC-UA has been already adapted to work in different environments and sectors: oil and gas exploration, smart building and automation, energy production are only some of the field in which it is currently employed. OPC-UA is considered a common base to establish communications among sensors, but it does not impose anything regarding the interfaces they expose or the formats of the exchanged data.

Using the **ETSI-M2M** standard as a reference, the **Open Machine to Machine (OM2M)** platform [1] offers a set of open source services for M2M interoperability. OM2M relies on REST to make the development of services and applications completely independent of the particular network used to connect them, and provides a modular architecture which can be extended through plugins. Also, it incorporates the SmartM2M [7] standard. Consortia are fully engaged in the creation of new standards for sensors' interoperability. A cross-industry consortium going under the name of **AllSeen Alliance**[1] has been dedicated to the definition of a framework, namely **AllJoyn**, to enable interoperability of devices, services and applications for the Internet of Things. The framework offers open and secure programmable interfaces, which enable software and services connectivity. In particular, it allows companies to create interoperable products that can connect, interact and discover other products implemented by following the AllJoyn framework. Applications and/or sensors developed using the AllJoyn framework are independent of the transport layer, platforms, brand and Operative Systems used. Another example of consortium engaged in the definition of a standard communication framework is represented by the Open Interconnect Consortium (**OIC**), which is focused on industry standards to enable wireless connections among IoT sensors and devices, and thus to manage the flow of data among them. As a result of the efforts of the consortium, the **IoTivity** Project[2] has been carried out, to build an open source software framework for device-to-device connectivity.

In February 2016 the Open Interconnect Consortium changed its name to **Open Connectivity Foundation (OCF)**[3] and in October 2016 it has merged with the AllSeen Alliance, thus creating a unique consortium working on IoT interoperability.

Among research projects we find **INTER-IoT**,[4] which is supported by the Horizon 2020 European program. The objective of the project is to create an interoperable IoT framework, comprehending tools and methodologies for the integration of heterogeneous IoT Platforms [11], and applicable to different domains.

OMG has also contributed to the realization of standards for the interoperable connection of devices, by creating the **Data-Distribution Service for Real-Time Systems (DDS)** [17]. DDS consists in a network protocol implementing the classical Observer Design Pattern, which regulates the transmission of data, acts on events and exchange of commands among devices belonging to the network topology. DDS

---

[1] AllSeen Alliance. https://allseenalliance.org/. Accessed on 20 March 2016.

[2] IoTivity. https://www.iotivity.org/. Accessed on 20 March 2016.

[3] Open Connectivity Foundation—https://openconnectivity.org/. Accessed on March 2020.

[4] Inter-IoT project. http://www.inter-iot-project.eu/.

defines a virtual Global Data Space where applications freely read and write data by using a name and key pair, which are used to uniquely identify them and make them discoverable by interested services.

2.1 Semantic Technologies

Semantic technologies are mostly based on the definition of **Ontologies** to provide a common reference for terms and concepts. In particular, an Ontology can be defined as a formalized representation of knowledge, a sort of standardly presented vocabulary composed of concepts and meaningful entities that are in relation with each other. One could be tempted to compare them with taxonomies, because of their innate hierarchical structures, but they differ because ontologies allow for the definition of complex relationships among concepts, which greatly surpasses a merely hierarchical categorization. Terms used in an ontology have a specific meaning, at least in the domain in which the ontology itself has been developed. That is why two different kinds of ontologies can be taken in consideration.

Upper Ontologies provide a more general description of knowledge, in which the represented terms, concepts and relations can be applied to several domains, without losing meaning. Information contained in Upper Ontologies can thus be applied in different context, and used as a common reference.

Domain Ontologies focus on a specific domain, and the concepts they represent can have different meaning from a context to another. It is common to have the same term representing different information in different Domain Ontologies.

Through the years, ontologies have been expressed through a series of different standards, generally based on XML, such as RDF and more recently **OWL** [4], which became the most used. Based on OWL several ontologies have been defined to support data and services interoperability. As an instance, OWL-S is a complex ontology framework used to describe services, their relations and compositions [13]. Through the use of a **Service** class, OWL-S defines a reference to a specific semantic service, which is in turn composed of three other parts:

- the **Profile** class, essentially meant to be human readable, offers a natural language based description of what the specific service offers. The Profile is connected to a series of properties which emphasize the service's name, complete description, quality of services, context applicability and possible limitations.
- the **Model** class offer a precise description of the interfaces of the Service, including its input and output parameters, message formats, adopted communication protocols or port numbers.
- the **Grounding** class allows the mapping between the abstract descriptions offered by Model and Profile to concrete implementation of the service, exposed via standard languages, namely WSDL.

All the connections between the Service class and the Grounding, Model and Profile Classes are obtained via specific properties.

One of the main characteristics of OWL-S, which makes it particularly adapt to describe Services and compositions thereof, is its capability to describe a service's behaviour, considering all the possible interactions that a user is allowed to have with it.

2.1.1 Application of Semantic Technologies to the IoT World

The use of Semantic Technologies in the IoT field is well documented, and several ontologies have been developed in past years to efficiently describe sensors, in terms of their capabilities and requirements, and of the data they exchange.

The work presented in [3] can be considered one of the first attempts to properly apply semantics to sensor networks, by using an ontology-driven approach to increase the network's adaptability.

Among the first standard ontologies developed to tackle interoperability and adaptability issues in sensors networks, the one presented in [6] is surely worth citing. The **Semantic Sensor Network** (SSN) ontology, developed by the W3C Semantic Sensor Network Incubator Group is freely available[5] can be considered an Upper Ontology, as it describes general sensor concepts, such as measurements, observations and sensors' relations, but it does not provide any domain-related concepts. The work of the incubation group also introduces the concept of Stimulus-Sensor-Observation (SSO) pattern in their ontology description, trying to reduce all the described concepts to three basic terms: *Stimuli*, that are detectable changes in the observed environment; *Sensors*, in charge of the actual observation and of the transformation of the physical stimulus to its digital representation; *Observation* representing the transformation of the stimulus into the sensor's output.

Other ontologies have been developed on top of SSN, such as the **Sensor, Observation, Sample, and Actuator Ontology (SOSA)** [10][6] which can be considered an lightweight update of the SSN, introducing a simplification of the original SSO pattern and providing support to modern sensors descriptions.

Other recent standards include **MyOntoSens** [16], an ontology which describes whole sensor measurement process including concepts like inputs, outputs, calibration, drift, latency, and precision. The ontology is being standardized as a Technical Specification (TS) within the SmartBAN Technical Committee of the ETSI (European Telecommunications Standards Institute).[7]

Since modern Smartphones contain numerous sensors that can be used to build complex sensor networks, especially with the development of Edge and Fog Computing, specialized ontologies have been built to provide formal description thereof. The work described in [2] presents the **SmartOntoSensor** ontology, based on previously

[5] Semantic Sensor Network Ontology—https://www.w3.org/2005/Incubator/ssn/ssnx/ssn.

[6] SOSA—https://www.w3.org/TR/vocab-ssn/.

[7] Smart Body Area Networks (SmartBAN); Service and application standardized enablers and interfaces, APIs and infrastructure for interoperability management—https://www.etsi.org/deliver/etsi_ts/103300_103399/103327/01.01.01_60/ts_103327v010101p.pdf.

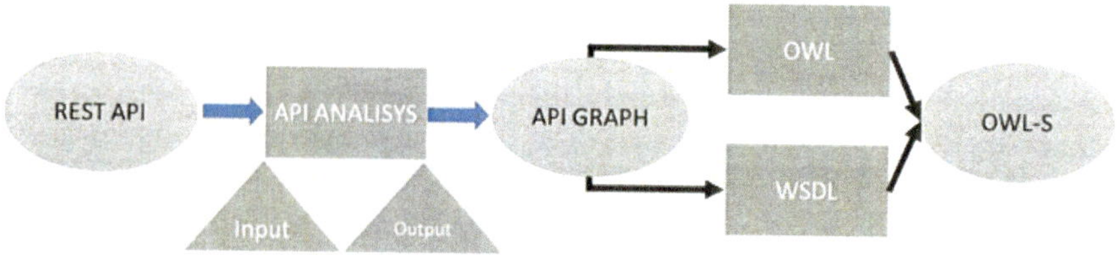

Fig. 1 The applied methodology

published standards such as SSN and SensorML,[8] which describes both smartphones and sensors from different points of view, including the specific platforms they are supported by, their measurements capabilities and observations, taking also in consideration their context model.

3 The Methodology

The approach applied in the development of the tool is composed of a series of consecutive steps, which are graphically summarized in Fig. 1. The first fundamental step is represented by the analysis of the sensors' API, which represents the base to populate an ontology of the sensors' functionalities, inputs and outputs. This step is essential, as without a shared machine-readable knowledge base, it would be impossible to actually match sensors and try to build new topologies. The analysis must take in consideration both input and output parameters of the sensors' API, as they will represent the "glue" between different types of devices. The result of the analysis is represented by an ontological representation, in OWL, which contains the information extracted from the API in a machine-readable format, and organized according to a structured graph, the so called API Graph. The information contained in the graph are complemented with a WSDL representation of the API, which is used as the grounding of an OWL-S representation. Both the API ontology and WSDL are obtained via the API analysis, whereas the OWL-S representation is created afterwards, and only if the representation of API requires dynamic information regarding, for an instance, the flow of steps needed to operate the sensors. The OWL-S representation becomes essential when sensors are composed, in order to manage their orchestration.

[8]Sensor Model Language (SensorML)—https://www.ogc.org/standards/sensorml.

4 The API Analysis

Analysing the API exposed by sensors' vendors is not only the first, but also the most delicate step. Indeed, understanding how the APIs can be called, through the accepted input parameters, and interpret the results of the call by reading the output, can be quite complicated when it comes to automatically doing it. Sensors' input parameters and output responses are generally described either in Json (which is the most common representation) or XML. In both cases, such parameters are very well structured, but such a structure varies greatly when comparing different providers. Also, most of the modern APIs are represented by REST calls. Starting with these assumptions, we are experimenting with three different possible approaches:

1. A completely automatic analysis of the REST APIs, obtained via commercial or open-source tools, such as IBM Appscan Security, ZAP and SoapUI. Such tools generally address security and quality of service issues, by analyzing the input parameters accepted by a REST call and the returned output.
2. The exploitation of text parsers, which can analyze the online documentation provided by manufacturers and identify the description of input and output parameters. This kind of approach requires the application of natural language processing (NLP) techniques, which are not always reliable, especially when the structure and semantics of the documentation vary from a source to another.
3. A completely manual analysis of the API and population of the API ontology. This is the most effective approach, as the possibility to make errors and create ambiguities is extremely reduced (provide that the human operator is skilled enough to understand the sensors' descriptions). However, it is highly inefficient, as such analysis would require time, and would also be impractical for frequent updates of the sensors' description.

Considering the practical difficulties of applying only one of the three approaches, we have considered a combination thereof: a human operator always acts as a supervisor for the automatic analysis obtained via REST tools and web parsers, solving ambiguities and possible misinterpretations. However, once the API ontology description has been consolidated, automatic tools are able to operate almost independently, as the possibility to introduce errors is reduced dramatically with a stable ontology. So, eventual updates of the API can be handled by automatic analysis tools, with sporadic intervention from a human operator required. In order to start the analysis of an API, the user can use the main control panel of the proposed tool, shown in Fig. 2. There are two main options available: the user can import existing services definitions (she has already analyzed an API before) or start a new analysis. In the first case, after the import, the user will be presented with a set of Services, which she can then select to proceed with the semantic annotation, as shown in Fig. 3. Otherwise, she will be asked to insert the URL of the sensor's REST service and she will be asked to choose the input parameters (which will be automatically retrieved) to use to probe the service and analyze the output.

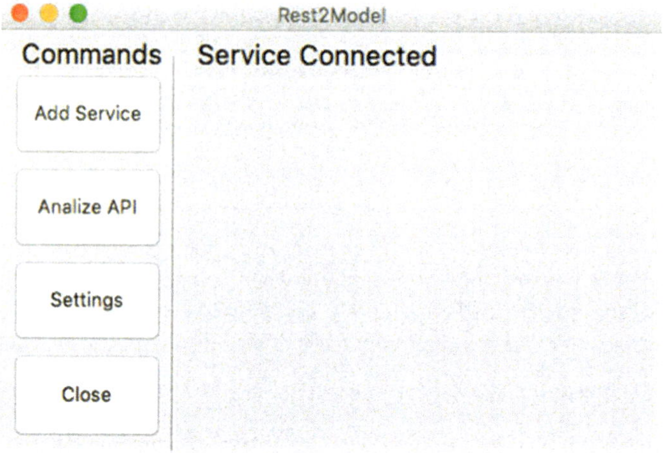

Fig. 2 The starting panel of the tool

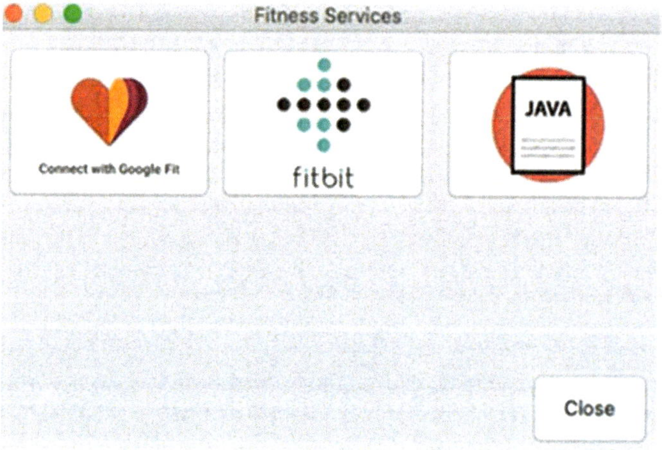

Fig. 3 Selection of imported services

Figure 4 shows the panels presented to the user in the first and latter cases. Three different possible inputs are identified: the user needs to select at least one of the proposed parameters in order to start the output analysis. An aspect to be necessarily considered is the eventual permissions needed to access the REST API. As an instance, Google services require explicit authorization: the tool will allow pop-ups to deal with this kind of situations.

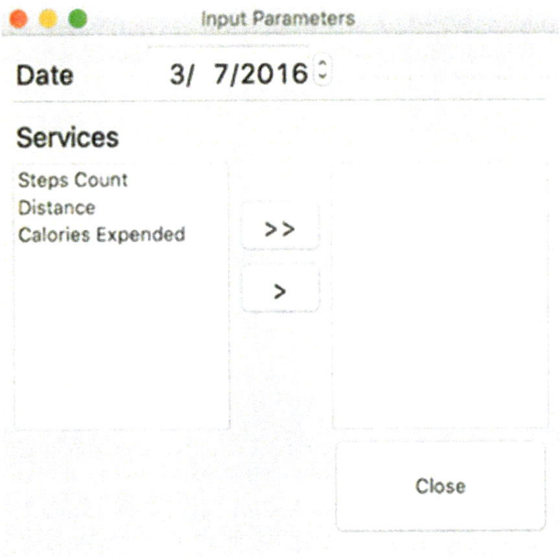

Fig. 4 Selecting the input parameters to probe the service

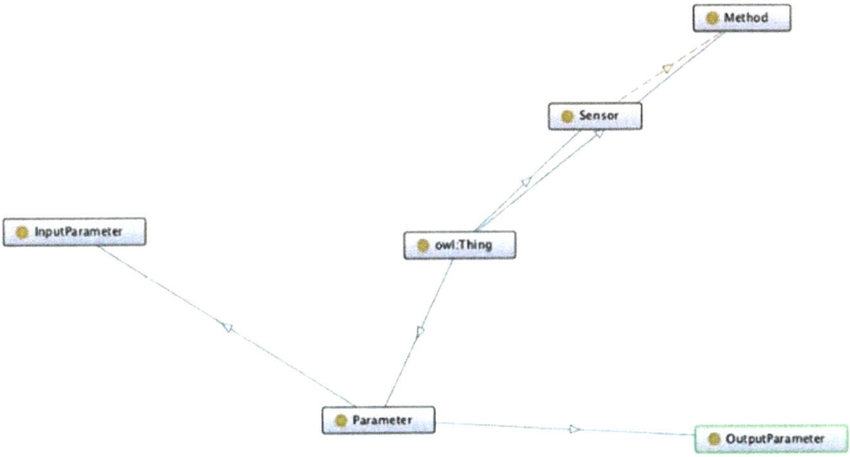

Fig. 5 The base API ontology

5 The API Graph

The information coming from the API analysis are used to populate a base ontology which represents the API graph. Such an ontology is, at the very beginning, completely empty, consisting only of a set of Classes and Relationships (Object properties in our case). The base ontology, reported in Fig. 5, is composed of three classes: (1)

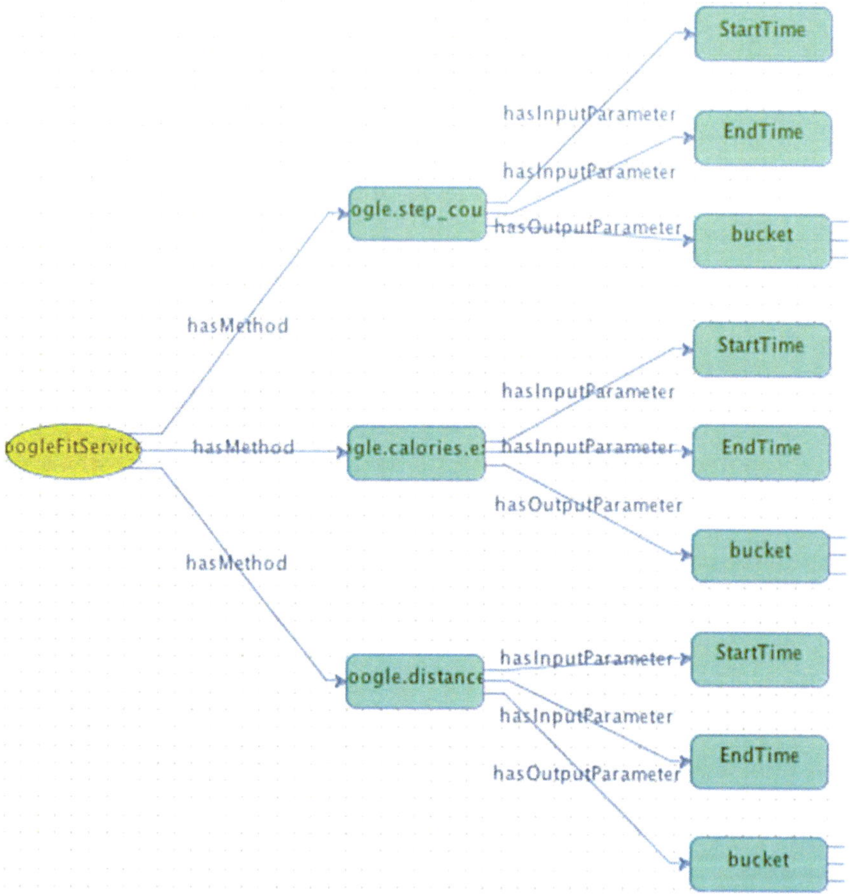

Fig. 6 The Google Fit Service representation

Sensor, to describe all possible sensors, not considering their specific functionality. (2) Parameter, which is subclassed by InputParameter and OutputParameter. Such classes represent input and output of the sensor. (3) Method, representing the particular function that can be called on the sensors.

The aforementioned classes are connected via specific object properties: (1) hasElement/ElementOf are converse properties that connect complex and simple parameters where structured ones exist. (2) hasParameters is used to link parameters to specific methods (3) Exposed is used to declare that a sensor has a specific functionality To express equivalences between methods, sensors and parameters no further elements are needed, as OWL provides native structures to identify equal concepts. Figure 6 provides a simple example of the API graph produced after analyzing the REST calls of the Google Fit Service.

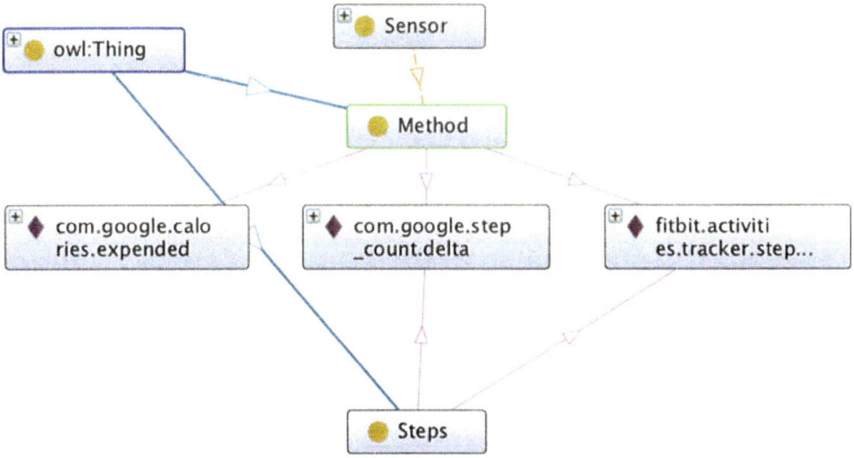

Fig. 7 FitBit API graph

The API graph contains, in this specific case, one Sensor (Google Fit Service), three methods (step_count, calories_estimate and distance_calculation) and three parameters (Start and EndTime, which are simple Input Parameters, and Bucket, which is a complex Output Parameter). Whereas this kind of representation is immediately readable by machines, it still lacks proper contextualization, that is a clear disambiguation of the meaning of the elements composing it. For this reason, a Domain Ontology is needed. Domain ontologies provide the semantics necessary to properly describe terms and entities. Several domain ontologies exist, especially in the medical field, but a good number of IoT ontologies have also been populated, as reported in [5]. If we consider a domain ontology, the API Graph can be enriched with further information: Fig. 7 reports the API Graph of a FitBit sensor, with the basic methods and parameters elements, but enriched via an external domain ontology with the "Steps" concepts, use to annotate the "delta" and "step" parameters belonging to the FitBit. By annotating the entities of the sensors' ontology, it is possible to create connections among elements and exploit the reasoning capabilities provided by semantic technologies. As an instance, if two methods are annotated with the same concepts, it is likely that they are interrelated and that, in a sensor network, they could serve the same objective and could be used together. Also, this allows the discovery of sensors and services used to interact with them. The main drawback is that annotation is currently done manually, via a semantic annotation window which is part of the tool. It is still very difficult to avoid human intervention in this phase, as parameters and methods' names are not always (if ever) meaningful. Figure 8 shows the annotation window which can be used to specify connections between elements of two different ontologies. The tool can operate on whatever ontology it is loaded with, so it is more of a generic semantic annotator than a sensor specific one. It can be used to connect individuals of the Parameter, Sensor and Method classes with each other, or with entities defined in domain ontologies. The tool support the free

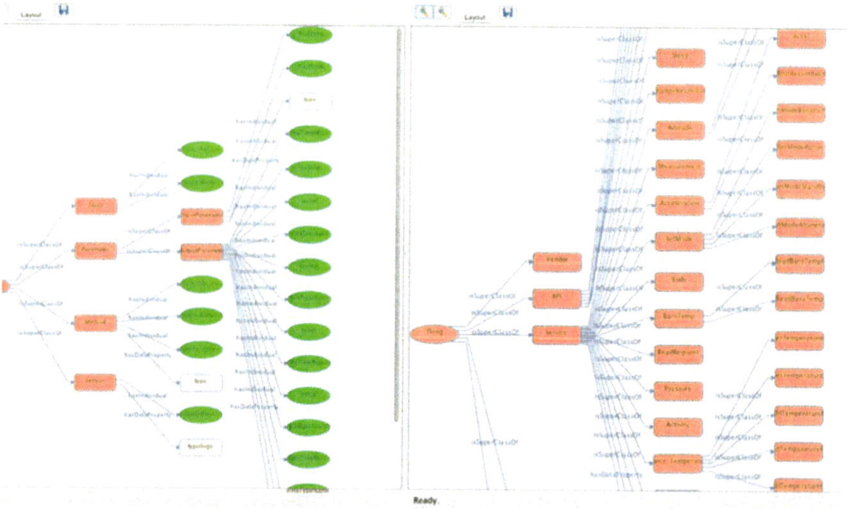

Fig. 8 The annotation window

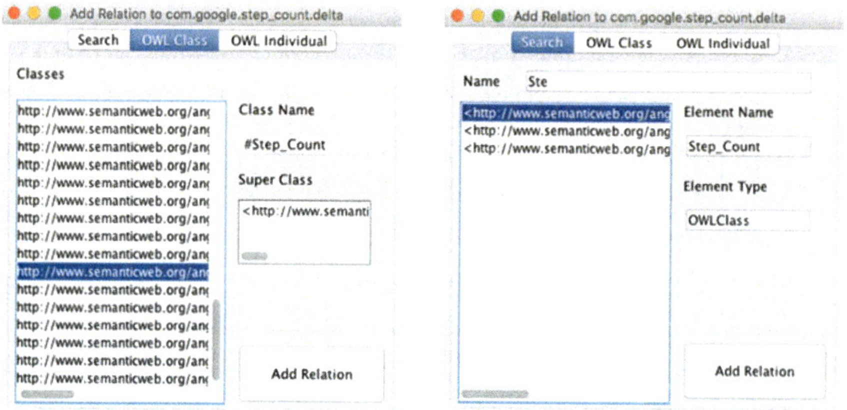

Fig. 9 The annotation panels

navigation of the loaded ontologies. In particular, it provides search capabilities to easily identify classes and their individuals. Figure 9 shows the two panels which can be used to create new relations by using the tool. On the right, the search box which allows to find classes, individuals and general elements of the ontology, and then to create relationships with the entities selected in the annotation Window of Fig. 8. On the left, the OWL Class search box which allows to find all individuals of a selected class.

```
<wsdl:interface name="com.google.calories.
    ↪ expended_Interface">
  <wsdl:operation name ="com.google.calories.
      ↪ expended_Operation">
    <wsdl:output messageLabel="OUT" element="
        ↪ types:ParameterOutput">
    </wsdl:output>
    <wsdl:input messageLabel="IN" element="
        ↪ types:ParameterInput">
    </wsdl:input>
  </wsdl:operation>
</wsdl:interface>
<wsdl:binding name="com.google.calories.
    ↪ expended_Binding" interface="tns:com.google.
    ↪ calories.expendend_Interface"
  type="http://www-w3-org/ns/wsdl/http">
   <wsdl:operation ref="tns:com.google.calories.
       ↪ expendend_Operation"/>
</wsdl:binding>
<wsdl:service name="com.google.calories.
    ↪ expended_Service"  interface="tns:com.google.
    ↪ calories.expendend_Interface">
<wsdl:endpoint name="com.google.calories.
    ↪ expended_ENDPOINT"  binding="tns:com.google.
    ↪ calories.expendend_Binding"
address="https://www.googleapis.com/fitness/v1/
    ↪ users/me/dataSources/derived:com.google.
    ↪ calories.expended:com.google.android.
    ↪ gms:aggregated">
</wsdl:endpoiTn>
</wsdl:service>
```

Listing 10.1 Example of WSDL representation

```
<rdf:RDF
  xmlns:owl=           "http://www.w3.org/2002/07/
    ↪ owl#"
  xmlns:rdfs=          "http://www.w3.org/2000/01/
    ↪ rdf-schema#"
  xmlns:swrl=          "http://www.w3.org/2003/11/
    ↪ swrl#"
  xmlns:j.0=           "http://www.semanticweb.org/
    ↪ katia/ontologies/2016/4/APIOWL3.owl#"
  xmlns=               "http://127.0.0.1/owl/com.
    ↪ google.calories.expended_Service.owls#"
  xmlns:list=          "http://www.daml.org/services
    ↪ /owl-s/1.2/generic/ObjectList.owl#"
  xmlns:expr=          "http://www.daml.org/services
    ↪ /owl-s/1.2/generic/Expression.owl#"
  xmlns:service=       "http://www.daml.org/services
    ↪ /owl-s/1.2/Service.owl#"
  xmlns:process=       "http://www.daml.org/services
    ↪ /owl-s/1.2/Process.owl#"
  xmlns:profile=       "http://www.daml.org/services
    ↪ /owl-s/1.2/Profile.owl#"
  xmlns:grounding =    "http://www.daml.org/services
    ↪ /owl-s/1.2/Grounding.owl#"
  xmlns:xsd=           "http://www.w3.org/2001/
    ↪ XMLSchema"
  xml:base=            "http://127.0.0.1/owl/com.
    ↪ google.calories.expended_Service.owls">

<owl:Ontology rdf:about="">
<owl:imports rdf:resource="http://www.semanticweb.
   ↪ org/katia/ontologies/2016/4/APIOWL3.owl#" />
</owl:Ontology>
```

Listing 10.2 Example of OWL-S representation

6 The WSDL and OWL-S Representations

Once the analysis of the API has terminated and the corresponding ontology has been created, building the WSDL representation becomes a trivial matter. Indeed, WSDL has a very precise structure, which can be easily replicated with the information contained in the API ontology. For our purposes, WSDL is useful in different ways. First of all, it provides a means to automatically build calls to the sensors' interfaces, using a standard representation which can be easily interpreted by external tools. Second, it is the standard language used to describe the grounding of OWL-S which is fundamental in our approach for the description of sensors' compositions. Listing 1 reports an example of WSDL representation, referring to the GoogleFit service. OWL-S representations at this stage are not very meaningful, as sensors do not have complex workflows to describe, unless they are part of a sensor network. In this case, OWL-S can be exploited to describe the net process and data workflows, by using the semantic annotations provided by the ontology itself. At this stage, however, we have no network configurations to represent, and OWL-S becomes a mere interconnection between WSDL parameters and ontological concepts, as shown in Fig. 2 where only the main ontologies and schemas are imported and connected.

7 Conclusions and Future Work

In this paper, a prototypal tool for the automatic analysis of sensors' REST API has been presented. The tool aims not only at automatically analyze the APIs, but also to derive a semantic representation of them, which can be then used to support the manual annotation with external domain ontologies. The main steps followed by the tool have been described: the critical analysis phase, in which a combination of automatic and manual approaches lead to the production of a semantic API Grap; the manual annotation phase, where the semantic representation is enriched with connections to domain ontologies; the WSDL and OWL-S production phase. At the moment, our research is focused on two aspects: to improve the analysis phase, in order to reduce the toll on the human operator; to actually build and represent sensor network using OWL-S and exploit it for orchestration.

Acknowledgements This work has received funding from the University of Campania Luigi Vanvitelli V:ALERE research programme under the SSCeGov (Semantic, Secure and Law Compliant e-Government Processes) project.

References

1. Alaya, M.B., Banouar, Y., Monteil, T., Chassot, C., Drira, K.: OM2M: extensible ETSI-compliant M2M service platform with self-configuration capability. Procedia Comput. Sci. **32**, 1079–1086 (2014)
2. Ali, S., Khusro, S., Ullah, I., Khan, A., Khan, I.: SmartOntoSensor: ontology for semantic interpretation of smartphone sensors data for context-aware applications. J. Sens. **2017** (2017)
3. Avancha, S., Patel, C., Joshi, A.: Ontology-driven adaptive sensor networks. In: The First Annual International Conference on Mobile and Ubiquitous Systems: Networking and Services, 2004. MOBIQUITOUS 2004, pp. 194–202 (2004)
4. Antoniou, G., Van Harmelen, F.: Web ontology language: OWL. In: Handbook on Ontologies, pp. 67–92. Springer, Heidelberg (2004)
5. Bajaj, G., Agarwal, R., Singh, P., Georgantas, N., Issarny, V.: A study of existing ontologies in the IoT-domain. arXiv preprint arXiv:1707.00112 (2017)
6. Compton, M., Barnaghi, P., Bermudez, L., GarcíA-Castro, R., Corcho, O., Cox, S., Graybeal, J., Hauswirth, M., Henson, C., Herzog, A., et al.: The SSN ontology of the W3C semantic sensor network incubator group. J. Web Semant. **17**, 25–32 (2012)
7. Datta, S.K., Bonnet, C.: Smart M2M gateway based architecture for M2M device and endpoint management. In: 2014 IEEE International Conference on Internet of Things (iThings), and IEEE Green Computing and Communications (GreenCom) and IEEE Cyber, Physical and Social Computing (CPSCom), pp. 61–68. IEEE (2014)
8. Di Martino, B., Esposito, A., Cretella, G.: Semantic representation of cloud patterns and services with automated reasoning to support cloud application portability. IEEE Trans. Cloud Comput. **5**(4), 765–779 (2015)
9. Di Martino, B., Esposito, A., Maisto, S.A., Nacchia, S.: A semantic IoT framework to support RESTful devices' API interoperability. In: 2017 IEEE 14th International Conference on Networking, Sensing and Control (ICNSC), pp. 78–83. IEEE (2017)
10. Janowicz, K., Haller, A., Cox, S.J.D., Le Phuoc, D., Lefrançois, M.: SOSA: a lightweight ontology for sensors, observations, samples, and actuators. J. Web Semant. **56**, 1–10 (2019)
11. Kubler, S., Främling, K., Zaslavsky, A.: IoT platforms initiative. Digitising the Industry Internet of Things Connecting the Physical, Digital and Virtual Worlds, pp. 265–292 (2016)
12. Leitner, S.-H., Mahnke, W.: OPC UA-service-oriented architecture for industrial applications. ABB Corp. Res. Center **48**, 61–66 (2006)
13. Martin, D., Burstein, M., Hobbs, J., Lassila, O., McDermott, D., McIlraith, S., Narayanan, S., Paolucci, M., Parsia, B., Payne, T., et al.: OWL-S: semantic markup for web services. W3C Member Submission **22**(4) (2004)
14. Mahnke, W., Leitner, S.-H., Damm, M.: OPC Unified Architecture. Springer, Berlin, Heidelberg (2009)
15. The European (EC-FP7) ICT strep project "mOSAIC—open-source API and platform for multiple clouds". Call: FP7-ICT-2009-5 Objective: ICT-2009.1.2 (Software Services and Cloud)
16. Nachabe, L., Girod-Genet, M., El Hassan, B.: Unified data model for wireless sensor network. IEEE Sens. J. **15**(7), 3657–3667 (2015)
17. Pardo-Castellote, G.: OMG data-distribution service: architectural overview. In: 23rd International Conference on Distributed Computing Systems Workshops, 2003. Proceedings, pp. 200–206. IEEE (2003)

Semantic IoT Interoperability and Data Analytics Using Machine Learning in Healthcare Sector

Pratiyush Guleria and Manu Sood

Abstract With the exponential growth of data in electronic form, it becomes a complex and tedious task to extract meaningful information. The vast collection of data has resulted in big data that may be in indeterminate form. The challenge is to extract meaningful data from internet sources that are spreading across multiple domains and to enable consistent resource sharing, interoperability on multiple IoT platforms. The use of emerging technologies like Machine Learning and IoT is realized on multiple platforms, systems, and service applications. The introduction of predefined libraries on Natural Language Processing in Machine learning platforms has emphasized on the semantic web technologies and its IoT future directions. In this chapter, authors have discussed the role of the semantic web, three layered framework for IoT interoperability, and have framed a web ontology structure for semantic interoperability in IoT for the healthcare sector. Authors have also proposed the text analytics model for the healthcare sector and performed semantic data classification on synthesized healthcare dataset to predict the patient diagnosis using Machine learning techniques.

Keywords Healthcare · IoT · Learning · Machine · Python · Semantic · Structured

1 Introduction

The semantic technologies involve technologies from Artificial Intelligence to Natural Language Processing. The NLP field involves the linked data and linguistic web. In the Semantic web, information is connected and linked from one source to another in the form of linked lists. Semantic technologies result in the evolution of Internet technologies. Machine Learning is the emerging field where one focus of research is on Natural Language Processing and predicting the results through past

P. Guleria (✉)
NIELIT Shimla, Shimla, Himachal Pradesh, India
e-mail: pratiyushguleria@gmail.com

M. Sood
Department of Computer Science, Himachal Pradesh University, Shimla, India

R. Pandey et al. (eds.), *Semantic IoT: Theory and Applications*, Studies in Computational Intelligence 941, https://doi.org/10.1007/978-3-030-64619-6_11

and historical events. In the semantic web, knowledge is the source for intelligent systems and ontologies support the tasks for it. The ontology is conceptualization i.e. knowledge base of a particular domain. In Artificial Intelligence, an agent performs the work of communication for which it uses the structure from some ontology. The Semantic web uses the term ontology which is in a simple term known as vocabulary. Ontologies help in organizing knowledge and the inference techniques are used on the Semantic web, which involves relationships. These techniques play an important role in the healthcare sector where patient symptoms, diagnosis, and treatments are prescribed using ontologies. The knowledge acquired using these terms helps in building intelligent decision support systems to predict the diagnosis. Another area of research includes social networking websites like language use on Facebook, Twitter, and Instagram, etc. The ontologies use linked data techniques to create an Intelligent Semantic system. In the semantic web, there are 4 techniques of ontologies defined. These are as follows: (a) RDF Schemes, (b) Simple knowledge organization system (SKOS), (c) Web ontology language (OWL), (d) Rule Interchange Format (RIF).

IoT semantic web uses ontologies which help in the medical field. E.g. the ontologies generate the vocabulary for problems like blood pressure and its correlated medicine i.e. aspirin to take immediate action in such a situation. In the present scenario, the IoT enabled devices are ontology processed. The technology is not limited only to smartphones, smart watches; this place is now taken by IoT. IoT is just like networking which connects gadgets and electronic devices for information dissemination and utilization. IoT is a concept where all devices which run in on and off state, now after IoT works by medium of the Internet. Here the information dissemination and intelligence come with web ontologies. The IoT technology is also known as non-screen computing where devices work like computers but with no screen in front of them. In IoT, there are smart objects with sensing competence, embedded recognition through RFID tags. With the help of semantic IoT, there is a unification of sensors, RFID tags, and communication. In Semantic IoT, the resource description framework provides semantic inoperability using different IoT devices. The semantic interoperability helps in effective and meaningful communication which is economical and ensures faster decision-making [1]. As the IoT devices generate heterogeneous data, there is a need for such information that contains unambiguous information of IoT resources. The information obtained facilitates data access, semantic interpretation, and knowledge extraction [2].

The introduction of the chapter is followed with the literature review in Sect. 2. The layered framework of interoperability in IoT is discussed in Sect. 3. Section 4 covers the methodology for classifying semantic data using Machine learning. The paper is concluded in Sect. 5 followed with references at the end.

2 Literature Review

Authors in [3] have discussed in detail about the semantic web. The semantic web provides proper format to the relevant information of web pages with the help of agents. Here, the information is having a decisive meaning. The two important technologies that have been developed for the semantic web are as follows: (a) XML, (b) RDF (Resource Description Framework). The Semantic web enables the machine to understand the semantic data. In the semantic web, another role is played by ontologies that find the synonym vocabulary which is having similar meaning e.g. "address" can also have synonym "location" in databases. Apart from the Semantic Web, there is need to develop knowledge representation techniques using artificial techniques such that meaningful information can be extracted from large data. On the contrary, knowledge representation and annotation languages have been developed using web infrastructure like HTML. The tags used in HTML gives the shape to domain-specific language to a machine processed form. The RDF Frameworks uses the ontology for providing semantic structure to an application [4]. In [5], authors have discussed ontology learning for the web which structures the data into a machine-understandable form. The semantic analysis helps in information synthesis whereas ontology results in knowledge representation. The main motive of the Semantic web is to give machine accessible information. In the Semantic web, there is a need for knowledge management but along with it the limitations which need to be handled are as follows: (a) keyword-based search, (b) extracting meaningful information and its maintenance, (c) deliver the information in the user and human-friendly manner, (d) effective use of ontologies for a shared understanding of a particular domain, (e) removal of ambiguities in terminologies i.e. to specify the terminology having similar understanding and meaning [6].

Authors [7] have explored the semantics and sentics related to healthcare for bridging the gap between structured and unstructured data. The framework proposed by the authors considers the patients' opinion on the web and health care providers to deliver improved services to the end-user. The innovative semantics-based methods help in addressing the healthcare problems which includes the effective use of biomedical vocabulary, semantic web technologies to extract meaningful information in biomedical and health data. The biomedical ontologies provide efficient domain knowledge to support data similarity and interoperability in a variety of healthcare information systems such as EHR, healthcare administration, and clinical decision support [8]. In [9], authors have stressed the role of agent-based systems and ontologies in the web world. With the help of agents and ontologies, there will be an effective use of programs to perform tasks with less human intervention. The author [10] has proposed a framework for handling clinical models through Semantic Web technology. The data source used for the framework is Electronic Health Records. The Analytical capability of the framework is ontology building through OWL.

In the semantic web, the patient medical records, history, medicines prescribed are represented in Web ontology language. The data in web ontology language form will be helpful for data-driven computing aid for personalized health maintenance. With the help of Intelligent Semantic Analytics, the medical records of patients can be retrieved taking symptoms in the user input query. Rahman et al. [11] have proposed a framework to maintain confidentiality in RFID based healthcare systems. In this framework, the data is generated from RFID tags. Semantic Web Analytics can help in the classification of Healthcare data for handling unstructured data. Authors [12] have proposed a framework for voice signals based pathology assessment. In Machine Learning, there are techniques like Association Rule Mining, Neural Network-based classification techniques that help in personalized healthcare services. The Web Ontology Language data sources of Healthcare services involves (a) Electronic Health care, (b) Sensor data in devices, (c) Epidemics data about infections, (d) Symptoms, and Diagnosis data. Semantic web uses an ontology to represent the terms meaning in web documents and applications. Semantic web works on the recommendations of W3C which involves (a) XML, (b) XML Schema, (c) RDF, (d) RDF Schema. The RDF and RDF Schema provide the vocabulary for properties and classes of RDF [13]. OWL [14] not only presents information to humans but also process meaningful information. The sublanguages of OWL are (a) OWL Lite, (b) OWL DL, and (c) OWL Full.

The RDF (Research description framework) consists of metadata which classifies the data on the WWW. The application of RDF has revolutionized the e-commerce sector like in Amazon.com, the products are categorized.

3 Layered Framework of Interoperability in IoT

The term Interoperability means to enable information systems to interpret, understand, and share the information without ambiguity.

In order to achieve this, the devices in IoT need to work with each other on cross-domain platforms. In semantic interoperability, devices need to confine to a common agreement on syntax and semantics [15]. In Fig. 1, the IoT Interoperability framework is shown. In this framework, the IoT Interoperability is divided into three layers which are as follows: (a) Device Interoperability, (b) Syntactic Interoperability and (c) Semantic Interoperability. In device Interoperability, the devices should be agreed to work seamlessly on the standard format and on cross-domain platforms

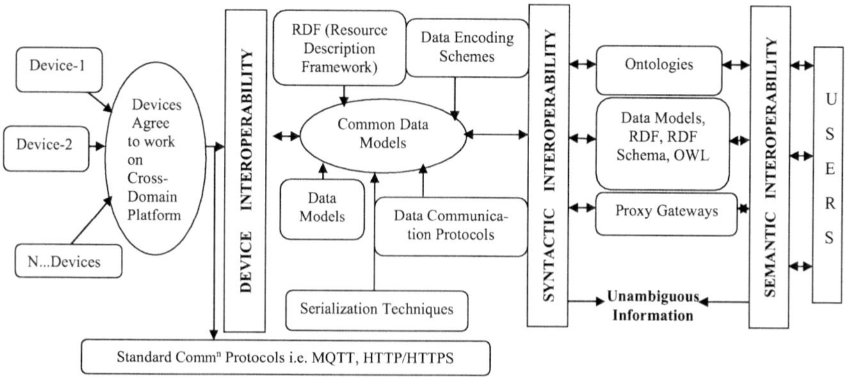

Fig. 1 Three layered Framework for IoT Interoperability

for effective and economical performance. In Syntactic Interoperability, common data models and formats need to be followed whereas in Semantic Interoperability, the information is interpreted and shared without ambiguity for information having similar meanings.

3.1 IoT Interoperability Challenges

The major IoT interoperability challenge is to enable consistent resource allocation and to ease the interoperability between different IoT platforms. The inconsistency between IoT platforms results in platform dependency and applications unable to run on multiple platforms. In such a scenario, there is a need for standardization in certain areas to find a solution to the problems related to IoT interoperability [16]. The certain challenges of IoT interoperability are as follows:

(a) Variation in IoT Infrastructure.
(b) Variation in devices and their libraries.
(c) Different data formats.
(d) Applications developed are unable to work on cross-platform or on different domains.
(e) There is no proper streamlining in IoT resource sharing.
(f) Inconsistency between different IoT platforms.

3.2 Web Ontology Framework for Semantic Interoperability in IoT

The framework has been proposed for the semantic web in the medical sector shown in Fig. 2. In the proposed framework, the patient query is monitored by the medical specialist in the specialized area. The symptoms of patients are preprocessed which involves the (a) patient past history, (b) medical prescriptions already diagnosed, (c) Clinical Records, (d) lab tests, etc.

In order to know the symptoms of the patient, the doctor monitors the same remotely using IoT devices. The interoperability in IoT needs to resolve the issues related to devices, network, schema format, etc. The semantic interoperability enables multiple platforms, applications to exchange information in a meaningful way on the web-enabled platform. The knowledge base of the semantic web involves the (a) Frame structures, (b) XML, (c) Predicate Logic, (d) UML modeling technique, (d) Logic Rules, (e) RDF (Resource Description Framework), etc. for uniformity in schema, syntax, and semantics.

The vocabulary related to patient symptoms is being checked in the knowledge base for finding the prognosis having similar meaning or terminology and finally after gathering the desired information, the web-based diagnostic results may be generated. The ontologies can be expressed in different forms which are as follows:

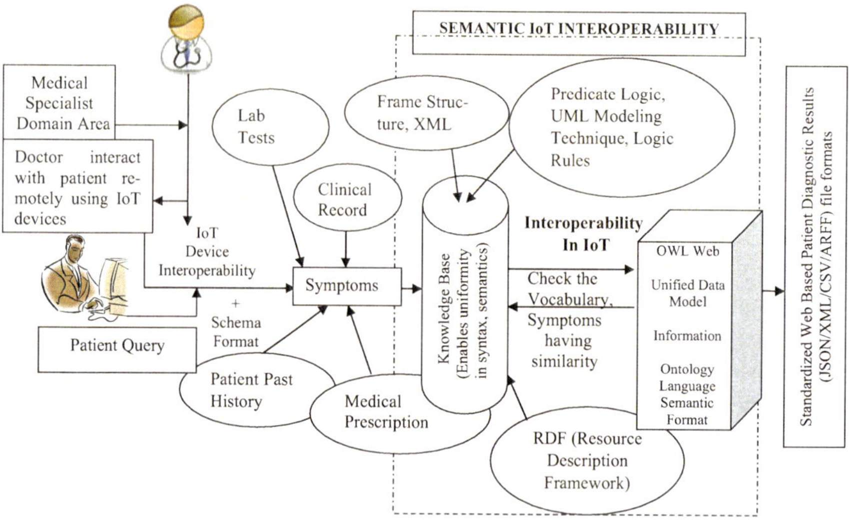

Fig. 2 Proposed framework for patient-centric web ontology based intelligent semantic analytics

(a) informal, when expressed in the form of natural language,
(b) semi-informal, when set in a defined and planned form of natural language,
(c) semi-formal when setting in a simulated and formerly defined language.

There are techniques like frames and first-order predicate logic followed in Artificial Intelligence techniques for representing ontologies. The web ontology represents (a) classes and relationship between classes, (b) inference rules that form a knowledge base. The Semantic web retrieves the information in the ontology and infers the new information on applying the logic rules.

DAML and OIL are the web markup languages developed for (a) content description, (b) web-enabled ontologies i.e. XML, RDF, RDF Schema, etc. The DAML means DARPA Agent Markup Language from DAML Programme and OIL represents the Ontology Inference Layer [17]. The ontology-based knowledge management systems help to formalize the sharing of knowledge [18].

3.3 Conceptual Graphs

The semantic net is expressed in the form of conceptual graphs. The conceptual graphs represent the semantics using Predicate logic. The relationship between Doctor and Patient in the form of a conceptual graph is shown in Fig. 3.

The relationship between doctor and patient is "diagnosis" which is shown in the form of an ellipse. The graphical representation shown in the Figure is represented in First-order predicate logic (FOPL) form as below:

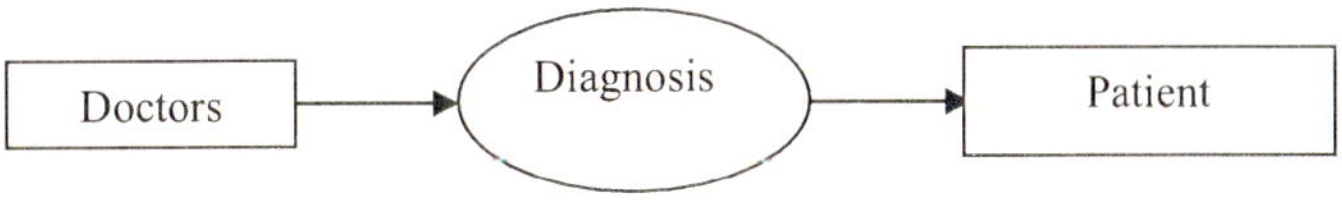

Fig. 3 Representation of doctor-patient centric conceptual graph

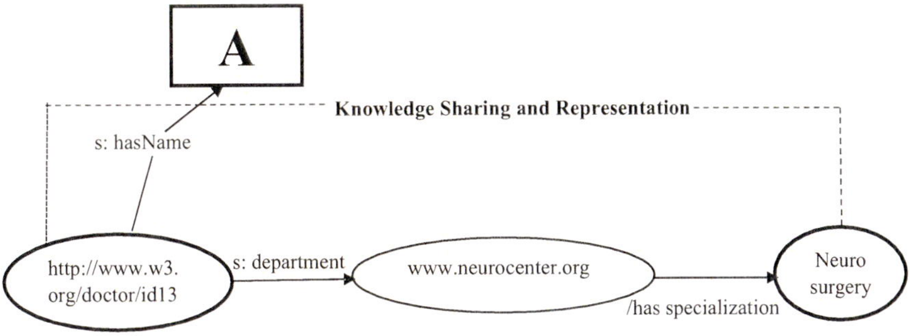

Fig. 4 Doctor-centric RDF graph

$$\exists x, y : \text{Doctor}(x) \wedge \text{Patient}(x) \wedge \text{diagnosis}(x, y)$$

The symbol ∃ is a quantifier symbol which means "there exists", the symbol ∧ means logical connective and is for "conjunction".

Web-based knowledge sharing, processing and reuse between applications is an important part of the ontologies [19]. The RDF and XML provide the foundation for the semantic web. The doctor centric representation in the form of the RDF graph is shown in Fig. 4.

The XML representation of the RDF Graph shown in Fig. 4 is the knowledge representation in the form of frame language.

```
<class-def>
        <class-name = "doctor"/>
        <subclass-of>
            <class-name= "surgeon">
        </subclass-of>
</class-def>
<class-def>
        <class-name = "specialization"/>
            <slot-constraint>
            <slot name = "is-part-of"/>
            <has-value>
                  <class name = "neurosurgery">
            </has-value>
        </slot-constraint>
</class-def>
```

The doctor, surgeon, specialization and neurosurgery are the class names whereas the slot represents the relations. The "has-value" defines the restriction class.

The ontologies are used in many applications which are as follows:

(a) Agent based systems, (b) Knowledge management systems, (c) Electronic commerce platforms, (d) Natural Language Processing, (e) Intelligent Information Systems [20]. The language stack in ontology is shown in Fig. 5.

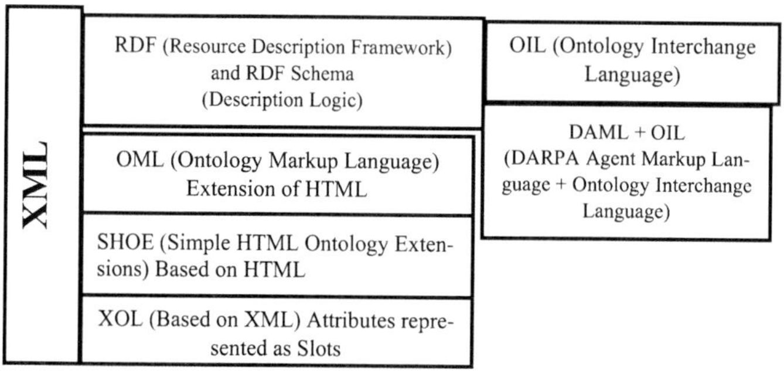

Fig. 5 Language stack in ontology

4 Methodology for Classifying Semantic Data Using Machine Learning

The methodology adopted for semantic data classification of Healthcare data is divided into four parts as follows: (a) the first part is collection of vocabulary associated with patient symptoms, (b) collection of synthesized or semi-synthesized data of healthcare sector for experimentation, (c) classification of data using Machine Learning algorithms in software like WEKA, MATLAB or Python programming, etc. and (d) finally the semantic analysis of these symptoms for improved results. The proposed model for text analytics in the Healthcare sector is shown in Fig. 6.

4.1 Vocabulary Associated with Healthcare Perspective

In the first part, it is necessary to construct the vocabulary representing patients' prognosis like (stroke, cold, attack, depression, fatigue, illness, etc.) The vocabulary of symptoms can be further categorized to get more precision.

4.2 Data Collection and Structure

The synthesized dataset for the experimentation purpose is shown in Table 1. The datasets available on the web resources have different attributes and the dataset related to pathological diagnosis of patients, as per our requirement is not available. Therefore, for the semantic analysis of the symptoms vocabulary and the prediction of the patient diagnosis, a dataset is synthesized. The dataset have 8 input features and the associated task related to the dataset is classification.

The attributes of the dataset are some of the symptoms like fatigue, restlessness, fever, sweating, cough, congestion, symptom description etc. The values of the dataset are both categorical and continuous.

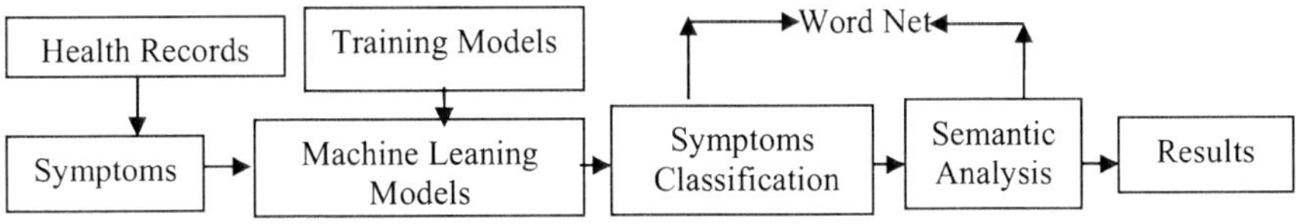

Fig. 6 Text analytics model for healthcare sector

Table 1 Sample dataset

Symptoms	Fatigue	Restlessness	Fever	Sweating	Cough	Congestion	Symptom description
"chills"	1	1	1	0	0	0	"The chills are there due to fever and sore throat"
"cold"	1	1	1	0	1	0	"The cold symptoms are running nose, cough, sneezing, bodyache, headache"
"fatigue"	1	1	0	0	0	1	"Fatigue causes due to symptoms of common cold"
"tired"	1	1	1	1	1	1	"bodyache and tiredness due to common cold"
"tension"	0	1	1	1	0	0	"prolonged illness"
"illness"	1	1	1	0	0	1	"chest congestion with cough occur in cold"
"cold"	1	0	0	0	1	1	"The cold symptoms are running nose, bodyache, headache"
"chills"	1	1	1	0	1	1	"Muscle aches are there due to fever. There is need of antibiotic"
"bitter"	1	1	1	0	1	0	"Watery eyes are there when cold is there"

(continued)

Table 1 (continued)

Symptoms	Fatigue	Restlessness	Fever	Sweating	Cough	Congestion	Symptom description
"icy"	1	1	1	0	1	0	"chest congestion occurs and it results in sinus infection"
"cough"	0	0	1	1	1	1	"chest congestion, soreness in throat"
"sneeze"	1	0	0	0	0	1	"common cold"
—	—	—	—	—	—	—	—
"rhinorrhea"	1	1	1	1	1	1	"common cold is also known as rhinorrhea"

4.3 *Classification*

In the third phase, the classification of vocabulary is performed using Machine learning techniques on the dataset. The ML approaches i.e. supervised learning has been applied on the dataset to predict the classification model for text analysis in patient prognosis. Supervised learning performs the classification on input training set with the desired output label whereas in Unsupervised learning there is clustering instead of classification. In unsupervised learning, the dataset is without input and desired output labels. The text analysis has been performed using n-gram frequency counts and a bag-of-n-grams model for analyzing text data.

4.4 *Semantic Analysis*

In the fourth phase, after applying the Machine learning algorithm, the semantic analysis is performed. The semantic analysis is performed on a healthcare dataset where the terms close to each other are semantically similar like "cold", "chills", "bitter" shows similarity in symptoms. Semantic data analytics extracts meaningful information from the large datasets [21].

The semantic analysis for text data using Machine Learning is processed as follows:

(a) Tokenize the text

(b) Lemmatize the words
(c) Erase Punctuation
(d) Remove stop words like "the", "and".

4.5 Results and Discussions

The result shown in Table 2 is obtained using the input dataset shown in Table 1 whereas the results of the test data are shown in Table 3. The test data sample i.e. symptoms description attribute containing the data for training and analytical purpose is shown below:

testdata = "The reasons for common cold may result into fever in kids and also results into chest congestion"

The Fig. 7 displays the input raw data and the preprocessed, cleaned data.

Table 2 Tokenization of dataset

ans = 8 × 1 tokenizedDocument	
5 tokens	chill due fever sore throat
7 tokens	cold symptom running nose cough sneeze bodyache headache
6 tokens	fatigue cause due symptom common cold
6 tokens	bodyache tire ness due common cold
2 tokens	prolong illness
5 tokens	chest congestion cough occur cold
6 tokens	cold symptom run nose bodyache headache
6 tokens	muscle ache due fever need antibiotic
cleanedBag = bagOfWords with properties	
Counts: Vocabulary NumWords NumDocuments	[13 × 37 double] [1 × 37 string] 37 13
cleanedBag = bagOfWords with properties	
Counts: Vocabulary: NumWords: NumDocuments:	[13 × 7 double] [1 × 7 string] 7 13

Table 3 Tokenization of testdata

newDocuments = tokenizedDocument	
9 tokens:	reason common cold result fever kid results into chest congestion
bagOfWords with properties	
Counts	[13 × 54 double]
Vocabulary:	[1 × 54 string]
NumWords:	54
NumDocuments	13
the reduction in data	0.8704

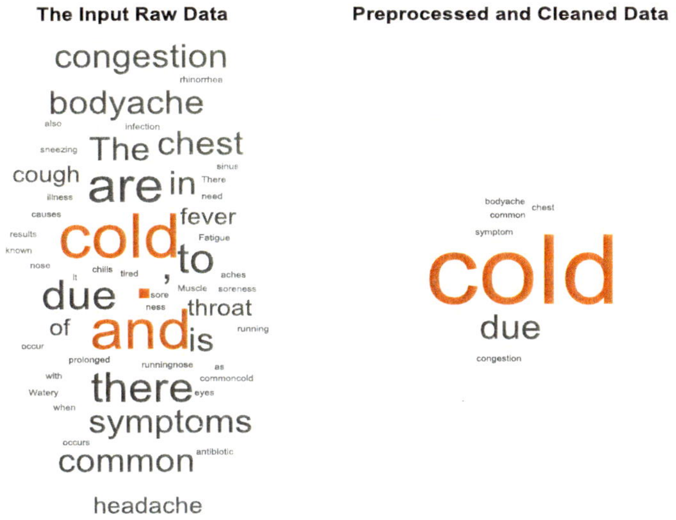

Fig. 7 Input raw data and preprocessed data

4.6 *Analyze and Visualize Text Using N-Gram Frequency Counts*

A Latent Dirichlet Allocation (LDA) model is used to find the ontology in a dataset shown in Table 1 and infers the probability of the ontology occurrence. The LDA model is an example of a topic model. This model is a statistical model used in Natural Language Processing for finding homogeneous values in a collected document. The LDA model is used for information retrieval, semantic analysis, and classification of words in a document. This model also finds the probability of word occurrence in a particular set of topics.

With LDA Model, the symptoms probability in the symptoms description attributes are inferred. The results obtained are shown in Table 4. The patient-centric trigrams and preprocessed bigrams are shown in Figs. 8 and 9. The bag-of-n-grams

Table 4 bagOfNgrams on input dataset

ans = 5 × 1 tokenizedDocument	
5 tokens	chill due fever sore throat
7 tokens	cold symptom running nose cough sneeze bodyache headache
6 tokens	fatigue cause due symptom common cold
6 tokens	bodyache tiredness due common cold
2 tokens	prolong illness
bag = bagOfNgrams with properties	
Counts Vocabulary Ngrams NgramLengths NumNgrams NumDocuments	[13 × 41 double] [1 × 37 string] [41 × 2 string] 2 41 13
Initial topic assignments sampled in 0.111156 s.	

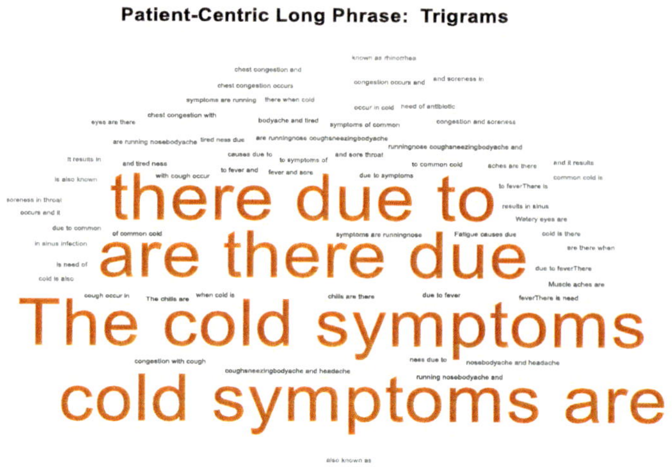

Fig. 8 Patient-centric Trigrams

models using a word cloud are visualized in Fig. 11 whereas the common n-grams of length 3 are shown in Table 5.

The semantic IoT interoperability framework for Patient-centric diagnosis is shown in Fig. 10. The results obtained in Fig. 9 are implemented in the framework for diagnosis by doctor. The semantic interoperability understands the patients' symptoms unambiguously and facilitates the exchange of meaningful information across

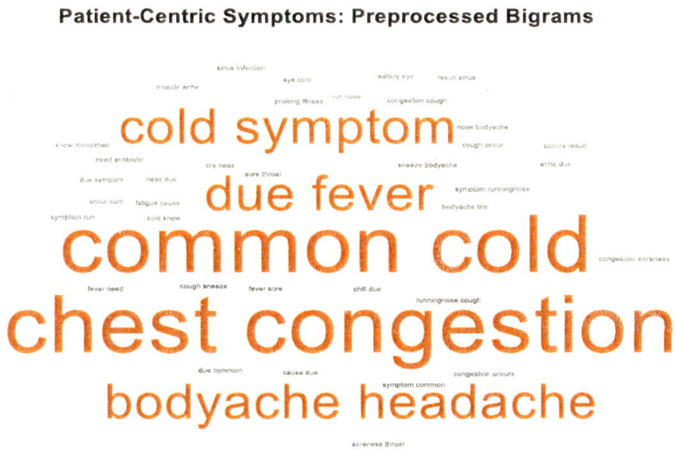

Fig. 9 Patient-centric Symptoms: Preprocessed Bigrams

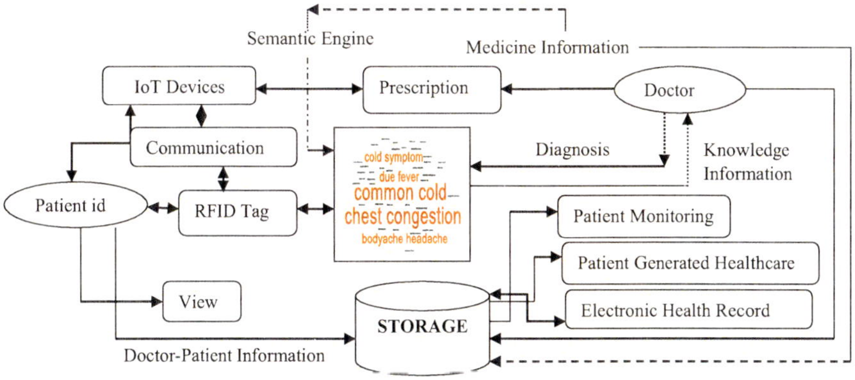

Fig. 10 Semantic IoT Interoperability framework for patient diagnosis

the different domains. The ontologies help to inter-relate the patient's symptoms which become vital information for machine learning and automation systems.

5 Conclusion

In the present scenario, where different technologies like data mining, artificial intelligence, and machine learning are combined to work on machine processed applications, the semantic web importance comes into force. The semantic web emphasis on the dissemination of meaningful information from the data stored in databases and on the internet. The intelligent semantic analysis works on machine learning

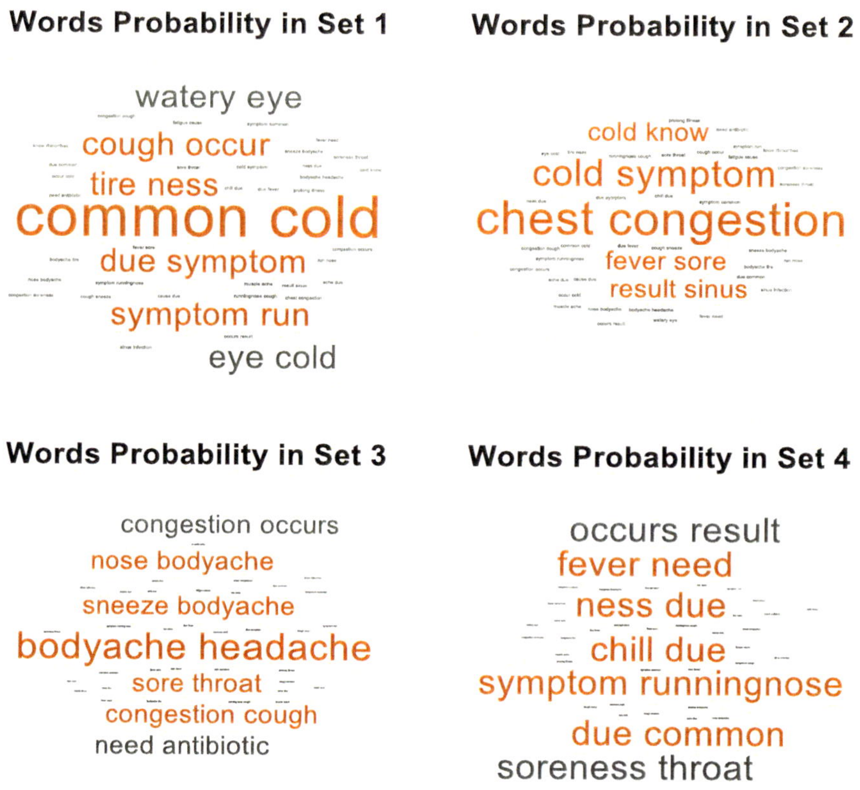

Fig. 11 bag-of-n-grams word cloud model

Table 5 Common n-grams of length 3

Ngram	Count	NgramLength
"are" "there" "due"	2	3
"there" "due" "to"	2	3
"The" "cold" "symptoms"	2	3
"cold" "symptoms" "are"	2	3
"The" "chills" "are"	1	3
"chills" "are" "there"	1	3
"chills" "are" "there"	1	3
"due" "to" "fever"	1	3
"to" "fever" "and"	1	3
"fever" "and" "sore"	1	3
"and" "sore" "throat"	1	3

with less human intervention for understanding the semantics in a right and unambiguous manner. There is a lot of future scope in the semantic web to automate the routine tasks. It helps the machine to understand those statements which are having similar meanings in databases where the data is stored. The ontologies developed using semantic web represent the knowledge structure for web science which helps in handling the unstructured data, machine-readable ontologies, developing IoT enabled devices for different domains, and use of semantic markup in query interfaces.

The IoT based semantic interoperability works on heterogeneous data capable of removing ambiguity in shared meaning and meaningful dissemination of data across multiple domains. The semantic interoperability in the IoT domain enhances its potential for offering value-added services in different sectors of society, especially in the healthcare sector. With the help of semantic interoperability in the IoT field, the meaning of the data will be interpreted correctly like a prescription of medicine to the patient based on the symptoms and exchange of medical history of patients using IoT devices on different platforms. The semantic web devices complete the work as per the necessity which saves time and is capable in handling multiple tasks, but in the present scenario, there is a lot of crime on the internet and connecting each device with the internet is a big question of security which needs to be carefully handled.

References

1. Jabbar, S., Ullah, F., Khalid, S., Khan, M., Han, K.: Semantic interoperability in heterogeneous IoT infrastructure for healthcare. Wirel. Commun. Mob. Comput. (2017)
2. Gomes, P., Cavalcante, E., Batista, T., Taconet, C., Conan, D., Chabridon, S., Pires, P.F.: A semantic-based discovery service for the Internet of Things. J. Internet Serv. Appl. **10**(1), 10 (2019)
3. Berners-Lee, T., Hendler, J., Lassila, O.: The semantic web. Sci. Am. **284**(5), 28–37 (2001)
4. Van Ossenbruggen, J., Hardman, L., Rutledge, L.: Hypermedia and the semantic web: a research agenda. J. Dig. Inform. **3**(1) (2002)
5. Maedche, A., Staab, S.: Ontology learning for the semantic web. IEEE Intell. Syst. **16**(2), 72–79 (2001)
6. Antoniou, G., Van Harmelen, F.: A Semantic Web Primer. MIT Press (2004)
7. Cambria, E., Hussain, A., Eckl, C.: Bridging the gap between structured and unstructured healthcare data through semantics and sentics (2011)
8. He, Z., Tao, C., Bian, J., Dumontier, M., Hogan, W.R.: Semantics-powered healthcare engineering and data analytics (2017)
9. Hendler, J.: Agents and the semantic web. IEEE Intell. Syst. **16**(2), 30–37 (2001)
10. Del Carmen Legaz-García, M., Martínez-Costa, C., Menárguez-Tortosa, M., Fernández-Breis, J.T.: A semantic web based framework for the interoperability and exploitation of clinical models and EHR data. Knowl. Based Syst. **105**, 175–189 (2016)
11. Rahman, F., Bhuiyan, M.Z.A., Ahamed, S.I.: A privacy preserving framework for RFID based healthcare systems. Future Gener. Comput. Syst. **72**, 339–352 (2017)
12. Hossain, M.S., Muhammad, G.: Healthcare big data voice pathology assessment framework. IEEE Access **4**, 7806–7815 (2016)
13. OWL Working Group: OWL 2 web ontology language document overview: W3C recommendation 27 October 2009

14. McGuinness, D.L., Van Harmelen, F.: OWL web ontology language overview, W3C Recommendation **10**(10) (2004)
15. Shah, S.S.A.: Semantic interoperability in Internet of Things (2018)
16. Noura, M., Atiquzzaman, M., Gaedke, M.: Interoperability in Internet of Things: taxonomies and open challenges. Mob. Netw. Appl. **24**(3), 796–809 (2019)
17. Horrocks, I., Patel-Schneider, P.F., Van Harmelen, F.: Reviewing the design of DAML + OIL: an ontology language for the semantic web. In: *AAAI/IAAI*, pp. 792–797
18. Staab, S., Studer, R., Schnurr, H.P., Sure, Y.: Knowledge processes and ontologies. IEEE Intell. Syst. **16**(1), 26–34 (2001)
19. Decker, S., Melnik, S., Van Harmelen, F., Fensel, D., Klein, M., Broekstra, J., Horrocks, I.: The semantic web: the roles of XML and RDF. IEEE Internet Comput. **4**(5), 63–73 (2000)
20. Gómez-Pérez, A., Corcho, O.: Ontology languages for the semantic web. IEEE Intell. Syst. **17**(1), 54–60 (2002)
21. Ullah, F., Habib, M.A., Farhan, M., Khalid, S., Durrani, M.Y., Jabbar, S.: Semantic interoperability for big-data in heterogeneous IoT infrastructure for healthcare. Sustain. Cities Soc. **34**, 90–96 (2017)

Domain-Specific Applications

SAGRO-Lite: A Light Weight Agent Based Semantic Model for the Internet of Things for Smart Agriculture in Developing Countries

Gaurav Kant Shankhdhar, Richa Sharma, and Manuj Darbari

Abstract The recent advancement of the Internet of Things (IoT) has led to the possibilities to process a large number of sensor data streams built upon large-scale IoT platforms. In developed countries IoT is already emerged successfully as a reasonable technique assuring the goal of self-complacency, hybrid and advanced decisions and computerization in the horticulture industry. Instant adoption of IoT in farming is impractical in developing nations because of less literacy, hesitance towards technology, smaller farm sizes and high cost of IoT farming solutions. Through a light weight IOT specifically focused on farming style of developing countries like India, farmers can increase their quality of farming by the use of this technology. The authors have developed a semantically enriched agent based model called Agent Based Semantic Model for Smart Agriculture, ABSMSA which uses SAGRO-Lite, a light weight ontology designed by the authors for specific farming characteristics in developing countries. The system uses two more ontologies the IoT-Lite and Complex Event Service Ontology (CESO) for semantic sensing and event recognition and handling.

G. K. Shankhdhar
Department of Basic Sciences, BabuBanarasi Das University, Lucknow, India

R. Sharma
Department of Information Technology, BabuBanarasi Das National Institute of Technology and Management, Lucknow, India

M. Darbari (✉)
Department of Computer Science, BabuBanarasi Das University, Lucknow, India
e-mail: manuj_darbari@acm.org

R. Pandey et al. (eds.), *Semantic IoT: Theory and Applications*, Studies in Computational Intelligence 941, https://doi.org/10.1007/978-3-030-64619-6_12

1 Introduction

1.1 Popularity of IoT

IoT devices and the framework, in developed countries are extraneously used and deployed to gather, worked upon and IoT frameworks are used to collect, process and examine data flow in dynamic environment and expedite facilities of clever solutions for better decision making [1]. IoT technologies are already being used in developed countries like U.S and are becoming increasingly popular in the developing countries. An IoT system comprises devices that catch signals through the sensors fixed at the domain sites. Actuators are built to function on the receiving of some signal. These readings from the sensors are monitored on the dedicated devices for that purpose. IoT devices can mutually interact with other devices or applications. They can also gather data and process it both locally and globally with the help of decentralized servers [2].

The benefits of using IoT in agriculture include [3–5]

- The incurred cost is reduced as the wastage is diminished.
- With the help of better monitoring of the agricultural fields, there is a significant decrease in the diseases that directly affects the produce and revenue generated.
- Reduce of water wastage and water is life.
- The overall maintenance of the agricultural activities is better done with the use of IoT.

India, like many other developing countries desperately looks to the government and its application programs of IoT in order to maximize profits in horticulture [6–8]. India is country in which 70% of the whole population depends on agriculture. 75% of the population resides in villages. India is an agro-based economy and is the classic realm for smart agriculture [9, 10]. The time has come for the government and the industry to boot strap the agro-IoT journey for rural upliftment and direct the country towards agro-economic development [11, 12].

1.2 Indian Agriculture and Farmers—Problems and Reforms

A significant problem of the people in India faced in the domain of agriculture is that this profession is looked below standard by the people both in the field of agriculture and even more by the non-professionals. If the farmer is a father then he never wants his sons to be the farmer anymore. This amounts to the less money generated in this field [13, 14]. The guardians want their children to be educated and get employed in worthier avenues except farming. In olden days as the population was less, agriculture methods were eco-friendly and the agriculturist was seen idealistic. The farmer was also looked upon with respect. But now, the whole scenario is changed and getting worse with time. The number of suicides committed by the farmers is also increasing at an alarming rate [15]. The reasons for the farmers committing suicide are as obvious as unavailability of water [16, 17]. We do use modern techniques to withstand the needs of the farmers like increased use of fertilizers and pesticides. But what we also face are the newer problems of toxic soils and poisoned animals [18, 19].

We also see the so called privileged farmers that are one in hundreds who utilize modern agricultural techniques like precision farming or smart agriculture and generate greater amounts of money. But, the problem with the masses is that most of them are illiterate or nerds when it comes to technology and this problem amounts to inability to perform even the basics of technology handling like a simple click. So, there has to be a training program for the farmers to get them acquainted with the required technological paraphernalia [20, 21].

1.3 Importance of Agriculture in Indian Economy

India is regarded mostly an agriculture country. The most significant occupation found in India is Farming. And this goes well in support of other developing countries. In India, agriculture contributed to 16% of GDP. And more to say, 10% of the total export from India are the agricultural products. 60% of India's land is suitable for agriculture [22]. The main crops grown in India are rice, sugar-stick, wheat, potato, onion, mangoes, tomato, beans, cotton, and more.

Sustenance is basic forever. We rely upon horticultural yields for our nourishment prerequisites. India creates enormous amount of food grains, for example, millets, oats, beats, and so forth. A noteworthy segment of the sustenance stuffs delivered is expended inside the nation. Our ranchers work day and night to sustain our populace that checks over 1.21 billion [23]. Other than farming with a business inclination, subsistence agribusiness with its accentuation on the generation of sustenance for the cultivator's family is boundless. Generally, Agriculture is pursued as the most straightforward technique for acquiring nourishment for the family. Farming in India is increasingly a 'lifestyle' than a 'method of business'.

In 2013, India sent out rural items esteeming around 39 billion dollars. Horticulture is the fundamental occupation for greater part of principle—laborers in India. An enormous number of country ladies are additionally occupied with farming. As indicated by 2001 registration, over 56.6% of the principle laborers in India are occupied with rural and unified exercises. Various businesses are agro-based enterprises, for example, jute, cotton, sugar, tobacco, and so on. Crude materials for such enterprises are provided from agrarian produce. Green transformation started in India with a target to give more prominent accentuation on Agriculture. The period of Green unrest that started in 1960s saw noteworthy increment in the creation of nourishment crops. The presentation of improved strategies for farming and high yielding assortments (HYV) seeds, for the most part wheat, had come about into amazing improvement in agrarian yields. The profitability of land expanded massively giving colossal monetary lift to the country.

1.4 Characteristics and Problems of Indian Agriculture

As stated above, Indian economy is pivoted on agriculture. The socioeconomic status of the people, the national polity and the gamut of life of the people is directly controlled by agriculture. The Indian agriculture, however, has its own characteristics [24].

Some of the important characteristics and problems of Indian agriculture have been described briefly in the following section:

1. **Exploding population**: The Indian agriculture is characterized by heavy pressure of population. About 70% of the total population of the country is directly or indirectly dependent on agriculture. The fast growth of population industrialization and urbanization are putting enormous pressure on arable land [24, 25].
2. **Food Grains as the most basic crops**: In both the Kharif (summer) and the Rabi(winter) seasons, grain crops occupy the greater proportion of the cropped area. Maize, rice, millets, ragi and bajra, and pulses are the dominant crops in the kharif season, and wheat, gram and barley occupy over three-fourth of the total cropped area in the Rabi season.
3. **Jhuming Cultivation**: In the rain-fed areas of the country, mixed cropping is a common practice. The farmers mix millets, maize and pulses in the kharif season and wheat, gram and barley in the Rabi season. In the areas of Jhuming (shifting cultivation), ten to sixteen crops are mixed and sown in the same field. The idea behind mixing crops is to get reasonable agricultural return. In case the monsoon is good, the rice crop will give better production and in case of failure of monsoon, the less water requiring crops like maize, millets, bajra and pulses will give good harvest.
4. **Lack of Technology**: A substantial number of the farmers of the country, especially in the rain-fed areas, still use draught animals (bullocks, male buffaloes and camels) for ploughing and other agricultural operations. India is a labor based agricultural enterprise in which most of the agricultural operations, like ploughing, leveling, sowing, weeding, spraying, sprinkling, harvesting, and threshing are carried on mainly by human hands. The use of machinery is still confined only to the rich fainters of Punjab, Haryana, Western Uttar Pradesh, plains of Uttarakhand, Bihar, Madhya Pradesh, Gujarat, and Maharashtra.
5. **Agriculture dependence on rain**: As mentioned before the problem in water supply for irrigation purposes is extremely high and 56% of whole nation depends on rainfall and summer monsoon. This dependency over monsoon is too optimistic in belief as summer monsoon in India is also not uniforms, thus amounting to the farmers' problems.
6. **Less farming of nitrogen fixing crops**: The nitrogen fixing crops like pulses are getting less area under their cultivation. Consequently, the natural fertility of the soil is depleting and the soils are losing their resilience characteristics [26].

7. **Low soil fertility**: One of the main problems of Indian agriculture is its low productivity. Indian agricultural products are among the lowest in the world. The main cause of low yield per hectare is the low fertility of soil and less care to replenish it through green-manure, fertilizers, fallowing, and scientific rotation of crops [27].
8. **Industrialization and Urbanization**: After the First Five Year Plan, Indian agriculture got a step-motherly treatment. The farming community has been ignored, while there has been more emphasis on industrialization and urbanization. The growth rate of agriculture is only about 2.5%, while the overall growth rate of the country is about 9% (2010) [28]. The severe drought of 2009 over greater part of the country has increased the miseries of the farmers, which is a set-back in the revival of Indian economy [29, 30].
9. **Low status of Agriculture in society**: Gone are the days when agriculture was regarded as a sign of dignity. This belief has gone a tumble as nowadays, farming is the occupation possessed by the weaker sections of the society. So, farming is looked down as a profession.
10. **Farmers are poor and in debts**: Although cultivators' indebtedness is universal in subsistent farming, its impact is perhaps nowhere as crushing as in India. Unfortunately, over 85% of all the cultivating families are under debt. It is because of heavy indebtedness that several thousand farmers in Andhra Pradesh, Karnataka, Tamil Nadu, Maharashtra, Orissa, Gujarat, Punjab, and Uttar Pradesh have committed suicide during the last ten years [30].
11. **Specialized training is required**: For the diffusion of agricultural innovations both in the irrigated and rain-fed areas, a team of skilled village level workers is required. A lot has to be done in this respect. The workers if trained can capitalize the opportunities of farmers to modernize their agriculture.
12. **Lack of agricultural research and poor education**: Though enough progress has been made in the field of agricultural research, there is no co-ordination between the farm and research laboratories in the different agro-climatic regions of the country. Hence, gains of new agricultural researches are not reaching the common cultivators, especially the marginal and small farmers. Very little attention is being paid for educating and training farmers for the adoption of new agricultural innovations and techniques to increase their agricultural production.
13. **Other Problems**: There are numerous other problems also which are affecting the agricultural production and rural economy and society adversely. For example, unscientific methods of agriculture, inadequate irrigation facilities, less use of chemical fertilizers, insecticides, pesticides, less remunerative prices of agricultural products, poverty, hunger, and malnutrition of farmers and lack of infrastructural facilities like roads, water, irrigation, electricity, credit, banking, and crop-insurance [31, 32].

1.5 Solution to Problems of Farmers

1.5.1 Farming Corporatization

The idea of corporate cultivating is that given a corporate a chance to possess a cultivating of a town, let him take the fields on rent for some time or 10 years relying on his venture. The corporate can give the rent sum either quarterly or month to month to the land proprietors. At that point he can utilize the ranchers (land proprietors or not) for his expected yield contingent on his arranging. The corporate would give every one of the advances, frameworks (like cold stockpiles, sustenance preparing units and so forth). At long last the corporate would reclaim every one of the yields. In the event that there is more benefit, he can appropriate reward among land proprietors and ranch workers [33].

This would end the issue of speculation, innovation and different issues which government can't manage. This will improve horticulture profitability, end worry among ranchers (land proprietor or not) and address the homestead suicide viably. There are a few issues too. The above thought can work in the zones having a few frameworks like availability, water system and other reasonableness. India's enormous ranch land needs availability, disregard water system.

Lawfulness is additionally a worry in numerous zones where corporate wouldn't chance its venture. Here and there corporate segment may need an assurance for gathering his reap. There would be some issue in labor law related issues.

Then again as we witness the way Sugar industrial facility cartel hassle the sugarcane ranchers (not paying the expense of sugarcane and at times paying less rate) as in UP and Maharashtra; same thing can create via cartel of corporates as well.

1.5.2 Precision, Smart and Digital Agriculture

India and Smart Farming

This is a high tech age where even some of our electronic gadgets are frequently more intelligent than us. In regards to horticulture we use same age old practices and we have never experienced a dawn or at least not made use of the existing technologies in farming [34, 35]. Strangled with obligations, extremely low salary and ever increasing costs, these farmers may never know the smart farming technologies (SFT).

Features of 'Smart Farming'

Nearly 80% of the farmers in US are in some way or the other exploiting the IoT techniques for reaping the benefits involved with it. More than 24% of the Europeans are involved with some kind of technological aids related to farming [36].

Information and Communication Technologies (ICT), for example, (accuracy hardware, the Internet of Things (IoT), sensors and actuators, geo-positioning systems, Big Data, Drones), robots, and so on are used to comprehend the need of agriculturists.

Some of the most useful measures used in IoT are moisture, nitrogen, fertility of the soil, phosphorous, soil impedance rate and water retention. These readings including the vegetation, characteristics and measurements taken at some day provide a good look up at some other date.

Adoption of Smart Farming Technologies in India

All the businesses involving agriculture provides for 18% of the GDP and employ 58% of the people throughout India. When there were 71.9% farmers in 1951, the population of farmers in India has shrunk to 45.1% in 2011. According to the Economic Survey 2018, the agronomists will be reduced to 25.7% by 2050 [37]. The high cost per agriculture, low production and lack of proper soil management has disinterested the farmers and there is a subsistent reduction in the next generation farmers. The time has essentially come for the farmers to take the advantage of the IoT in agricultural domain to reap its benefits.

Is the Farmer Ready for Smart Agriculture?

The gap between the introduction of a new technology its implementation and its adoption is very big. This becomes all the more aloof when the target users are the farmers of and undeveloped country. Most of the farmers in India are novice to even smart phones and their lack of interest and hesitation towards learning a new technology makes the task of adoption difficult by many times. The technical words like ‘transaction’ is probably not even heard by them [38, 39].

The Indian farms are small in size with nearly 80% of them even less than 2 ha and 30% irrigated by natural sources.

Including drones and satellite images, sensors, the horticulture industry is changing quite noticeable way [40]. Smart farming [41], Precision and digital farming [42] though related terms do have a finer difference in their meaning.

1.5.3 Precision Agriculture

McKinsey and Company characterize precision agriculture as: “a technology-enabled approach to farming management that observes, measures, and analyzes the needs of individual fields and crops” [43]. 2016 report apprises on how useful information collected over years will modify the global food chain, According to McKinsey, the graph of precision agriculture is governed by two trends: “big-data and advanced-analytics capabilities on the one hand, and robotics—aerial imagery, sensors, sophisticated local weather forecasts—on the other" [44].

1.5.4 Smart Farming

In order to optimize the difficult and complex task of farming system, the application of information technologies and devices such as sensors, receivers, hubs and wireless computing is known as smart farming [41].

The information can be data regarding condition of soil and plants, graphs of the terrain, readings of the climate, weather, statistics of resource usage, use of manpower, funding, etc., can be taken up by the Smart Farmers by use of hand-held devices such as smart phones or sometimes tablets. A farmer armed by the IoT tools will get all the information needed to make informed decisions based on statistical data prepared by smart phenomenon, rather than just anticipation.

1.5.5 Digital Farming

The idea of digital farming lies in extracting and mining value from data. The concept of actionable intelligence and semantic added value is attached with the data which is much more than mere availability of data.

Digital farming is amalgamation of both concepts—precision farming and smart farming [42].

2 IoT and Its Potential for Agriculture

2.1 IoT Functional Blocks

The most acceptable definition of IoT given by Smith says that a dynamic global network infrastructure with self-configuring capabilities based on standard and interoperable communication protocols where physical and virtual "things" have identities, physical attributes, and virtual personalities and use intelligent interfaces, and are seamlessly integrated into the information network, often communicate data associated with users and their environments [45].

An IoT system consists of many functional blocks to provide for sensing, identification, actuation, communication, and management. Figure 1 presents these functional blocks as described below.

- **Device**: The devices in IoT serve to provide sensing, manifestation, control, monitoring and practical implementation activities. They can exchange data with peer devices and applications, or collect data from devices and process the data either locally or with centralized servers even, cloud back-ends.

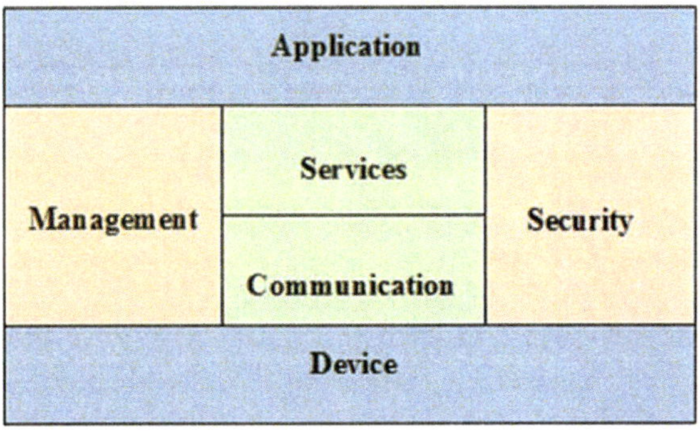

Fig. 1 IoT functional blocks

- **Communication**: The task of message exchange between devices is carried by the communication block.
- **Services**: The services provided by the IoT include device discovery, service allocation, etc.
- **Management**: Different functions are provided by the Management block to govern an IoT system.
- **Security**: Provides authentication, authorization, privacy, message integrity, content integrity, and data security.
- **Application**: Acts like an interface to the user and the IoT system.

2.2 *IoT Agriculture Framework*

This part discusses a comprehensive framework to accommodate full-fledged agro-solutions using Internet of Things (Fig. 2). The presented framework depicts a six layered concept including hardware facilitics, internet and third party communication technologies, IoT middle ware, cloud services, big data analytics, and the experience of the agriculturists providing full support [46].

- **Physical Layer**: This layer talks about the lowest abstraction of the IoT including sensors, actuators, microcontrollers, network equipment like routers, gateways and switches.
- **Network Layer**: This layer comprises of Wi-Fi, GSM, CDMA, LTE (4G) technologies
- **Middleware Layer**: The functions performed by the middleware layer include device operation, context awareness, co-operation, portability, and system security.
- **Service Layer**: This layer is cloud assisted. It is used by the other layers up the stack for storage on the cloud and SaaS. It deals with Sensor data, equipment recognition, store house of plant disease, and heuristics.
- **Analytics Layer**: Here prediction is done using Big Data. Prediction can be for measuring probabilistic chance of productivity of crops in upcoming season.

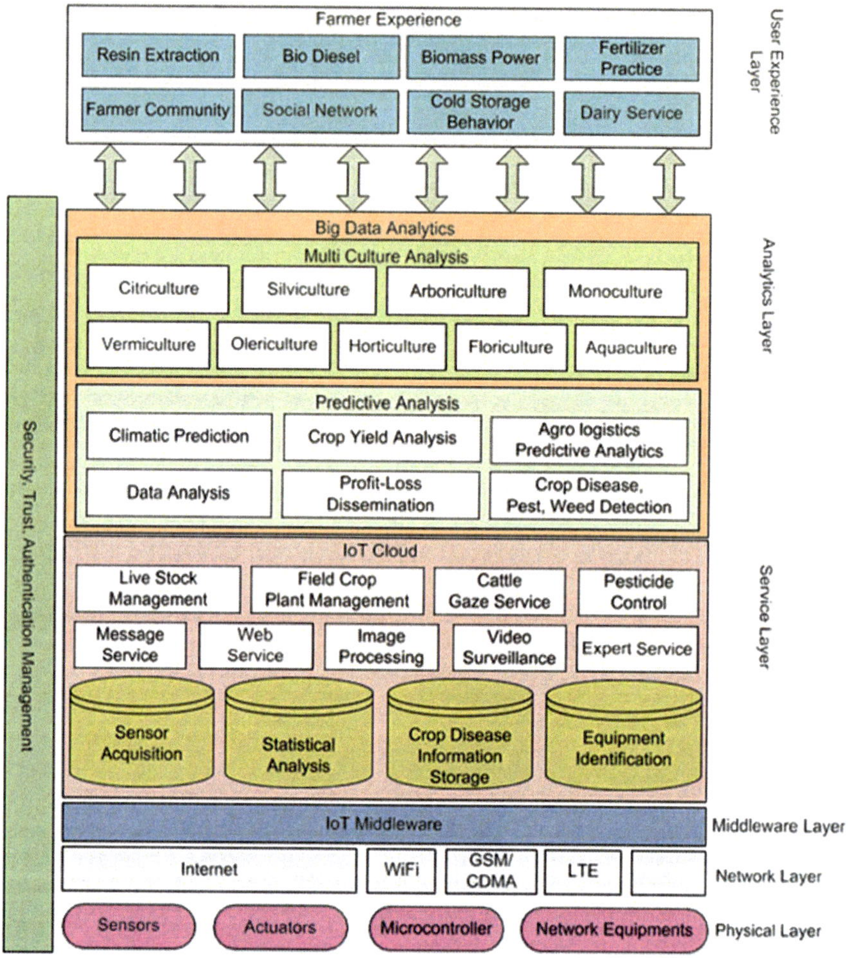

Fig. 2 IoT based agricultural framework (adapted from [46])

- **User Experience Layer**: This application layer is designed for the agriculturists. This is the top most layer and through this layer the farmer communicates with the other members using social network to exchange their views.

2.3 *IoT Based Agriculture Applications*

This section throws light upon the various possible IoT applications including agricultural, farming and related applications.

2.3.1 Irrigation Management System

Modern agribusiness requires an improved irrigation management system along with executives' framework to enhance the water use in cultivating and related exercises. Four elements are popularly being utilized in smart irrigation system framework, for example, integration of real-time weather forecast data, control of farmer's framework from anyplace on the planet utilizing home, empowering WiFi and Ethernet connection, including synchronization with dampness or moisture sensors introduced in farmer's yard, and decreasing farmer's month to month bills while monitoring constrained water assets. IoT is always getting fame in water and irrigation management related system around the globe.

2.3.2 Pest and Disease Control

With the help of IoT devices and actuators a constant vigilance over the farming area can be kept and through the use of the drones held cameras the agriculturist is able to keep his eye on crop quality. He can also figure out by the images sent to him that whether the crop has caught some disease. He can also have the probability of the occurrence of pest in harvests and can arrange the quantity of pesticides beforehand. If one farmer goes through an experience of pests and diseases then through the social apps, he can also share them with his peers and counterparts.

2.3.3 Cattle Movement Monitoring

Constant observing of any dairy cattle is additionally accomplished through the IoT. A lot of text is available showing smart technology based ecological framework for pig and other cattle [47].

2.3.4 Dairy Monitoring

Connecterra as an example is being currently used to manage dairy in following the line of the IoT [48]. The animal behavior is studied and analyzed using this technique. This knowledge is then used to predict animal reproductive system. By constant vigilance over the animal habitat, the predictions can be streamlined even for animal diseases and cure.

2.3.5 Water Quality Monitoring

We can measure temperature, pH value, salts, and turbidity of water by water quality observing. This is done by placing wireless sensors near water spouts.

2.3.6 Greenhouse Condition Monitoring

A related term with agriculture is Greenhouse that affects the former. Temperature rise is attributed to the ozone layer depletion and this has direct effects on the crops.

2.3.7 Soil Monitoring

The crop quality directly depends on the soil. The knowledge of soil property is crucial for agricultural paradigm. Measuring soil quality and constant measures of the soil sent to the agriculturists is seen in 6LoWPAN and its benefits are quite obvious.

2.3.8 Precision Agriculture by UAV

Drones and the UAVs have revolutionized the way technology has assisted mankind and precision farming and smart agriculture is no exception. With the help of drones and UAVs the agriculturists keep constant eye over the farms. This helps in diagnosing diseases and weather and wind prediction. The farmer is able to take decisions based on statistical tools that forecast the weather, rain, humidity and presence of pests and other troublesome animals that may ruin the farm or cause damage.

2.3.9 Production Supply Chain Management

IoT has direct effect on supply chain management due to better quality of crops, timely harvest and lesser losses incurred due to many reasons.

3 IoT Equipments

3.1 Existing IoT Products for Smart Farming

Next few paragraphs are devoted to highlight some recent progression in Smart Farming.

- **Network**: In an ongoing study by Sinha, creators found that LoRa is the best contender for savvy agribusiness applications [49]. The creators of Lukas structured a long-extend water level checking framework for troughs utilizing a Wireless Sensor Network (WSN) dependent on LoRa handsets, enabling the cattleman to watch water accessibility for domesticated animals notwithstanding when the horse shelter was 1 or 3 km away. In an alternate application, proposed an IoT structure to add to rustic advancement executing rural applications bolstered by open-source equipment and long-extend specialized gadgets. The main arrangement of this arrangement utilized LoRa handsets since provincial towns were situated in remote regions and it was helpful to have a minimal effort and non-exclusive foundation [50].

- **Efficient management of Energy**: One of the primary necessities for gadgets utilized in IoT tasks is that they should be vitality proficient. This is especially significant for inescapable arrangements conveyed outside that can't be powered from the electric grid nor routinely kept up in light of the fact that they are introduced in troublesome or remote conditions. In WSN situations, the present test is to create multi-source vitality reapers and ultra-proficient sensors to make sans battery arrangements [51]. These contemplations are significant for IoT answers for agro industrial and natural issues and environment problems as recharging batteries isn't practical and ambient energy sources are generally accessible. As far as savvy vitality control for IoT ventures, proposed a novel vitality the executives methodology for sun based controlled gadgets that expect to control the heap straightforwardly from the sun powered cell, dodging power converters and vitality stockpiling components that add to vitality misfortunes, more prominent weight/volume proportion, and more significant expense [52].
- **Control**: Ongoing natural monitoring solutions are presently offering extra capacities regarding decision making and management. For instance, the authors in [53] proposed a uniquely designed land slide hazard checking framework dependent on a WSN that permits quick arrangements in antagonistic conditions without human mediation in light of the fact that the framework can manage hub disappointment and low quality correspondence connections revamping the system independent from anyone else. Wong and Kerkez exhibited a Web administration and constant information engineering that incorporates a versatile controller that updates the parameters of each detecting hub inside a WSN dependent on a recently characterized approach [54]. Edwards-Murphy presented a bee sanctuary observing framework that gathers inward and outer information to depict the status of the honey bee province from a lot of potential states utilizing a characterization calculation dependent on decision trees [55, 56].
- **Orchestration**: Some of the agri-business areas of IoT that have emerged are concerned with food quality with no compromise to supply best food to the customer and channelizing production with the demand [57, 58].

3.2 *Multi-agent System Architecture*

The use of ontology works to provide a thorough knowledgebase in support of the multi-agent system and extends the functioning of the MAS by introducing the concepts like rules and reasoning or inferencing. Ontology is a term taken from Meta-Physics that gives the foundations of being or existence of an organism in ecosystem. In this chapter the authors have used ontologoies along with MAS to simulate the actual manual working of some of the existing and some novel systems like agents for fertilizers, pesticides, sprinklers, soil measures, moisture and disease monitoring by use of UAVs and drones. Smart agents function as autonomous and situated intelligent objects of farms, crops, fertilizers, manures, soil, weather, pests, etc. [59, 60].

In order to develop agents to work in our domain firstly, O-MaSE methodology for agent development is utilized. The primal work here is to first identify the goals that are needed to be met in order to completely design a MAS for smart agriculture. When the goals have been identified then the roles are acknowledged that function for fulfilling the goals. Then the agents are identified based on the grouping of roles [61]. For team goals, organizations do allow agents to work together by using best cooperation among agents that work in collaboration even with previously unknown agents resulting in mutual clubbing of each agent for joint functioning. In order to develop a MAS, we need to follow a Agent Oriented Systems Engineering (AOSE) technique. The development strategy followed here in this research is O-MaSE. There are numerous benefits of O-MaSE that make it powerful [62].

3.2.1 Organizational Multi-Agent Systems Engineering (O-MaSE)

The authors have considered O-MaSE for designing the MAS for smart agriculture and this is done in order to provide medium level projects that can be extended to complex and high level projects as well. A thorough study of O-MaSE is provided in [62].

O-MaSE is a meta-model and is tailored to match the project's needs.

Twelve method-roles have been identified as part of the O-MaSE methodology: requirements engineer, goal modeler, domain modeler, organization modeler, role modeler, agent class modeler, protocol modeler, policy modeler, plan modeler, capabilities modeler, action modeler, and programmer. Each O-MaSE method-role is responsible for carrying out the tasks by applying the appropriate techniques to produce the end result. The chief constituent products needed in ABSMSA are shown in Fig. 3.

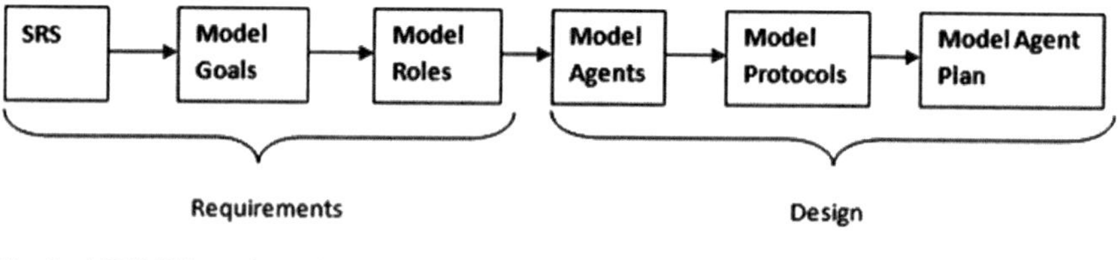

Fig. 3 ABSMSA work products

4 Proposed Model

4.1 *Agent Based Semantic Model for Smart Agriculture (ABSMSA)*

A digital ecosystem is a distributed, adaptive, open socio-technical system with properties of self-organization, scalability and sustainability inspired from natural ecosystems. Digital eco-system models are inspired by knowledge of natural ecosystems especially for aspects related to competition and collaboration among diverse entities [63, 64].

MAS is decentralized and its functioning in accordance with the cloud systems forms the foundation for the heterogeneous style of horticulture. Even each and every stakeholder who is associated with agriculture including buyers, sellers, and also third party people like pesticide, fertilizers, seeds and farming equipment suppliers will benefit.

The constituents of ABSMSA are discussed below.

4.1.1 Cloud Based Platform

The services provided here are through intelligent agents that are autonomous and can react to events. The cloud based platform provides for securely saving the data in the form of big data and conduct analytics and prediction through negotiations via enterprise service bus.

4.1.2 Ontologies in Smart Agronomics

Is formed by the entities recognized in the domain of smart horticulture like all the information needed about the seeds, crops, diseases, pests, seasons, weathers, soil, machines, etc. and the relationships between these entities in the form of predicate logic. These ontologies are designed by the authors keeping in mind the requirements of agriculture by the agronomists of developing countries. Here careful consideration is made in order to eliminate too extensive ontologies and focused selection of terms and attributes are made to help the farmer and for better performance. Three separate knowledge bases are used. Existing Light weight ontology called IoT-Lite is user for the purpose of sensory data in heterogeneous IoT platforms [16]. Another ontology, named Complex Event Service Ontology (CESO) is used to trigger and manage farm events, like immediate need to irrigate, detection of sick animals or pests in crops [65]. This also aids in better decision power by the agronomist.

4.1.3 Smart Constituents

The designed agents' services for pesticides, fertilizers, crop, soil, water and relater machines like sprinklers, etc. expose functionalities which monitor the state of the objects they are bound to. Agents develop and elect plans and generate recommendations. Most importantly they negotiate results with other agent services and users.

4.1.4 Service of "Virtual Discussion Room" for Coordinated Decision Making

Software agents are organized in a Virtual Discussion Room (VDR) for coordinated and synchronized decision-making and to provide consultation for farmers. It is a regular happening that new unpredictable situations arise in the realm of the ecosystem. A re-balancing of goals, their related plans, beliefs of the agents need to be done for new scenarios emerging from the current situations. This will give rise to new vents that are triggered and this cycle will go on which will constantly require discussion among the agents. Negotiator agent, in case of a conversation with other agents invites all the agents to the discussion. After receiving acknowledgement from other agents, the Negotiator agent demands for proposals from other agents. Now the other agents analyze the problem scope and they submit the proposal to the Negotiator agent. Now the Negotiator re-calculates the indicators (checks that there is no more in-homogeneity in the system). As a result either the proposal is accepted or rejected. This goes on for a fixed number of iterations or a fixed time stamp until the conflict is resolved or finally rolled over and the system is restored to a previous state.

4.1.5 Smart Agent of Agronomist Simulator in Mobile Phone

A mobile phone or a tablet preferably a smart phone with internet is required by the agriculturist to receive relevant information, news, happenings through events that are triggered on the farm or anywhere that is linked to the farmer. He can also be a part of some discussions. Advices from professionals can also be made use of by him. All this is done by the use of a service called Agronomist Simulator through which nearly all the task of the agronomist comes to his finger tips or we can say that he can operate upon all the functions on the field by sitting at home.

The Fig. 4 shows the ABSMSA model where respective agents are exposing services to carry out the intelligent agro-automation process. The agents responsible for the fields and the crops expose the farm service. The supervisor agent controls the satellites and drones for vigilance over the agriculture land. These services are grouped under 'monitoring services'. The Machine Agent, PPP agent and Manure_FertAgent provide the 'resource management services'.

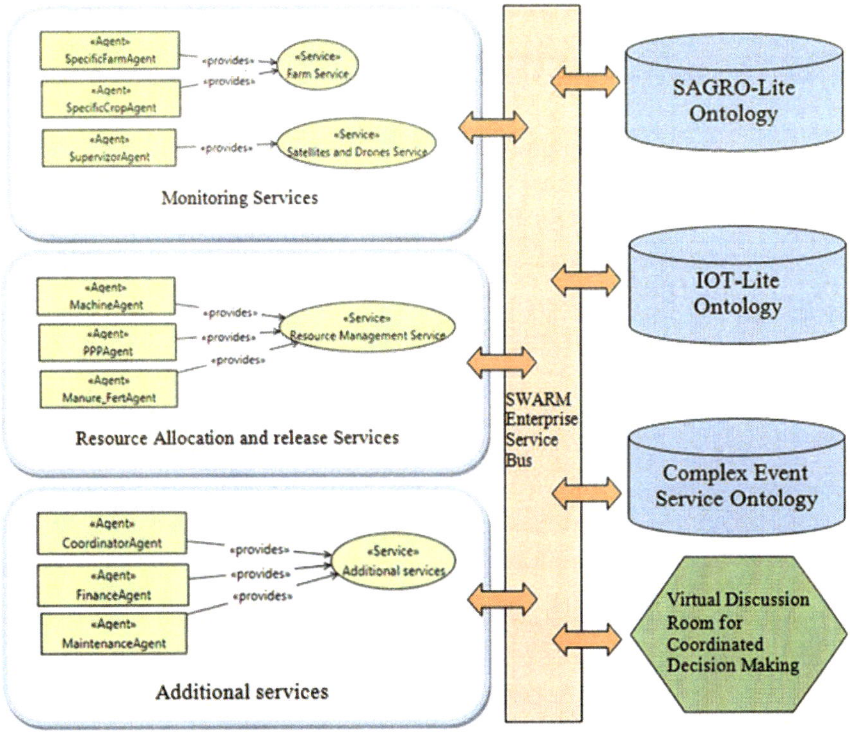

Fig. 4 ABSMSA model

Additional services include day to day plan implementation, finance and maintenance conducted by coordinator agent, finance agent and maintenance agent respectively.

The Swarm ESB acts as a virtual medium for communication between agent services, the three knowledge bases and the enabler for the coordinated decision making in the virtual discussion room.

5 Design and Development of ABSMSA

5.1 Goal Diagram

O-MaSE methodology is a meta-model and its various work-products are available that are chosen to suite a particular organization's needs. In Sect. 3.2.1 the authors have already discussed the reason to choose the particular work-products [66]. In goal modeling, a hierarchy of goals is constructed that comprehensively include all the functionality of the O-MASE [67] (Fig. 5).

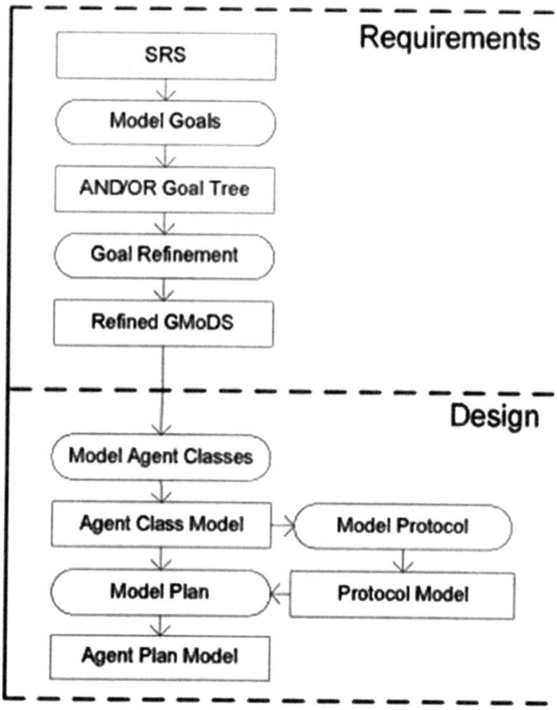

Fig. 5 O-MaSE process

Ten goals have been identified as part of the O-MaSE methodology as shown in Fig. 6.

5.2 Role Diagram

In order to model roles we assign different goals identified during design of goal model to the most suitable roles or actors in the system. Role model is shown in Fig. 7. Each leaf goal is assigned either in entirety or partially to the roles. More than one goal can also be assigned to the role. After proper role assignment, the allotted roles will then be mapped to the agents. This also defines the capability of the agents [68, 69].

5.3 Agent Diagram

The agents derived from the roles in our proposed system 'ABSMSA' can be shown in Fig. 8. Each agent is capable of performing certain functions. This is depicted by the capabilities it possesses. For Example the SimulatedFarmerAgent

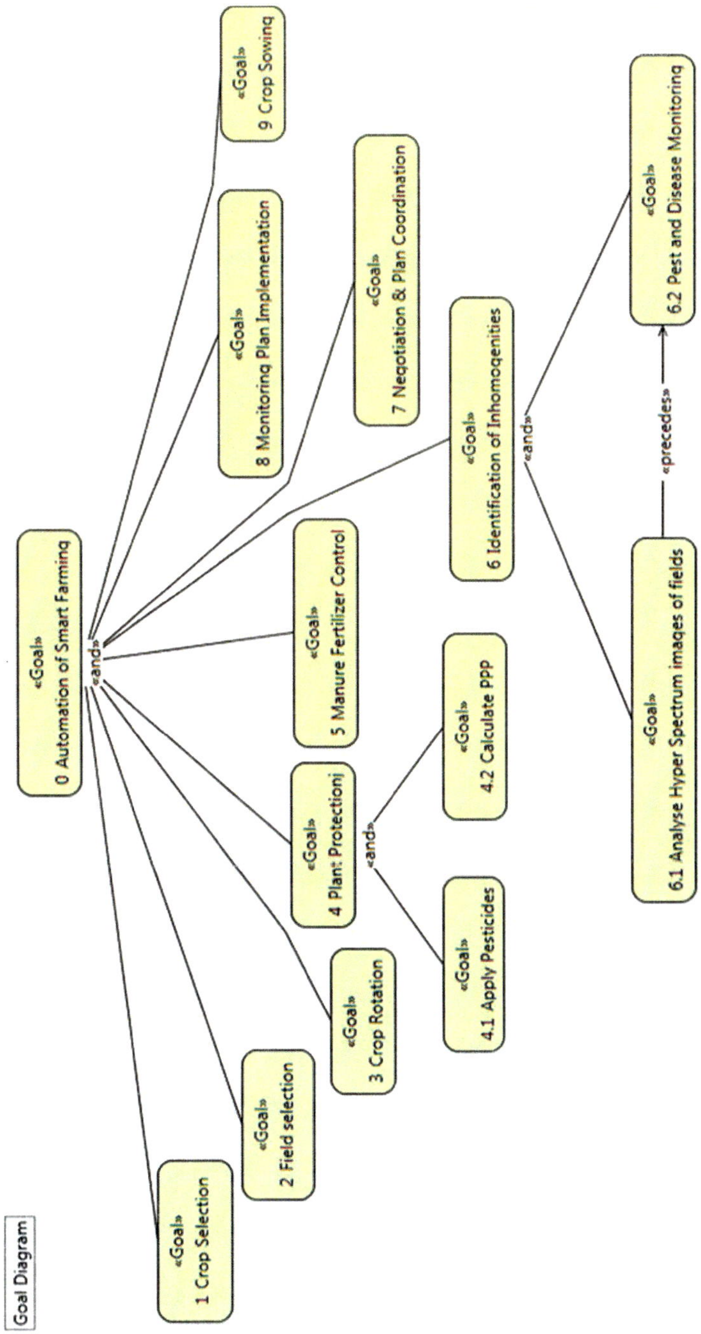

Fig. 6 Goal diagram for ABSMSA

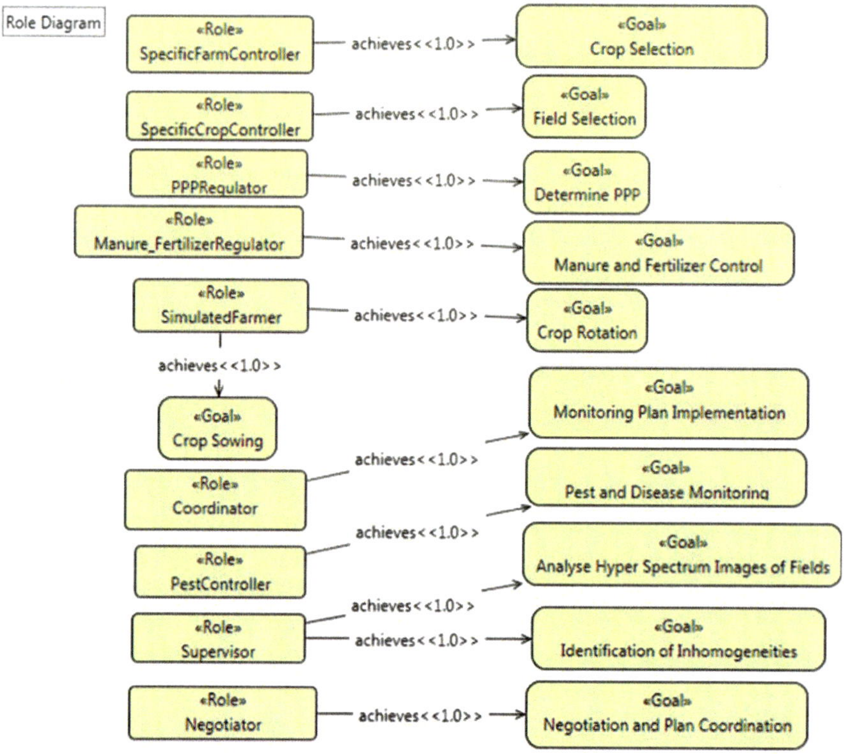

Fig. 7 Role diagram for ABSMSA

possesses the capabilities namely, CropRotation_PLAN, cropProcessing_PLAN and TechnologyOperation_PLAN [70].

5.3.1 SpecificCropAgent

The task of this agent is to intelligently select the optimum crop taking into consideration the weather, soil, moisture, etc. The SpecificCropAgent intrigues to query the SAGRO-Lite ontology to determine the most suited crop for the existing climatic condition. It also takes into account the level of rainfall, type of soil, temperature, etc. in order to make its selection.

5.3.2 SpecificFarmAgent

SpecificFarmAgent gives the intelligent decision making power of the finest field for the plant variety and then the sequence of undertakings for cultivation and control of the whole phenomenon. The SpecificFarmAgent also generates tasks for agents

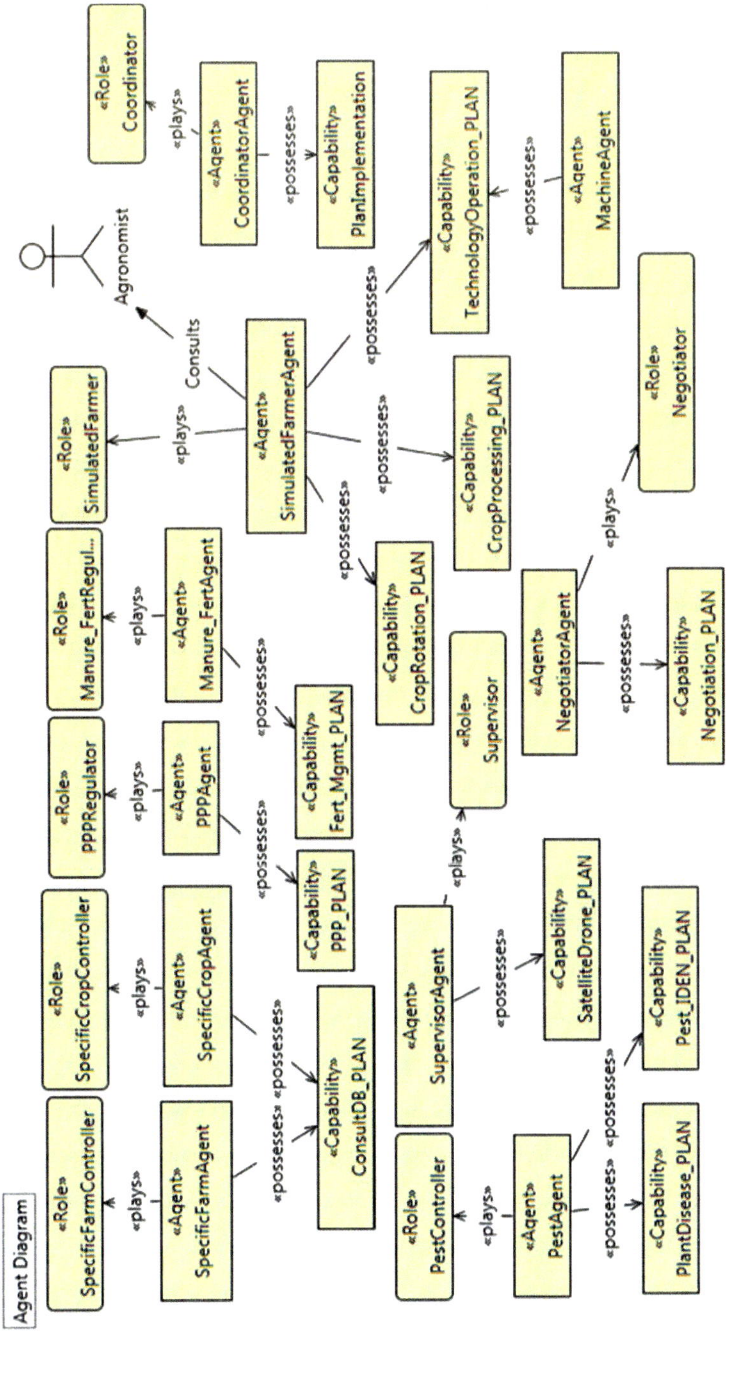

Fig. 8 Agent diagram representing agents in ABSMSA

of machinery, brigades, and monitoring. In other words everything related to a farm or corporatized farms starting from accepting signals of in-homogeneity detected by the satellites through SupervisorAgent, activating the drones to capture images of the infected area and finally sending the images to the farmer or agronomist to take future action. SpecificFarmAgent is also responsible for initiating the MachineAgent for sprinkling pesticides over the infection diagnosed area of the fields.

5.3.3 SimulatedFarmerAgent

We can say that the SimulatedFarmerAgent acts on behalf of the farmer and is always connected with the farmer by means of an application programming interface through a mobile device. In cases, where something has to be done on the fields like watering, ploughing, etc., the farmer instructs this agent by the mobile device and with multi-agent collaboration between MachineAgent, SpecificFarmAgent and Manure_FertAgent the task is accomplished.

5.3.4 PPPAgent

PPPAgent decides the usage of varied types of machines for plant curing products. This agent typically plans technological proceedings for containing the flow of applied pesticides. PPPAgent also governs the barest minimum application of pesticides in order not to diminish the vitality of the crop.

5.3.5 Manure_FertAgent

The horticulturist has full control over this agent in terms of the fertilizers and nutrients for better crop production. This agent works in tandem with the PPPAgent to control the pesticide and fertilizer usage in the farm.

5.3.6 PestAgent

Smart agriculture gives the agronomist deep decision making assistance, be it about the disease or contamination by the pests. And all this is done by considering the state of the farm. The PestAgent by taking information from UAVs and drones apprises the agronomist about possible threats of the pests in the farm.

5.3.7 MachineAgent

Machine Agent is hub that manages all the hardware devices and tools used by the smart farming industry. These can range from ploughs to sprinklers, carriers and brigades to daily used equipments.

5.3.8 CoordinatorAgent

Coordinator agent looks for fine execution of the plans generated by the SimulatedFarmerAgent, SpecificFarmAgent, SpecificCropAgent andSupervisorAgent and coordinates their mutual tasks on shift-daily basis.

5.3.9 SupervisorAgent

This agent is the smart worker that clicks images through drones and UAVs, transfers them to the agronomist holding the smart phone through the satellites. This agent is also capable of diagnosing an infected crop by comparing the images of leafs or flowers and checking them against a huge database or ontology logistics. Firstly, some homogeneity is identified from the satellite for the whole farms coloration. Then further investigation is done by requesting the drone. And if there is some alarming condition then the information is sent to the agronomist along with the possible solutions, if any.

5.3.10 NegotiatorAgent

Moderates the "discussion room" for coordinated decision making, controls indicators and is also the think tank of the whole eco-system. It is also responsible for conflict resolution between agents if some deadlock is reached, for proper functioning of the multi-agent system. Whenever there is a problem at hand, this agent invites solutions from all the participators and by negotiation decides the best solution under existing situation.

5.4 Protocol Diagram

The process of inter-agent communication is governed by the protocols designed for the system. Two or more agents interact by the use of a protocol. Messages are exchanged by the rules governed by the protocols. In order to diagrammatically represent the protocols before implementing them we need to show all possible message exchanges and sessions among the agents via a protocol model. The protocol model is also the message verification system that has to be abided during a course

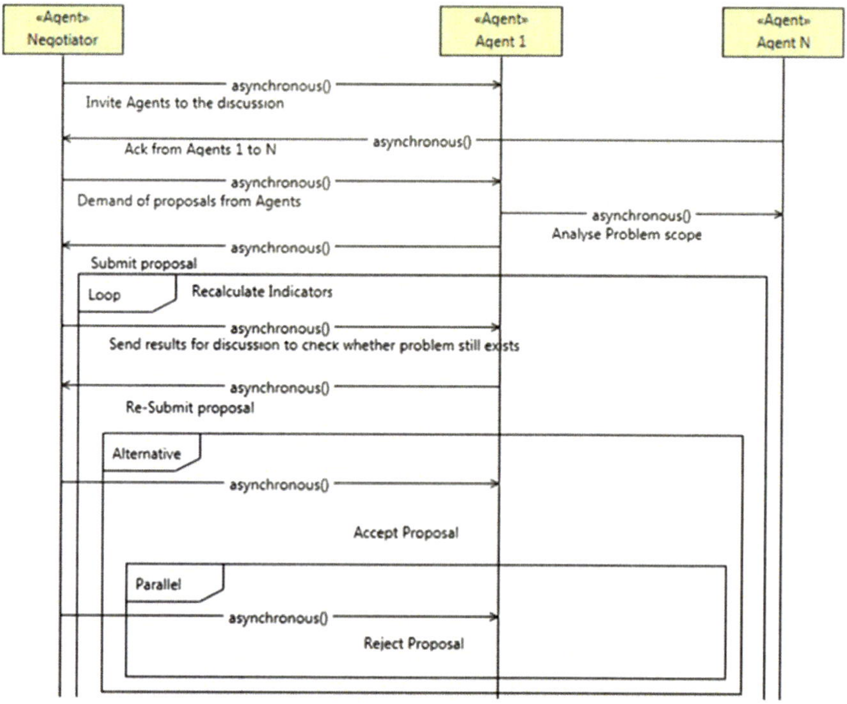

Fig. 9 Virtual discussion room protocol/interaction diagram

of agent communication. The policy rules define the contents of the protocol model. The interaction diagram for virtual discussion room is given in Fig. 9.

6 Ontologies Used in ABSMSA

6.1 IOT-Lite

In order to exclude unimportant details of existing monolithic ontologies for the Internet of Things for varied applications, IoT-Lite was proposed [71], shown in Fig. 10, a launch of the semantic sensor network (SSN) ontology to depict key IoT ideas permitting interoperability and revelation of sensory data in heterogeneous IoT stages by a lightweight semantics. IoT-Lite is the result of an exploratory research that spotlights on revelation and looks for the base ideas and connections that can give answers to the vast majority of the end client questions within time and complexity constraints [72]. This ontology is incorporated in ABSMSA for it being light weight and small, only holding the knowledge that is sufficient for smart farming in countries that are developing and less varied in agricultural crops and conditions like India [16].

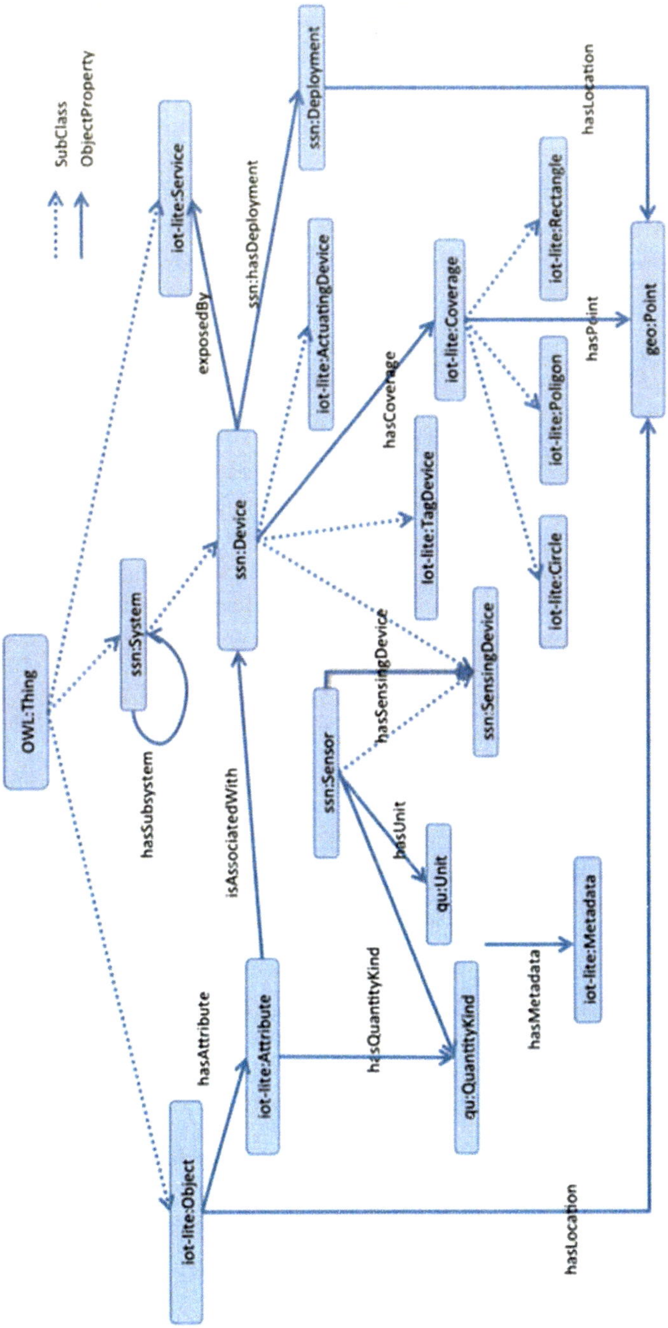

Fig. 10 IoT-lite ontology

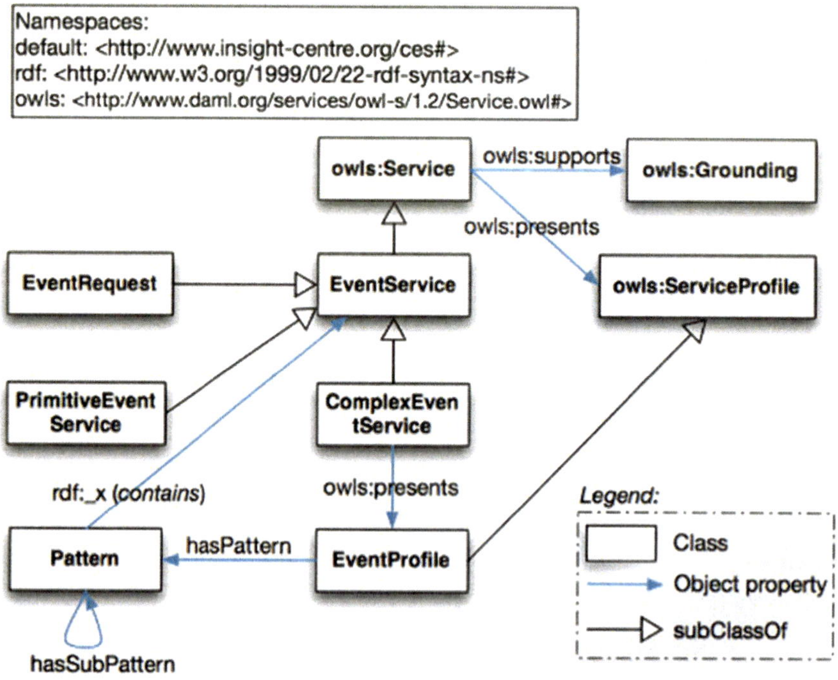

Fig. 11 Common event service ontology

6.2 *Complex Event Service Ontology*

It provides devices to prepare and function upon aggregated data streams to collect farm events, for example, requirement for irrigation system, diseased animals or vermin/pests identification in crops. This ontology mainly deals with all types of sensor generated events triggered by smart devices. The advantage of using this ontology is that it can function in collaboration with IoT-Lite ontology already mentioned. Where, IoT-Lite nearly has all the knowledge regarding the usage pattern of majority of IoT applications, it ignores the less used classes and relationships of SSN, making it fast and spontaneous. On one hand, where the IoT-Lite functions to provide quick solutions to most of the common queries, CESO provides the event triggering and handling functionality [65]. CESO is shown in Fig. 11.

6.3 *SAGRO-Lite*

During development for this precise and explicit ontology intended for the domain of horticultural in developing nations, the following points were considered:

- What are the reasonable harvests or crops that are to be grown?
- The choice of crops.
- Selection of manures.
- Selection of time to use manure.
- Identification of pests and plant infections.
- Disease preventive measures.
- Appropriate techniques to counter particular disease.
- Side effects of a particular disease.
- Significant steps to keep up nature of crop harvestings.
- Consideration for post harvest methods.
- Harvests developed by different ranchers and quantity.

The next task is to identify broad areas of cultivation to figure out the details regarding the above identified points. These include nurseries, harvests and post harvests considerations including pest control, fertilizers and common control tasks.

To semantically annotate data streams, SAGRO-Lite works in collaboration with IOT-Lite, a light-weight information model developed on top of SSN [73–76].

By providing an RDF-based representation of heterogeneous streams, C-SPARQL solves the challenge of giving reasoners an access protocol for heterogeneous streams [77, 78]. As RDF is the most accepted format to feed information to reasoners, C-SPARQL allows existing reasoning mechanisms to be further extended in order to support continuous reasoning over data streams and rich background knowledge. C-SPARQL is excellent to be used in complex and multi-stream queries while CQELS is primarily used in queries requiring static data.

The logical idea behind designing this ontology was that the Indian farmer needs knowledge regarding crops, climate, humidity, soil condition, pests and diseases only under the boundary of his agricultural needs and cultivation. So monolithic datasets describing about useless crops, yields, soil, optimum moisture levels, temperature and knowledge about related paraphernalia clearly is not needed. This also increases the functioning and performance of the eco-system.

The SAGRO-Lite ontology is derived from the Generic Crop Knowledge Module shown in Fig. 12. Centered on crop, its related entities are basic characteristics, climate, fertilizers, disease, symptoms, cure, harvesting, marketing and economics.

An extract of this ontology developed in Protégé 5.2 in the form of an OWL-Viz diagram is shown in Fig. 13. The entity crop is expanded to clearly depict crop type, common name, scientific name, etc. shown in Fig. 14. Then to another level, the ontology is further expanded where we can see the details and relationships of the attributes of 'Wheat' crop with hasMinTemp, hasMaxTemp, hasMinRainfall, hasMaxRainfall, hasSoilType, hasDisease, etc. shown in Fig. 15.

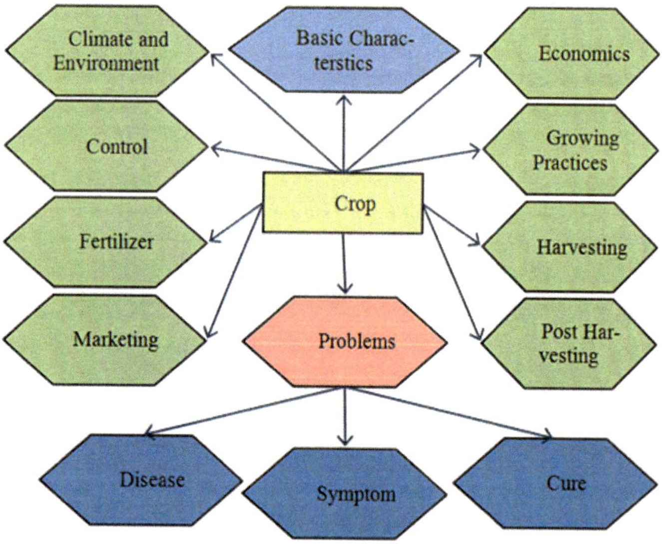

Fig. 12 Generic crop knowledge module

7 Scenario

A "virtual discussion room" in ABSMSA as pointed out earlier is shown with an example scenario using smart services is considered below:

- The satellite identifies that one of the farms of the horticulturist is abnormal triggered by Supervisor Agent, having a new in-homogeneity by reading the pattern of the field photographs as shown in figure and triggering the SpecificFarmAgent (Fig. 16).
- The SpecificFarmAgent comes to action and activates the drone agent to determine causes of the problem by closely investigating the place of issue and constructs an execution plan to deal with the newly investigated problem.
- The drone and UAVagent preliminarily plans shooting. This is done over the problem area using hyper spectral cameras as shown in Fig. 1. The whole process and the related problem and solution is recorded in the Farm Database and also the experience is registered in the knowledge base (Fig. 17).
- The SpecificFarmAgent sends warning messages to alert the agronomist along with the photographs taken by the drone (Fig. 18).
- The image recognition API initiates the task of comparing these images with the images stored in the databases and other approachable knowledge bases. Accordingly, PPPAgents are activated (Fig. 19).
- By consulting the knowledge base through SAGRO-Lite ontology interface, the likely reasons of the problem situation are anticipated by coordinated decision

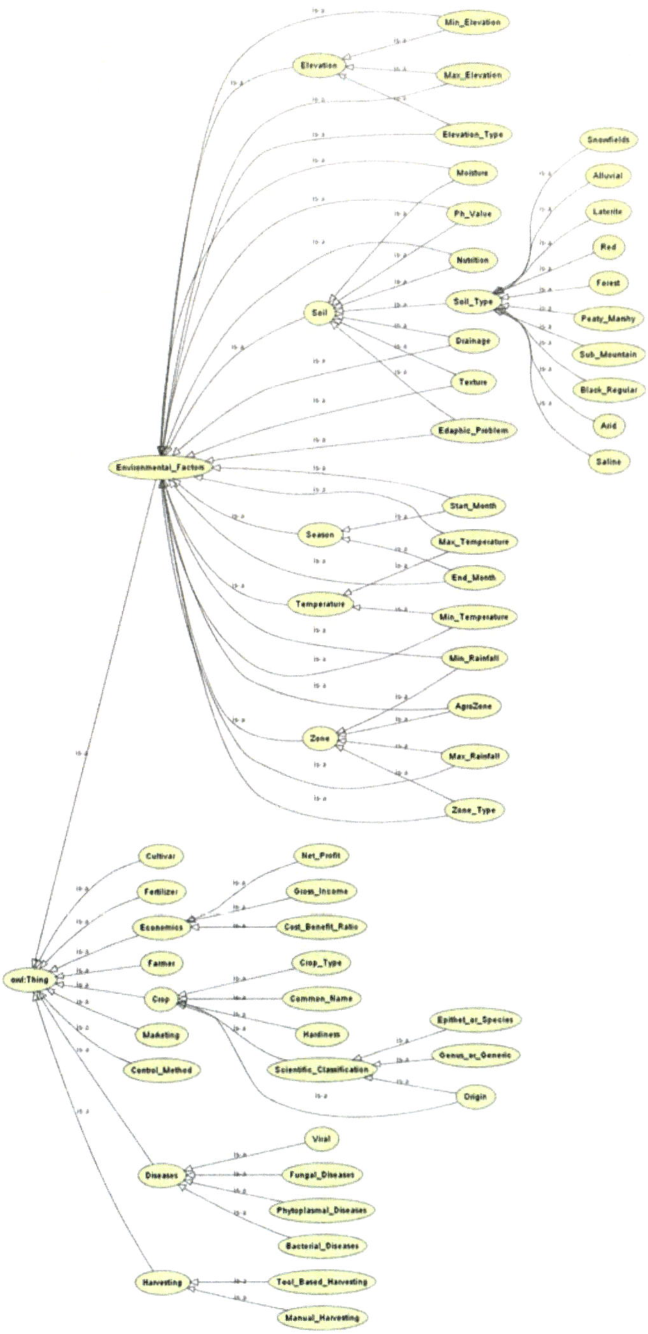

Fig. 13 SAGRO-lite ontology

Fig. 14 SAGRO-lite ontology: crop class

making. The likely reasons are found to be the disease 'wheat leaf rust' due to the pests or under-feeding.

- The SimulatedFarmerAgent plans the horticulturist's trip to the problem area to observe and find if everything works under control and if the expected problem is the actual problem.

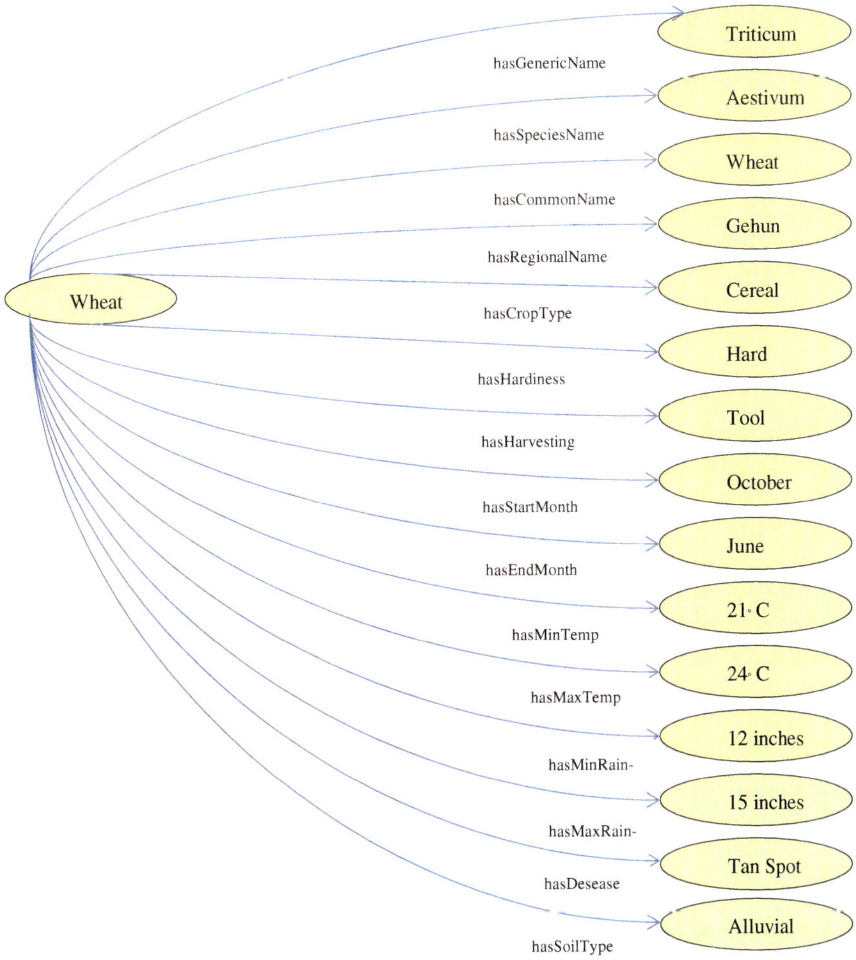

Fig. 15 SAGRO-lite ontology: wheat as individual

- The actual cause of the problem is found to be the wheat leaf rust disease. By investigating further he also figures out that the real cause is pests.
- After inputting the required information about the problem in the smart app, the pesticide sprinklers are activated to be used to the infected area of the farm.
- The daily tasks of the MachineAgents and Machine Operators are scheduled.
- SpecificFarmAgent logs causes and solutions of the problem in order to be more confident to diagnose a similar problem in future.
- If more inspection is required then the SpecificFarmAgent provides greater control and tracks any changes that have been observed.

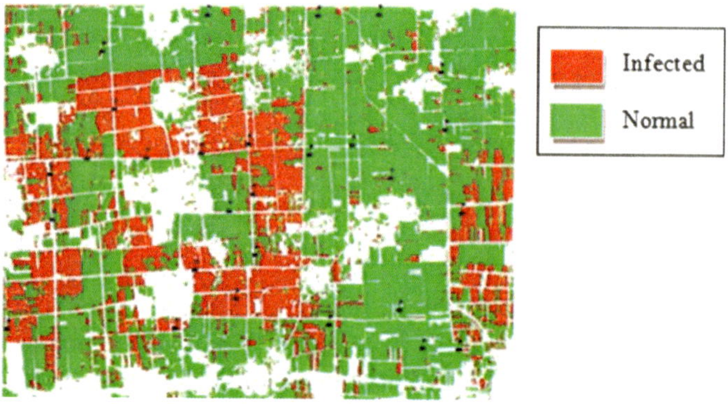

Fig. 16 View of the wheat field from satellite

Fig. 17 Drone for close monitoring of the field

Fig. 18 Diseased wheat leaf through drone photography

Fig. 19 Image of a diseased wheat leaf sent to the agronomist

8 Future Trends and Conclusion

Despite high dependency on agriculture, India is going through a deep agrarian crisis. The survey of 5000 farm households across 18 states says that 76% farmers would prefer to do some work other than farming. The difference in pace at which the technologies develop and at which they are adopted and consumed until they become a standard is big and in case of developing countries it is all the more significant. So through a light weight IOT specifically focused on farming style of developing countries like India, farmers can increase their quality of farming and thereby have better production and returns by the use of this technology. But instant adoption of IoT in farming is impractical in developing nations due to following reasons: India cannot afford high prices so in order to be successful the IoT based horticulture solutions must be low cost. The farmers are less literate and less tech savvy. The farmers need some easy and straightforward solution to their queries and an easy interface. The present frameworks for Smart Farming prove to be too complicated to be practically implemented. The authors propose a light weight semantic agent based model SAGRO-Lite, that uses the IOT-Lite meta-model [79, 80].

The authors in this chapter present the work which is the outcome of a research effort that focuses on discovery and seeks for the minimum concepts and relationships that can provide answers to most of the end user queries in the farming domain. The authors have developed an ontological model, SAGRO-Lite, a part of the bigger Multi-Agent System, ABSMSA for use in Smart Farming typically for the challenges faced by the developing countries [29]. To avoid the delays, complexity and not so profitable features, ABSMSA utilizes the IoT-Lite Meta Model for performance improvements and also provides a smart agriculture ontology tailored for use by farmers in the developing countries like India. Semantic modeling is only the initial part of the whole design, SAGRO-Lite with its agents has to take into account how the models will be used; how the annotated data will be indexed and queried with real-time data; and how to make the publication suitable for constrained environments and large scale deployments when applications often require low latency and processing time. All events triggered by the sensors are handled by a dedicated ontology called are Complex Event Service Ontology discussed in earlier sections [81].

The farmers in India, and other developing countries have a common problem of lack of technological knowledge and even a greater concern arises from the farmers' non acceptance to technological changes and up-gradation. The whole effort of the smart agriculture will go in vain and incur huge costs if the farmers do not cooperate. While the framework provides constructive features and benefits to farmers in developing countries, this paper presents a research plan with an overarching goal to help ensure that the farmers in countries like India gain from the IoT [82].

The future study focuses on the potential impact of technological developments of smart farming on the (Indian) agriculture and food sector in the long term, transcending domain boundaries and disciplines, and thus offering a view on uncertainties and room for strategic decisions. By working with various methods of future studies we have tried to do justice to the many uncertainties that are intrinsic to the future of a complex domain such as agriculture and food. Despite the focus on technological developments this study also touches the area of social problems and solutions by reflecting on the scenarios and by looking at developments in a context of technological and non-technological trends. The future offers a wide view of the smart farming as well as the food sector in general in developing countries like India.

References

1. Atzori, L., Iera, A., Morabito, G.: The internet of things: a survey. Comput. Netw. **54**(15), 2787–2805 (2010)
2. Ray, P.P.: A survey on internet of things architectures. J. King Saud Univ.-Comput. Inf. Sci. **30**(3), 291–319 (2018)
3. Elijah, O., et al.: An overview of Internet of Things (IoT) and data analytics in agriculture: benefits and challenges. IEEE Internet Things J. **5**(5), 3758–3773 (2018)
4. Khanna, A., Kaur, S.: Evolution of Internet of Things (IoT) and its significant impact in the field of precision agriculture. Comput. Electron. Agric. **157**, 218–231 (2019)

5. Luthra, S., et al.: Internet of Things (IoT) in agriculture supply chain management: a developing country perspective. In: Emerging Markets from a Multidisciplinary Perspective, pp. 209–220. Springer, Cham (2018)
6. Chandra, A., McNamara, K.E., Dargusch, P.: Climate-smart agriculture: perspectives and framings. Clim. Policy **18**(4), 526–541 (2018)
7. Lipper, L., et al.: Climate smart agriculture. Nat. Resour. Manag. Policy **52**, 2018 (2018)
8. Salam, A., Shah, S.: Internet of things in smart agriculture: enabling technologies. (2019)
9. Vuran, M.C., et al.: Internet of underground things in precision agriculture: architecture and technology aspects. Ad Hoc Netw. **81**, 160–173 (2018)
10. Keswani, B., et al.: Adapting weather conditions based IoT enabled smart irrigation technique in precision agriculture mechanisms. Neural Comput. Appl. **31**(1), 277–292 (2019)
11. Agoramoorthy, G.: Can India meet the increasing food demand by 2020? Futures **40**(5), 503–506 (2008)
12. Reddy, D.N., Mishra, S. (eds.): Agrarian Crisis in India. Oxford University Press, Oxford (2010)
13. Walter, A., et al.: Opinion: smart farming is key to developing sustainable agriculture. Proc. Natl. Acad. Sci. **114**(24), 6148–6150 (2017)
14. Kpadonou, R.A.B., et al.: Advancing climate-smart-agriculture in developing drylands: joint analysis of the adoption of multiple on-farm soil and water conservation technologies in West African Sahel. Land Use Policy **61**, 196–207 (2017)
15. Lakhwani, K., et al.: Development of IoT for smart agriculture a review. In: Emerging Trends in Expert Applications and Security, pp. 425–432. Springer, Singapore (2019)
16. Bermudez-Edo, M., et al.: IoT-Lite: a lightweight semantic model for the internet of things. In: 2016 International IEEE Conferences on Ubiquitous Intelligence and Computing, Advanced and Trusted Computing, Scalable Computing and Communications, Cloud and Big Data Computing, Internet of People, and Smart World Congress (UIC/ATC/ScalCom/CBDCom/IoP/SmartWorld), IEEE (2016)
17. Jara, A.J., et al.: Semantic web of things: an analysis of the application semantics for the iot moving towards the iot convergence. Int. J. Web Grid Serv. **10**(2–3), 244–272 (2014)
18. Maliappis, M.T.: Applying an agricultural ontology to web-based applications. Int. J. Metadata Semant. Ontol. **4**(1-2), 133–140 (2009)
19. Beck, H.W., Kim, S., Hagan, D.: A crop-pest ontology for extension publications. Proceedings (2005)
20. Wang, Y., et al.: An ontology-based approach to integration of hilly citrus production knowledge. Comput. Electron. Agric. **113**, 24–43 (2015)
21. Xie, N., Wang, W., Yang, Y.: Ontology-based agricultural knowledge acquisition and application. In: International Conference on Computer and Computing Technologies in Agriculture. Springer, Boston, MA (2007)
22. Arjun, K.M.: Indian agriculture-status, importance and role in Indian economy. Int. J. Agric. Food Sci. Technol. **4**(4), 343–346 (2013)
23. Reich, D., et al.: Reconstructing Indian population history. Nature **461**(7263), 489 (2009)
24. Chaurasia, V.B., Singh, M.: Step towards the improvement of Indian agriculture. In: 14th Annual Conference, pp. 61 (2018)
25. Bhojani, S.H., Patel, A.R.: Information technology: an arising concept in agriculture sector. J. Comput. Technol. Appl. **4**(1), 23–27 (2019)
26. Kumar, Y., Singh, P.K.: To study the influence of insurance policy on the agriculture field and Indian economy: concept paper. In: Renewable Energy and its Innovative Technologies, pp. 13–24. Springer, Singapore (2019)
27. Verma, C., Pandey, R.: Big data representation for grade analysis through Hadoop framework. In: 2016 6th International Conference-Cloud System and Big Data Engineering (Confluence), IEEE (2016)
28. Kotwal, A., Ramaswami, B., Wadhwa, W.: Economic liberalization and Indian economic growth: what's the evidence?. J. Econ. Lit. **49**(4), 1152–99 (2011)

29. Pandey, R., Dwivedi, S.: Ontology description using owl to support semantic web applications. Int. J. Comput. Appl. **14**(4), 30–33 (2011)
30. Postel, S., et al.: Drip irrigation for small farmers: a new initiative to alleviate hunger and poverty. Water Int. **26**(1), 3–13 (2001)
31. Pandey, R., Dwivedi, S.: Interoperability between semantic web layers: a communicating agent approach. Int. J. Comput. Appl. **12**(3), 0975–8887 (2010)
32. Pandey, M., Pandey, R.: JSON and its use in semantic web. Int. J. Comput. Appl. **164**(11), 10–16 (2017)
33. Kuruvilla, A., Jacob, K.S.: Poverty, social stress and mental health. Indian J. Med. Res. **126**(4), 273 (2007)
34. Kumari, Sneha, et al. "Sparql: semantic information retrieval by embedding prepositions. Int. J. Netw. Secur. Appl. **6**(1), 49 (2014)
35. Pandey, R., Dwivedi, S.: RDF/RDF-S providing framework support to OWL ontologies. Int. J. Comput. Sci. Inf. Technol. **3**(4) (2012)
36. Jagannathan, S., Priyatharshini, R.: Smart farming system using sensors for agricultural task automation. In: 2015 IEEE Technological Innovation in ICT for Agriculture and Rural Development (TIAR), IEEE (2015)
37. Channe, H., Kothari, S., Kadam, D.: Multidisciplinary model for smart agriculture using internet-of-things (IoT), sensors, cloud-computing, mobile-computing and big-data analysis. Int. J. Comput. Technol. Appl. **6**(3), 374–382 (2015)
38. Khatri-Chhetri, A., et al.: Farmers' prioritization of climate-smart agriculture (CSA) technologies. Agric. Syst. **151**, 184–191 (2017)
39. Patil, A., et al.: Smart farming using Arduino and data mining. In: 2016 3rd International Conference on Computing for Sustainable Global Development (INDIACom), IEEE (2016)
40. Auernhammer, H.: Precision farming—the environmental challenge. Comput. Electron. Agric. **30**(1-3), 31–43 (2001)
41. Katyal, N., Pandian, B.J.: A comparative study of conventional and smart farming. In: Emerging Technologies for Agriculture and Environment, pp. 1–8. Springer, Singapore (2020)
42. Bronson, K.: Looking through a responsible innovation lens at uneven engagements with digital farming. NJAS-Wageningen J. Life Sci. (2019)
43. Carolan, M.: Publicising food: big data, precision agriculture, and co-experimental techniques of addition. Sociologia Ruralis **57**(2), 135–154 (2017)
44. Popović, T., et al.: Architecting an IoT-enabled platform for precision agriculture and ecological monitoring: a case study. Comput. Electron. Agric. **140**, 255–265 (2017)
45. Atzori, L., Iera, A., Morabito, G.: Understanding the Internet of Things: definition, potentials, and societal role of a fast evolving paradigm. Ad Hoc Netw. **56**, 122–140 (2017)
46. Kamilaris, A., et al.: Agri-IoT: a semantic framework for Internet of Things-enabled smart farming applications. In: 2016 IEEE 3rd World Forum on Internet of Things (WF-IoT), IEEE (2016)
47. Ilapakurti, A., Vuppalapati, C.: Building an IoT framework for connected dairy. In: 2015 IEEE First International Conference on Big Data Computing Service and Applications, IEEE (2015)
48. Madsen, S.L., et al.: Quantifying behaviour of dairy cows via multi-stage support vector machines: book of proceedings. In: 8th European Conference on Precision Livestock Farming (2017)
49. Sinha, R.S., Wei, Y., Hwang, S.-H.: A survey on LPWA technology: LoRa and NB-IoT. Ict Express **3**(1), 14–21 (2017)
50. Pham, C., Rahim, A., Cousin, P.: Low-cost, long-range open IoT for smarter rural African villages. In: 2016 IEEE International Smart Cities Conference (ISC2), IEEE (2016)
51. Shaikh, F.K., Zeadally, S.: Energy harvesting in wireless sensor networks: a comprehensive review. Renew. Sustain. Energy Rev. **55**, 1041–1054 (2016)
52. Wen, Z., et al.: Self-powered textile for wearable electronics by hybridizing fiber-shaped nanogenerators, solar cells, and supercapacitors. Sci. Adv. **2**(10), e1600097 (2016)
53. Francesco, A.,et al.: Combined finite–discrete numerical modeling of runout of the Torgiovannetto di Assisi rockslide in central Italy. Int. J. Geomech. **16**(6), 04016019 (2016)

54. Wong, B.P., Kerkez, B.: Real-time environmental sensor data: an application to water quality using web services. Environ. Model. Softw. **84**, 505–517 (2016)
55. Murphy, E., et al.: Diet of stoats at Okarito Kiwi Sanctuary, South Westland, New Zealand. N. Z. J. Ecol. 41–45 (2008)
56. Singh, H., Sarangi, S.C., Gupta, Y.K.: French Phase I clinical trial disaster: issues, learning points, and potential safety measures. J. Nat. Sci. Biol. Med. **9**(2), 106 (2018)
57. Ruan, J., Shi, Y.: Monitoring and assessing fruit freshness in IOT-based e-commerce delivery using scenario analysis and interval number approaches. Inf. Sci. **373**, 557–570 (2016)
58. Liu, Y., et al.: An Internet-of-Things solution for food safety and quality control: a pilot project in China. J. Ind. Inf. Integr. **3**, 1–7 (2016)
59. Kant, G.S., Singh, V.K., Darbari, M.: Legal semantic web-a recommendation system. IJAIS **7**(3) (2014)
60. Mishra, S.K., Singh, V.K., Shankhdhar, G.K.: Ontology development for wheat information system. IJRET-Int. J. Res. Eng. Technol. **04**(05) (2015)
61. Verma, A., Shankhdhar, G.K., Darbari, M.: Verified message exchange in providing security for cloud computing in heterogeneous and dynamic environment. Int. J. Appl. Inf. Syst. **11**(10), 15–18 (2017)
62. Garcia-Ojeda, J.C., et al.: O-MaSE: a customizable approach to developing multiagent development processes. In: International Workshop on Agent-Oriented Software Engineering. Springer, Berlin, Heidelberg (2007)
63. Shankhdhar, G.K., Verma, A., Singh, V.K., Darbari, M., Singh, V.: Application of IOT in electrical grid. IOSR J. Eng. ISSN (e): 2250–3021, ISSN (p): 2278-8719 **08**(4), 01–03 (2018)
64. Shankhdhar, G.K., Darbari, M.: Building custom, adaptive and heterogeneous multi-agent systems for semantic information retrieval using organizational-multi-agent systems engineering, O-MaSE. IEEE Explore, ISBN: 978-1-5090-3480-2 (2016)
65. Gao, F., Ali, M.I., Mileo, A.: Semantic discovery and integration of urban data streams⋆. Challenge **7**, 16 (2014)
66. Shankhdhar, G.K., Darbari, M.: Introducing two level verification model for reduction of uncertainty of message exchange in inter agent communication in organizational-multi-agent systems engineering, O-MaSE. Int. Organ. Sci. Res. (2017). https://doi.org/10.9790/0661-1904020818
67. Shankhdhar, G.K., Darbari, M.: Integrating COCOMO II model in O-MaSE methodology for estimating effort in building heterogeneous and dynamic multi-agent systems. Sci. Eng. Res. Support Soc. Int. J. Softw. Eng. Appl. 29–40
68. Shankhdhar, G.K., Darbari, M.: Implementation of validation of requirements in agent development by means of ontology. Int. J. Comput. Sci. Eng. **6**, 1129–1135 (2018). https://doi.org/10.26438/ijcse/v6i7.11291135
69. DeLoach, S.A., Garcia-Ojeda, J.C.: The o-masemethodology. In: Handbook on Agent-Oriented Design Processes, pp. 253–285. Springer, Berlin, Heidelberg (2014)
70. Garcia-Ojeda, J.C., DeLoach, S.A.: agentTool III: from process definition to code generation. In: Proceedings of the 8th International Conference on Autonomous Agents and Multiagent Systems-Volume 2. International Foundation for Autonomous Agents and Multiagent Systems (2009)
71. Agarwal, R., et al.: Unified IoT ontology to enable interoperability and federation of testbeds. In: 2016 IEEE 3rd World Forum on Internet of Things (WF-IoT), IEEE (2016)
72. Seydoux, N., et al.: IoT-O, a core-domain IoT ontology to represent connected devices networks. In: European Knowledge Acquisition Workshop. Springer, Cham (2016)
73. Compton, M., et al.: The SSN ontology of the W3C semantic sensor network incubator group. Web Semant. Sci. Serv. Agents World Wide Web **17**, 25–32 (2012)
74. Caracciolo, C., et al.: The AGROVOC linked dataset. Semant. Web **4**(3), 341–348 (2013)
75. Lauser, B., et al.: From AGROVOC to the agricultural ontology service/concept server. An OWL model for creating ontologies in the agricultural domain. In: Dublin Core Conference Proceedings. Dublin Core DCMI (2006)
76. Hu, S., et al.: AgOnt: ontology for agriculture internet of things. In: International Conference on Computer and Computing Technologies in Agriculture. Springer, Berlin, Heidelberg (2010)

77. Barbieri, D.F., et al.: C-SPARQL: SPARQL for continuous querying. In: Te 18th international conference on World wide web-WWW'09 (2009)
78. Dao-Tran, Minh, and Danh Le Phuoc. "Towards Enriching CQELS with Complex Event Processing and Path Navigation."*HiDeSt@ KI*. 2015
79. Fulton, M., Giannakas, K.: Organizational commitment in a mixed oligopoly: agricultural cooperatives and investor-owned firms. Am. J. Agric. Econ. **83**(5), 1258–1265 (2001)
80. Patnaik, U.: Unbalanced growth, tertiarization of the Indian economy and implications for mass living standards. In: Towards Progressive Fiscal Policy in India. Sage Publications, New Delhi, pp. 299–325 (2011)
81. Pandey, R., Saxena, P., Tripathi, S.: Data interpretation for social network using R API. In: 2018 8th International Conference on Communication Systems and Network Technologies (CSNT), IEEE (2018)
82. Verma, C., Pandey, R.: Mobile cloud computing integrating cloud, mobile computing, and networking services through virtualization. In: Design and Use of Virtualization Technology in Cloud Computing. IGI Global, 140–160 (2018)

How to Understand Better "Smart Vehicle"? Knowledge Extraction for the Automotive Sector Using Web of Things

Mahda Noura, Amélie Gyrard, Benjamin Klotz, Raphael Troncy, Soumya Kanti Datta, and Martin Gaedke

Abstract How to understand better the knowledge provided by Google results to build future "smart vehicle-centric" applications? What is the knowledge expertise required to build a smart vehicle application (e.g., driver assistance system)? Automotive companies (e.g., Toyota, BMW, Renault) are employing Internet of Things (IoT) and Semantic Web technologies to model the automotive sector. We aggregate this "common sense knowledge" in a automotive dataset which comprises 42 semantics-based projects between 2005 and 2019. The knowledge is already encoded with knowledge representation languages (e.g., RDF, RDFS, and OWL) and supported by the World Wide Web Consortium (W3C). However, only a subset of those projects share their expertise by publishing their ontologies online. For this reason, at the current time or writing, only 16 ontologies are processable. Our innovative Knowledge Extraction for the Automotive Sector (KEAS) methodology analyzes what are the most popular terms required to build a smart car, it provides: (1) a set of keyphrase that are synonyms to smart cars to find domain-specific knowledge, (2) synonyms are used to build a corpus of scientific publications to train the k-means

M. Noura · M. Gaedke
Technische Universitat Chemnitz, Chemnitz, Germany
e-mail: mahda.noura@informatik.tu-chemnitz.de

M. Gaedke
e-mail: martin.gaedke@informatik.tu-chemnitz.de

A. Gyrard (✉)
Kno.e.sis, Wright State University, Dayton, USA
e-mail: amelie@knoesis.org

B. Klotz · R. Troncy · S. K. Datta
EURECOM, Sophia Antipolis, Biot, France
e-mail: benjamin.klotz@eurecom.fr

R. Troncy
e-mail: raphael.troncy@eurecom.fr

S. K. Datta
e-mail: soumya-kanti.datta@eurecom.fr

A. Gyrard
Trialog, Paris, France

R. Pandey et al. (eds.), *Semantic IoT: Theory and Applications*, Studies in Computational Intelligence 941, https://doi.org/10.1007/978-3-030-64619-6_13

machine learning algorithm, (3) a dataset of smart car ontologies that we collected, is analyzed by the k-means algorithm, and (4) the extraction of the most common terms from the ontology dataset for the automotive sector. Our KEAS findings can be used as a starting point for further domain-specific investigations (e.g., Volvo willing to integrate semantic web) and for future information extraction from structured knowledge.

Keywords Internet of Things (IoT) · Knowledge directory service · Semantic ontology interoperability · Ontology validation · Reusability · Semantic Web of Things (SWoT) · Semantic web technologies · Reusable knowledge

1 Highlights

- Reusing knowledge already designed for knowledge-based smart car projects.
- Automatic knowledge extraction for the automotive sector based using the k-mean machine learning algorithm.

2 Introduction

How to understand better the results provided by Google to build the future "smart vehicle-centric" applications? What is the knowledge expertise required to build a smart vehicle application such as the driver assistance system? According to PC magazine,[1] a **smart car** is an automobile with advanced electronics. Microprocessors have been used in car engines since the late 1960s and have steadily increased in usage throughout the engine and drivetrain to improve stability, braking and general comfort. According to Gartner's 2018 prediction,[2] "IoT platforms", "Autonomous Driving Level 4", and "Knowledge Graphs" are the next challenges for the coming 5–10 years or even beyond.

Automotive companies (e.g., **Toyota**[3] [1], **BMW** [2–5], **Renault** [6]) are already employing Internet of Things (IoT) and Semantic Web technologies. BMW Autonomous Driving in the Internet of Cars Summer School[4] demonstrates interest in IoT technologies and even Semantic Web technologies [3]. BMW is designing the Vehicle Signal and Attribute (VSSO) ontology[5] [2] and the Vehicle Driving Context (VDC)

[1] http://bit.ly/2xMZQDv.
[2] https://gtnr.it/2SgUvOi.
[3] http://bit.ly/2Y3A1xL.
[4] http://www.bmwsummerschool.com/.
[5] http://automotive.eurecom.fr/vsso.

ontology.[6] **auto.schema.org**[7] defines 4 types, 20 properties and 3 enumeration values (in December 2018) which clearly shows that the knowledge could be extended. Volvo is investigating the integration of semantic web technologies (RDF, Linked Data, ontologies) for automomous cars.[8]

Acquiring knowledge about automotive (e.g., technological survey, reading scientific publications and staying updating with the latest progresses) is a time-consuming approach. The survey about transportation ontologies [7], published in 2018, can be easily enriched with numerous ontologies that we collected within the LOV4IoT ontology catalog for IoT and transport[9] that we designed. The survey [7] compares 11 ontologies according to 7 criteria: (1) Precision (relationship diversity, axiom complexity), (2) Evaluation, (3) Knowledge management services, (4) Generality, (5) Granularity, (6) Competence, and (7) Span.

We designed the "semantic-based IoT smart vehicle" LOV4IoT dataset thats collects common sense knowledge for the automative sector. We classified 42 projects between 2005 and 2019 since they claim that the knowledge is already encoded with knowledge representation languages (e.g., RDF, RDFS, and OWL) and supported by the World Wide Web Consortium (W3C). However, only a subset of those projects share their expertise by publishing their ontologies online. For this reason, at the current time or writing, our dataset comprises only 16 processable ontologies.

Motivation are as follows:

- M1: Why cannot we find the entire Ph.D. thesis, entitled "Using Ontologies and Intelligent Systems for Traffic Accident Assistance in Vehicular Environments" [8] published in 2014 relevant for smart car on the first page of Google results? It is provided on the third page on Google[10] whereas years of research and expertise are explained in the thesis.
- M2: How to find more knowledge than Google for a specific domain (e.g, smart vehicle)?
- M3: Why does the Google Knowledge Graph cannot provide results to handle the synonyms used for the automotive domain (e.g., smart car, smart vehicle, smart mobility)?

Research questions are as follows:

- RQ1: How to automatically analyze structured knowledge (e.g., ontologies) from existing projects? We found that numerous projects designed ontologies that are also explained within scientific publications can be analyzed.
- RQ2: What are the most used entities (e.g, concepts, instances) within those ontologies? Statistical methods can help to achieve this task.

[6]http://automotive.eurecom.fr/vdc.

[7]https://auto.schema.org/.

[8]https://twitter.com/olafhartig/status/1121539105924550661.

[9]http://lov4iot.appspot.com/?p=lov4iot-transport.

[10]"Smart car ontology" search on Google, December 2018.

Contributions are as follows: Our innovative Knowledge Extraction for the Automotive Sector (KEAS) methodology understands the "common sense knowledge" required to build smart vehicle applications which provides:

1. C1: A set of keyphrase synonyms for the smart vehicle domain to find domain-specific knowledge in past or current projects that published their results within scientific publications,
2. C2: Synonyms are used to build a corpus of scientific publications to train the k-means machine learning algorithm,
3. C3: A dataset of smart car ontologies is built and analyzed by the k-means algorithm to cluster knowledge, and,
4. C4: The extraction of the most common knowledge for the automotive sector. We refined a previous methodology [9] that we applied to the smart vehicle domain in this book chapter.

Structure of the Paper: Sect. 3 introduces the related work. Section 4 explains our Knowledge Extraction for the Automotive Sector (KEAS) methodology to find the relevant knowledge already implemented within ontologies. Section 5 evaluates our proposed approach. Section 6 concludes the paper and provides future work.

3 Background and Related Work

Toyota Motor Europe (TME)[11] uses auto.schema.org in their web site to describe cars to sell. For instance, 7000 URLs including the type "Car" from the TME web site have been encoded and indexed by Google.

auto.schema.org[12] defines 4 types (BusOrCoach, CarUsageType, Motorcycle, MotorizedBicycle), 20 properties (accelerationTime, acrissCode, bodyType, emissionsCO2, engineDisplacement, enginePower, engineType, fuelCapacity, meetsEmissionStandard, modelDate, payload, roofLoad, seatingCapacity, speed, tongueWeight, torque, trailerWeight, vehicleSpecialUsage, weightTotal, wheelbase) and 3 enumeration values (DrivingSchoolVehicleUsage, RentalVehicleUsage, TaxiVehicleUsage) (in December 2018). It clearly shows the the knowledge could be extended.

OpenSensingCity[13] references 12 ontology URLs relevant to mobility: Transport, travel domain, transportation networks, transport disruption, soft mobility, PASSIM, location concept for travel support system, route, ASK-IT, road, transit.

SAREF4AUTO is being specified and supported by the ETSI standard; the ontology code and specification cannot be found yet, only those slides can be investigated [10] at the time of this writing.

Conclusion: Ontology-based projects are introduced in Table 1 when ontologies are publicly available, that we analyze thanks to the KEAS methodology in Sect. 4.2.

---

[11] http://bit.ly/2Y3A1xL.

[12] https://auto.schema.org/.

[13] http://ci.emse.fr/opensensingcity/ns/result/domain/transportation/.

Table 1 Ontology-based IoT automotive projects used in the dataset that we analyzed

Authors	Year	Expertise	OA	Reasoning
Klotz et al. [2, 4] [5]	2018	BMW: vehicle signal and attribute	✓	✓ OWL restriction
OpenSensingCity	2018	Parking scenario	✓	×
		Bike scenario	✓	×
CityPulse [11, 12]	2016	Traffic analysis scenario	✓	–
Gyrard et al. [3] BMW summer school	2014	Transport ontology	✓	✓ Jena rules
Morignot et al. [13]	2013	Autonomous vehicle assistance	✓	✓foggy -> mode manual
Zhao et al. [1, 14–16]	2015	Toyota: safe autonomous driving	✓	–
Lecue et al. [17]	2014	STAR-CITY: transport ontology	✓	–
Ruta et al. [18–20]	2017	iDriveSafe ontology Mafalda projet (3 ontologies)	✓ ✓	✓ OWL restrictions
Maarala [21]	2017	Traffic ontology	✓	✓ 16 rules, OWL restrictions
Bermejo et al. [22]	2013	Road traffic management ontology	✓	✓ 77 rules/actions (SWRL DLSafe rule in the ontology)
Dardailler et al.	2012	W3C road accident ontology	✓	×
Corsar et al. [23]	2015	Transport disruption ontology	✓	✓ OWL restriction
Codescu et al. [24]	2011	Open street map and route planning	✓	×
Grausberg, Fuchs et al. [25, 26]	2008	Driver assistance system ontology	✓	✓ OWL restriction (rule speed max min)
Hepp et al.	–	W3C vehicle sales ontology	✓	× No owl: restriction

Legend Ontology Availability (OA)

Other projects related to the topic that cannot be used since ontologies are not shared (as depicted in Table 2). Although scientific publications were really interesting, those ontologies have been discarded since we cannot find their ontology online (see Table 2).

Table 2 Other ontology-based IoT automotive projects that cannot be employed since ontologies are not available (even upon request)

Authors	Year	Expertise	OA	Reasoning
Wetterwald et al. [10]	2019	SAREF4AUTO	×	×
Katsumi et al. [7]	2018	Survey – 11 ontology referenced	×	×
Fernandez et al. [27]	2016	Automatic traffic lights settings	×	×
Armand et al. [6]	2014	Renault: driving assitance	×	✓ 14 SWRL rules
Villalba et al. [8, 28, 29]	2014	VEACON: vehicular accident	×	–
		CAOVA: Car accident for VANETs	×	–
Stocker et al. [30] [31]	2014	Road vehicle classification	×	✓ Rule-based inference (vehicle type)
Ebers et al. [32]	2013	VANETs ontology	×	×
Mnasser, De Oliveria, Houda, Zidi et al. [33–36]	2013	Transportation ontology	×	✓ Jess engine for SWRL rules
Li et al. [37]	2012	Car ontology (in chinese)	×	×
Calavia et al. [38]	2012	Traffic ontology	×	✓ Semantic reasoning, SWRL rules
Madkour et al. [39]	2011	Ontology of transportation networks	×	×
Hamilton et al. [40]	2013	Ontology of transportation networks	×	✓ Pellet, SWRL, Jess
Feld, Muller et al. [41]	2011	Automative, distance between cars	×	×
Wang et al. [42]	2011	Traffic accident ontology	×	×
Hulsen et al. [43]	2011	Ontology for driver assistance	×	✓ RacerPro
Berdier et al. [44]	2011	Ontology for Urban mobility	×	×
Kannan et al. [45]	2010	Intelligent driver Assistance system for vehicle	×	✓ Pellet reasoner (consistency)
Baumgartner et al. [46]	2010	Ontology for situation awareness	×	✓ 10 rules
Liu et al. [47]	2010	Road surveillance system	×	✓ SWRL rules (inform, alert)
Niaraki et al. [48]	2009	personalized route planning	×	×
Yue et al. [49]	2009	Traffic accident	×	×
Zhai et al. [50]	2009	Traffic information	×	✓ (dryness, dampness)
Sun et al. [51]	2009	Smart car	×	×
Belhadef et al. [52]	2009	Urban geographical information system	×	×
Regele et al. [53]	2008	Autonomous driving system: trajectory planning, traffic coordination	×	×
Mair, Eigner et al. [54]	2008	Collision avoidance in VANETs Ontology (in German)	×	×
Cheng et al. [55]	2008	Transportation	×	✓ Ontology-based reasoning (Racer) rule-based reasoning (Jess, LISP, SWRLJessTab)
Hu et al. [56]	2007	Oil	×	×
Lorenz et al. [57]	2005	Ontology of transportation networks	×	×

Legend Ontology Availability (OA)

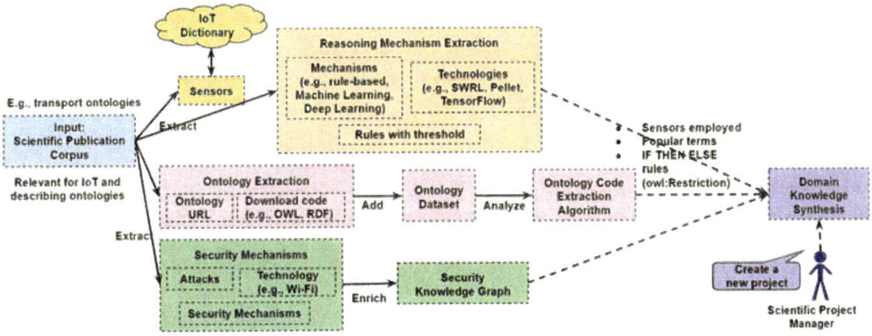

Fig. 1 Knowledge extraction from scientific publications and ontologies (Long-term vision)

4 Knowledge Extraction for the Automotive Sector (KEAS) Methodology

The long-term vision of the Knowledge Extraction from IoT-related ontologies project is depicted in Fig. 1. In this paper, we are focused on the *Ontology code extraction algorithm* and the *Ontology Dataset* components applied to the smart vehicle domain.

4.1 Survey Methodology to Collect Ontologies for Smart Vehicles

Scientific Publication Corpus and Ontology Dataset. We collected a total of 42 projects from 2005 to 2018 more or less related to smart vehicles. However, the aggregation of knowledge has been done since several years. The methodology has been search on Google and Google Scholar a set of specific keywords, as an example those keyphrases: (1) start with ontology-based, (2) finished with ontology, or (3) start with semantic-based. Keyphrases are as follows:

- Automotive, Automated vehicle, Autonomous vehicle, Car, Cars, Vehicle, Vehicles, Smart car.
- Transportation, Transportation networks, Transport, Public transportation.
- Road Traffic Management, Roads, Road system, Traffic Jam Avoidance.
- Personalized Route Planning, Route Planning.
- Car Driving Assistance, Driver Assistance Systems, Advanced Driver Assistance, Intelligent Driver Assistance System.
- Vehicular Ad Hoc Networks (VANETs), Wireless Vehicular Networks (VANETs), Vehicle-to-Vehicle (V2V), Vehicle-to-Vehicle Networks, Vehicular Networks.

- Road Accident, Vehicular Accident, Traffic Accidents, Road Safety, Car Accident Prevention, Accident Rescue Mission.
- Intersection Assistance.
- Vehicle Context-aware Services.
- Pedestrian Detection.
- Car Pooling Recommendation System.

For the set of scientific publications, we focused on the following criteria:

- Are ontology URLs available within the scientific article? Frequently, URLs are missing. Authors have been contacted to retrieve ontology code and we enriched the dataset when receiving positive answers. Table 1 summarizes the 16 ontologies that share their ontologies online, which is the smart vehicle ontology dataset later analyzed. Other ontology-based projects are referenced in Table 2, unfortunately, the ontologies cannot be processed yet since they are not accessible.
- Are sensors mentioned within the paper?
- Are there reasoning mechanisms and already defined rules to interpret data generated by the smart vehicle applications?
- Is the reference section provide more resources to investigate? We enrich our scientific publication dataset accordingly (e.g., LOV4IoT-transport knowledge repository).

The main difference between our survey and the existing ones, is that our survey is the result of a continuous enrichment of the LOV4IoT ontology catalog since 2012 and we provide tools to support the reuse of the survey outcome (e.g., dump of ontology code). Meanwhile, we are aware of Systematic Literature Review (SLR) guidelines such as [58–60].

4.2 Building the Corpus of Knowledge for the Transportation Domain

To train the dataset, we need to build a corpora of knowledge for the transportation domain. word2vec helps in transforming texts from either scientific publications or ontologies into vectors that can be processable by machine learning algorithms. **word2vec** performs the training of the term embeddings and the process of building a word2vec model for all identified unique words. The word2vec algorithm is based on neural networks and builds a vocabulary from a pre-training text model and attaches the vector representations to each word. Around 20 of the terms were not part of the pre-trained model thus we removed those terms from the list of words. The output of this step is the word embedding vector space representation. The genism python library is used to implement word2vec.
Transport Ontology Dataset: We have collected 16 ontologies that can be downloaded and analyzed (as depicted in Table 1): 2 ontologies are excluded since they

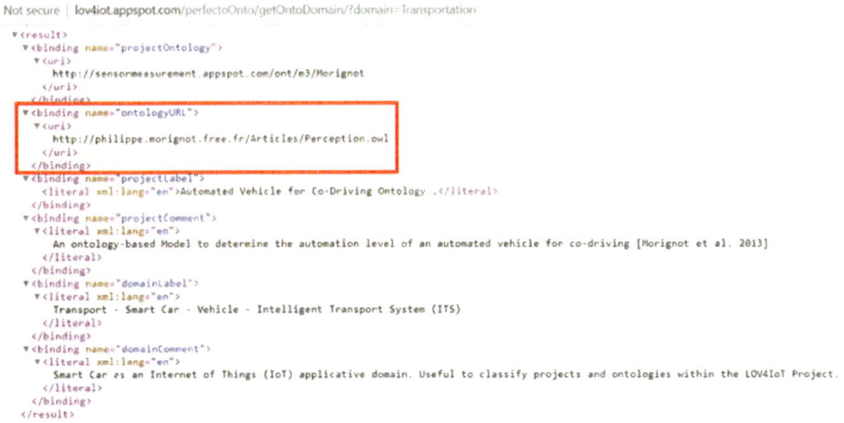

Fig. 2 Web service example to automatically retrieve ontology-based projects for the smart car domain: ontology URLs and scientic publications

were in Chinese or German. We released the list of ontology URLs within an online table.[14] For the convenience of the developers, we created a tutorial web page (http://lov4iot.appspot.com/?p=queryTransportOntologiesWS) to either use the web service or easily download the dump of the ontology code that we collected.

For instance, the developer can query the web service http://lov4iot.appspot.com/perfectoOnto/getOntoDomain/?domain=Transportation which returns the list of the projects relevant for the smart car domain that we collected within the LOV4IoT ontology catalog for transport,[15] it includes: the name of the project and the ontology, the ontology URL, and additional information such as the scientific publication describing the ontology (see Fig. 2). The web service is more up-to-date with latest ontologies collected, compared to the dump file. However, when the projects are not maintained anymore, the URLs can become dead links, which is the reason we store the ontology code within dump files.

5 Evaluation

Planned Evaluation: To identify the most popular concepts from smart car ontologies, the proposed KEAS methodology is evaluated in an empirical study which includes an analysis that gives a complete overview of the performance of the descriptiveness of the most popular concepts (in the same way it has been done in our Knowledge Extraction for the Web o Things (KE4WoT) work [61]). The objective of this experiment is to identify if the keywords provided by KE4WoT can sufficiently describe existing ontologies.

[14] http://shorturl.at/jEIQ7.

[15] http://lov4iot.appspot.com/?p=lov4iot-transport.

Ontology Selection: 16 smart car ontologies are collected from LOV4IoT for evaluation purposes (Table 1). The most important ontologies in each domain have been selected according to the following criteria:

- Citations of the scientific publications describing the ontology (e.g., the SSN ontology v1 [62] has more than 1000 citations): higher is the number, better the ontology might be. However, this criteria cannot be applied to recent publications.
- Journal impact factor and conference ranking: higher the ranking is, better would be the ontologies. Within the references section, the ranking is added for publications cited and classified within Tables 1 and 2.
- Recent publications increase the chance to have the authors maintaining the ontology and integrating previous ontologies.
- Ontologies disseminated in standardizations (e.g., W3C Web of Things ontology,[16] W3C SSN/SOSA ontology [63], ETSI M2M SAREF ontology [64]) can be considered as more reliable.
- Industrial partners involved, the project is considered more impactful, and the implementation is more reliable.
- Domain experts involved (not computer scientists) since they share their human expertise.
- Ontology code that can be downloaded, because, in science, the experiments should be replicable, following the FAIR (Findability, Accessibility, Interoperability, and Reuse) principles.

Ground Truth Dataset Design: Domain experts can participate in the questionnaire to design the ground truth (a similar questionnaire for smart cities, weather, and smart home is available online,[17] see [61]) for detailed information). Experts were either involved in developing smart car ontologies or are an open audience having the domain expertise **to describe each ontology using three keywords**. The participants' level of expertise in the automotive domain and knowledge engineering, is asked in a Likert scale of five levels, from 'totally disagree' to 'totally agree'. The experts are given the list of ontologies (through a series of figures from the ontology classes in Protege) in different domains to select the top three keywords that best describes that ontology in relation to the keywords that were obtained from the main concepts in the generated clusters. Domain experts chose keywords among a total number of keywords in our evaluation form.

6 Conclusion and Future Work

The Systematic Literature Survey (SLS) in any research topics is a time-consuming approach. Finding knowledge returned by Google results still require a huge work on learning, classification and summarizing. To ease this time-consuming task, we

[16]https://www.w3.org/TR/wot-thing-description/.

[17]https://bildungsportal.sachsen.de/survey/limesurvey/index.php/716626/lang-en.

built a "common sense knowledge" dataset for the automotive sector comprising 42 projects between 2005 and 2019. However, only 16 ontologies are processable and published online with knowledge representation standards. Our innovative Knowledge Extraction for the Automotive Sector (KEAS) methodology aims to analyze the most popular knowledge required to build smart vehicle applications by applying the k-means machine learning algorithm to a dataset of 16 ontologies that we collected.

This work highly encourages researchers to share their reproduceable experiments by publishing online their smart vehicle ontologies. As a future work, we would like to re-generate an ontology to aggregate and unify the knowledge from existing ontologies. Furthermore, we would like to automatically recognize the sensors mentioned within ontologies and scientific publications to maintain our IoT dictionnary, and reasoning mechanisms used to detect abnormal sensor data and execute actions.

Acknowledgements This work has partially received funding from the European Union's Horizon 2020 research and innovation programme under grant agreement No. 857237 (Interconnect). The opinions expressed are those of the authors and do not reflect those of the sponsors.

7 Appendix

7.1 Clustering Results

See Figs. 3, 4, 5, 6, 7, 8 and 9

```
[assessment, viewpoint, analysis, classification, investigation, survey]
[course, begins, advance, back, go, last, step, put, forward, ahead, end, initial,
first, start, meeting, date, point]
[new, several, multiple, relevant, previous, following, involved, additional, interest,
number]
[making, made, make, driven, maker, trees]
[manually, add, rule, search, update, meta, validity, observable, valid, static,
match, set, inference, predefined]
[asset, recovery, upgrading, installation, improvement, reconstruction, evacuation,
progress, preventative, cleaning, cycle, preparation, plan, repair, maintenance,
planning, inspection, construction, operation]
[held, manager, play, played, reports, participant, board, report]
[articulated, autonomous, automation, robot, navigation, adas, assistance, platooning,
automated, driving, cruise, cooperative]
[blizzard, ssn, record, sampling, reading, observation, description, measurement,
weather]
[aid, main, dedicated, common, includes, created, inter, based, shared, communications,
built, general, technical, communication, sharing, works, connected, establishment,
open, extension, involving, working, address, work, related, active]
[declares, org, info, title, uri, doi]
[publisher, guest, pub, ski, books, cert, email, comments, photo, fi, signed, creator]
[access, using, usage, offer, store, delivers, availability, storage, collection,
delivered, users, avail, used, use, available, consumption]
[checks, sender, message, check, massage]
[build, economy, urban, buildings, spaces, community, security, mobility, environment,
city, environmental, network, smart, transport, transportation, management, public,
building, infrastructure]
[history, attraction, adult, birth, young, currency, age]
[layout, riot, parameters, execute, steps, hardware, details, software, module,
model, execution, run, mode, configuration, version, setup, running]
[order, dealing, involves, problem, human, language, also, tools, parts, since,
part, possibility, kind]
[rescue, customs, prohibited, bank, reserve, administrative, airport, court, insurer,
company, owned, authority, agency, organization, rights, insurance, post, engaged,
police, site, facility]
```

Fig. 3 Cluster Results Part I

```
- [decision, stand, turn, u, action, right, left, drive, v]
- [done, specified, meets, fitness, standard, defined, selected, zone, modified]
- [note, minor, attributed, remaining, major, recognized, indirect, associated, subject,
  primary, secondary, cause, contributor]
- [registration, manual, fee, household]
- [school, office, informal, town, college, bureau, arts, chair, arena, turkish, department,
  theme, university, senior, tcs, greek, spanish, french, italian, official, centre,
  siemens]
- [relationship, belongs, vocabulary, element, named, property, document, attribute,
  relation, entity, item, restriction, operator, feature, object, category, class,
  event, name, agent, type]
- [strings, obj, slots, atomic, sequence, datatype, columns, rows, row, equivalence,
  matrix, timestamp, tree, block]
- [willingness, owners, owner, owns, man, people, child, person, contact, place, thing,
  body]
- [hard, carrying, restrict, protected, regular, covered, maintain, closure, allowed,
  'yes', non, pre, internal, temporary, local, free, direct, short, long]
- [library, identification, barcode, label, rfid, image, code, qr, bar, card, tagging,
  id]
- [reliability, performance, accuracy, uptime, precision, quality, visibility, success]
- [black, white, color, box, style]
- [attack, leak, anti, detection, fault]
- [march, demonstration, campaign, regional, spur, national, program, catalyst]
- [impossible, unknown, deviated, weigh, witness, disruptive, wrong, even, bad, behind,
  abnormal]
- [experience, behavior, situation, nature, dynamic, ability, self, phenomenon, pattern,
  characteristics, changing, behaviour, game, activity, conditions, context, term,
  perception, condition]
- [convenience, walking, outdoor, infotainment, occupant, occupants, motorized, pedestrian,
  disabled, health, living, care, pedestrians, safety, emergency]
- [spacial, descent, variant, fourier, sax, symbolic, wavelet, cosine, piecewise,
  hybrid, derivation, blinding, gauss, lm, var, approximation, linear, discrete, dimensionless]
- [crowd, music, mobile, video, remote, obd, media, multimedia, cctv, biometric, smartphone,
  touch, nfc, phone, computer, tv, electronic, machine, text, telephone, voice]
```

Fig. 4 Cluster Results Part II

```
- [xvalue, eclipse, juridical, vip, mopeds, bull, dop, timespan, van, cartography,
  topo, routier, stationery, pizza, butcher, bunker, hostel, flightpath, waterfall,
  baptist, hindu, burg, jain, basket, hut, aerodrome, disused, burger, christian,
  dentist, cave, jewelry, drama, jewish, heath, casino, bakery, kindergarten, marsh,
  orthodox, catholic, pier, fountain, chemist, cliff, desc, ascent, alt, kmeans, moto,
  aux, artist, wiping, wmi, tps, drivetrain, ucum, lumber, album, bomb, unspecified,
  slush, derivable, wheelbase, vin, storey, athletics, caravan, thai, landuse, wayside,
  denomination, bic, comedy, arabic, pascal, doa, milliliter, freezer, diastolic,
  insufficiently, hydrophone, milliliter, milliampere, okta, gsr, cupboard, crosswind,
  lx, candela, lumen, farad, hg, unclassified, horn, clutch, args, tennis, ruins,
  commanded, mil, tournament, cuisine, rab, radian, milliwatt, unsigned, osc, vehicule,
  microgram, sens]
- [spot, orientation, ego, moment, movement, trajectory, occupancy, along, segment,
  heading, tracks, proximity, moving, track, current, line, area, space, field, status,
  location, distance, direction, position, state]
- [parade, exhibition, concert, demolition, marathon, festival, rugby, cricket, flyover,
  archaeological, beach, beauty, motel, dog, memorial, pharmacy, dance, hiking, lockers,
  climbing, worship, hall, marina, boulevard, zoo, theatre, airfield, naval, cinema,
  skiing, mall, fortress, mexican, soccer, veterinary, camp, cafe, playground, courthouse,
  museum, stadium, monument, avenue, rearview, basketball, baseball, football, picnic,
  garden, bpm, hood, lorry, gallery, golf, castle]
- [directed, stationary, force, normal]
- [diversion, agricultural, metering, plant, drinking, wastewater, irrigation, forecasting,
  agriculture, underground, monitoring, hvac, water]
- [maximum, limit, range, full, wind, level]
- [obstructing, insects, ill, gusty, assault, fight, viaduct, procession, occasion,
  tongue, bicyclist, volcano, grave, halt, defensive, scrub, skating, woman, infant,
  tired, pregnant, declined, refuel, hockey, seafood, religion, smoking, emotion,
  tailwind, careless, overspeed, dusk, bolster, convoy, overturned, snowy, foggy,
  race]
- [squad, snowploughs, rolling, helicopter, aircraft, lorries, petrol, watercraft,
  ship, cargo, sailing, rickshaw, valet, yard, taxi, parked, tow, cabin, ambulance,
  abandoned, subway, ferry, taxis, docking, gates, clearance, rental, shop, tram,
  bicycles, truck, boat]
```

Fig. 5 Cluster Results Part III

```
[trade, japanese, opera, american, electronics, maritime, civil, international,
asian, nuclear, german, chinese, indian, european, military]
[cancellation, jumping, swing, overnight, cleared, rush, sweeping, jam, sunset,
fell, idle, stuck, fallen, shake, busy, night, sunny, stands]
[pipe, anchors, wall, pipeline, branch, sewer, concrete]
[cultural, gardening, commercial, hospital, fishing, shelter, hotel, recreation,
toilets, residential, leisure, swimming, amenity, tertiary, sport, historic, tourism,
places, park, food, sports, restaurant]
[herd, swarms, flower, clothes, musical, fish, laundry, shoes, supermarket, toys,
doctors, beverages, worn, fridge, washing, pedometer, manifold, animal, heart, animals,
coffee]
[de, g, n, e]
[raid, flash, slam, transformer, miniature, pet, antilock, vac, anpr, dish, rrc,
airbag, latch, ignition, gearbox, wiper, glove, esp, volt, abs, restraint, coolant]
[avoidance, violation, reverse, warning, brakes, decelerate, signs, drives, drivers,
counter, timing, maneuver, overtaking, alarm, braking, belt, accelerator, pedal,
beat, steering, alert, seat, brake, sign, throttle, wheel, collision, driver]
[storeys, millibar, tonne, celsius, kilo, kg, pound, ug, lo, fahrenheit, centimeter,
centimetre, kelvin, mmol, ampere, kilogram, rpm, maf, watt, litre, liter, gram,
milligram, cubic]
[entry, detour, origin, trip, path, journey, destination, alternate, departure,
arrival, route, stop, travel]
[damaging, theft, sabotage, terrorist, destruction, vulnerable, explosion, protection,
aggressive, soft, natural, excessive, dangerous, serious, wear, spillage, hazardous,
reactive, damaged]
[horse, motorcycles, gears, coach, automobile, chassis, tesla, revolutions, hertz,
trailer, motorcycle, licence, comment, brand, motor, license, gear]
[shoulder, strike, contraflow, narrow, widening, adjacent, sight, guard, exit, gate,
shadow, footway, ballet, pole, runway, ends, plane, roll, mirrors, curb, roadway,
blind, windshield, carriageway, ramp, marking, bridge, tunnel, broken, head, obstruction,
obstacle, front, rear, side, crossing, crosswalk, lane]
[esc, ecg, glucose, photosynthetically, gyrometer, coulomb, orp, ohm, voc, capacitance,
luminosity, lux, electrocardiogram, cholesterol, chill, wetness, luminous, decibel,
ion]
```

Fig. 6 Cluster Results Part IV

```
[patches, pavements, spray, chippings, gritting, carriageways, gantry, cables, film,
laying, resurfacing, litter, blasting, spun, packed, ball, seating, navigable, curbs,
rails, rocks, bowls, wash, ditch, wood, apron, drain, stone, cable, bench, hydrant,
furniture, cycleway, old, pad, pavement, grass, cutting, rock, beds, leafs, sloped,
drained, gravel, dome, collapsed, loose, holes, pulling, mud, sand, asphalt, washer,
cushion, trunk, fill]
[sensitivity, derived, index, ratio, relative, absolute, intensity, indicator, degree,
weight]
[axles, foot, axle, mm, miles, inch, kilometers, kilometres, diameter, travelled,
odometer, millimeter, millimetre, kilometre, kilometer, square, metre]
[delivery, offering, business, platform, marketplace, services, connection, cloud,
service]
[deceleration, speeds, drag, stress, traction, bending, burst, displacement, friction,
pitch, angular, vertical, torque, tension, rotational, velocity, instantaneous,
motion, longitudinal, lateral, sound, angle, acceleration, speed]
[fully, single, sub, scale, multi, tire, preferred, micro, second]
[min, dtc, evap, ref, ra, dist, temp, max, avg, dew]
[market, difficulty, barrier, political, obstacles, potential, barriers, gap]
[c, b, w, k]
[permanently, slide, roof, doors, outside, entrance, chamber, house, kiosk, closed,
opening, rooms, room, refrigerator, empty, curtain, locked, container, lock, door]
[cross, stability, boundary]
[sleet, flooding, tornadoes, storms, slippery, intermittent, floods, avalanches,
showers, overflow, weighing, thunderstorms, hurricane, smog, gusts, blowing, disturbance,
salting, frost, hail, shower, icy, storm, rainy, smoke, fire, heavy, dust, rain]
[earthquake, impacts, events, crash, landslips, damages, insured, severity, severe,
roadworks, extreme, risk, injury, closures, incident, damage, hazard, failure, accident,
danger]
[odometry, localization, compass, barometer, gyro, positioning, speedometer, yaw,
laser, microwave, seismometer, radar, optical, magnetometer, ultrasonic, gyroscope,
accelerometer, magnetic, clock, tilt, camera, gps, mirror, beam, detector, thermometer]
[diesel, production, fossil, recycling, renewable, goods, ethanol, gross, billion,
million, chains, oil, waste, fuel]
[undefined, leaves, sink, flux, leaf, stream, fog, source]
```

Fig. 7 Cluster Results Part V

```
- [glare, taxiway, shed, photovoltaic, illuminance, dimming, lighting, solar, window,
  lamp, shade, sun, lights, light]
- [pylon, sensors, deployed, sensing, monitor, actuation, information, data, actuating,
  device, sensor]
- [controlled, loops, carries, actuated, diagnostic, signals, terminal, appliance,
  controller, command, actuator, engine, control, switch, system, meter]
- [mountain, alpine, bay, glacier, rink, grassland, geological, valley, coastline,
  river, canal, reef, dam, wetland, forest, waterway, pan, delta]
- [duration, peak, hours, millisecond, instant, interval, year, times, day, volume,
  mileage, minute, hour, unit, time, per]
- [generator, generated, input, variable, interface]
- [dense, moderate, large, localized, isolated, clear, near, around, fair, poorly,
  medium, weak, strong, poor, presence, high, low]
- [deep, pool]
- [buses, transit, intersections, bikes, bike, vehicles, motorway, signalling, lanes,
  street, toll, highway, roadside, train, rail, passenger, railway, congestion, bicycle,
  car, bus, station, parking, traffic, road, vehicle]
- [noise, pollutant, co, emission, pollution, carbon, gas, energy]
- [difference, mean, equivalent, inflation, altitude, median, percent, elevation,
  density, variation, slope, height, error, width, length, values, average, value,
  rate, count]
- [oxide, monoxide, dioxide]
- [cold, avalanche, drifts, vegetation, patchy, subsidence, spring, uneven, freeze,
  warmer, snowfall, dry, rainfall, land, falling, melting, fall, cloudy, winds, cloudiness,
  cover, winter, wet, ground, ice, freezing, soil, precipitation, snow, surface]
- [heat, moisture, salinity, resistance, conductance, skin, ph, conductivity, ambient,
  blood, humidity, pressure, temperature]
- [toxic, fumes, fresh, coal, hydro, radioactive, methane, mg, oxidation, calcium,
  nitrate, ppm, gaseous, ammonium, salt, sodium, potassium, ammonia, combustion, ozone,
  oxygen, sulphur, alcohol, volatile, organic, nitrogen, concentration]
- [big, fast, pace, accelerate, smooth, transform, backward, slow]
- [artwork, points, coordinates, map, spatial, latitude, longitude, reference]
- [tidal, barometric, global, atm, radiation, atmospheric]
```

Fig. 8 Cluster Results Part VI labelfig

```
- [exhaust, injection, lift, chiller, conditioner, boiler, drop, cooler, conditioning,
  recirculation, tower, fan, ventilation, intake, heater, heating, fluid, tank, flow,
  air]
- [overhead, queuing, latency, cons, delay, throughput, handover]
- [filter, materials, particle, aggregate, particles, compound, material, mass, chemical,
  quantity]
- [charging, cell, backup, charge, electrical, capacity, battery, voltage, load, electric,
  power]
- [payload, cc, timer, pulse, wi, par, pc, packet, transmission, uplink, frequency,
  downlink, signal, link, mhz]
- [impact, reduced, influence, depends, result, increase, decrease, ect, distribution,
  negative, reduction, loss, change, total, show]
```

Fig. 9 Cluster Results Part VII

References

1. Zhao, L., Ichise, R., Mita, S., Sasaki, Y.: An ontology-based intelligent speed adaptation system for autonomous cars. In: Joint International Semantic Technology Conference (Conference rank not found). Springer (2014)
2. Klotz, B., Troncy, R., Wilms, D., Bonnet, C.: VSSo—a vehicle signal and attribute ontology (Short Paper). In: SSN Workshop at ISWC. CEUR Workshop Proceedings(2018)
3. Gyrard, A., Bonnet, C., Boudaoud, K.: Ontology-based intelligent transportation systems. In: BMW Summer School 2014, Autonomous Driving in the Internet of Cars (2014). [Online]. Available: http://sensormeasurement.appspot.com/publication/PosterBMW.pdf
4. Klotz, B., Troncy, R., Wilms, D., Bonnet, C.: Generating semantic trajectories using a car signal ontology. In: The Web Conference. WWW, A-rank Conference (2018)
5. Klotz, B., Datta, S.K., Wilms, D., Troncy, R., Bonnet, C.: A car as a semantic web thing: motivation and demonstration. In: Global IoT Summit GIoTS, colocated with the IoT Week (2018)
6. Armand, A., Filliat, D., Ibañez-Guzman, J.: Ontology-based context awareness for driving assistance systems. In: Intelligent Vehicles Symposium (IEEE IV, B-rank conference). IEEE (2014)
7. Katsumi, M., Fox, M.: Ontologies for transportation research: a survey. Elsevier Transp. Res. Part C Emerg. Technol. J. (IF: 5.775 in 2018) (2018)
8. Villalba, J.B.: Using Ontologies and Intelligent Systems for Traffic Accident Assistance in Vehicular Environments. Ph.D. dissertation (2014)
9. Noura, M., Gyrard, A., Heil, S., Gaedke, M.: Concept extraction from the web of things knowledge bases. In: International Conference WWW/Internet 2018. Elsevier, Outstanding Paper Award (2018)
10. Wetterwald, M.: Slides: towards a SAREF extension for automotive. In: W3C Workshop on Data Models for Transportation
11. Puiu, D., Barnaghi, P., Toenjes, R., Kümper, D., Ali, M.I., Mileo, A., Parreira, J.X., Fischer, M., Kolozali, S., Farajidavar, N., et al.: Citypulse: large scale data analytics framework for smart cities. IEEE Access (2016)
12. Kolozali, S., Bermudez-Edo, M., Puschmann, D., Ganz, F., Barnaghi, P.: A knowledge-based approach for real-time iot data stream annotation and processing. In: IEEE iThings Conference (2014)
13. Pollard, E., Morignot, P., Nashashibi, F.: An ontology-based model to determine the automation level of an automated vehicle for co-driving. In: International Conference on Information Fusion (2013)
14. Zhao, L., Ichise, R., Yoshikawa, T., Naito, T., Kakinami, T., Sasaki, Y.: Ontology-based decision making on uncontrolled intersections and narrow roads. In: IEEE Intelligent Vehicles Symposium (IV). IEEE (2015)
15. Zhao, L., Ichise, R., Mita, S., Sasaki, Y.: Core ontologies for safe autonomous driving. In: International Semantic Web Conference (Posters & Demos). ISWC, A-rank conference (2015)
16. Zhao, L., Ichise, R., Mita, S., Sasaki, Y.: Ontologies for advanced driver assistance systems. In: The 35th Semantic Web & Ontology Workshop (SWO) (2015)
17. Lécué, F., Tallevi-Diotallevi, S., Hayes, J., Tucker, R., Bicer, V., Sbodio, M.L., Tommasi, P.: Star-city: semantic traffic analytics and reasoning for city. In: Proceedings of the 19th international conference on intelligent user interfaces. ACM (2014)
18. Ruta, M., Scioscia, F., Gramegna, F., Di Sciascio, E.: A mobile knowledge-based system for on-board diagnostics and car driving assistance. In: International conference on mobile ubiquitous computing, systems, services and technologies (UBICOMM, B-rank conference). Citeseer (2010)
19. Ruta, M., Scioscia, F., Gramegna, F., Loseto, G., Di Sciascio, E.: Knowledge-based real-time car monitoring and driving assistance. In: SEBD. Citeseer (2012)

20. M. Ruta, F. Scioscia, G. Loseto, A. Pinto, and E. Di Sciascio, "Machine Learning in the Internet of Things: a Semantic-enhanced Approach," *Semantic Web Journal*, 2017
21. A. I. Maarala, X. Su, and J. Riekki, "Semantic reasoning for context-aware internet of things applications," *IEEE Internet of Things Journal*, 2017
22. Bermejo, A., Villadangos, J., Astrain, J., Cordoba, A.: Ontology based road traffic management. In: Intelligent Distributed Computing VI. Springer (2013)
23. Corsar, D., Markovic, M., Edwards, P., Nelson, J.D.: The transport disruption ontology. In: International Semantic Web Conference (ISWC, A-rank Conference). Springer (2015)
24. Codescu, M., Horsinka, G., Kutz, O., Mossakowski, T., Rau, R.: Osmonto—an Ontology of OpenStreetMap tags. In: State of the Map Europe (SOTM-EU) (2011)
25. Fuchs, S., Rass, S., Lamprecht, B., Kyamakya, K.: A model for ontology-based scene description for context-aware driver assistance systems. In: Proceedings of the 1st International Conference on Ambient Media and Systems (2008)
26. Fuchs, S., Rass, S., Kyamakya, K: Integration of ontological scene representation and logic-based reasoning for context-aware driver assistance systems. In: Electronic Communications of the EASST (2008)
27. Fernandez, S., Ito, T.: Using SSN ontology for automatic traffic light settings on intelligent transportation systems. In: IEEE International Conference on Agents (ICA). IEEE (2016)
28. Barrachina, J., Garrido, P., Fogue, M., Martinez, F.J., Cano, J.-C., Calafate, C.T., Manzoni, P.: CAOVA: a car accident ontology for VANETs. In: IEEE Wireless Communications and Networking Conference (WCNC, A-rank conference). IEEE (2012)
29. Barrachina, J., Garrido, P., Fogue, M., Martinez, F.J., et al.: VEACON: a vehicular accident ontology designed to improve safety on the roads. Elsevier J. Netw. Comput. Appl. (IF: 5.273 in 2018) (2012)
30. Stocker, M., Rönkkö, M., Kolehmainen, M.: Making Sense of Sensor Data Using Ontology: A Discussion for Road Vehicle Classification (2012)
31. Stocker, M., Rönkkö, M., et al.: Situational knowledge representation for traffic observed by a pavement vibration sensor network. Trans. Intell. Transp. Syst. (2014)
32. Ebers, S., Hellbuck, H., Pfisterer, D., Fischer, S.: Short paper: collaboration Between VANET applications based on open standards. In: Vehicular Networking Conference (VNC, B-rank conference). IEEE (2013)
33. De Oliveira, K.M., Bacha, F., Mnasser, H., Abed, M.: Transportation ontology definition and application for the content personalization of user interfaces. Elsevier Expert Syst. Appl. J. (IF: 4.292 in 2018) (2013)
34. Zidi, A., Abed, M.: A generalized framework for ontology-based information retrieval: application to a public-transportation system. In: International Conference on Advanced Logistics and Transport (ICALT, B-Rank Conference). IEEE (2013)
35. Mnasser, H., Gargouri, F., Abed, M.: Towards an intelligent information system of public transportation. In: International Conference on Advanced Logistics and Transport (ICALT, B-Rank Conference). IEEE (2013)
36. Houda, M., Khemaja, M., Oliveira, K., Abed, M.: A public transportation ontology to support user travel planning. In: International Conference on Research Challenges in Information Science (RCIS, B-Rank Conference). IEEE (2010)
37. Li, G., Ma, D., Loua, V.: Fuzzy ontology based knowledge reasoning framework design. In: International Conference on Software Engineering and Service Science (ICSESS, Ranking Not Found). IEEE (2012)
38. Calavia, L., Baladr0n, C., Aguiar, J.M., Carro, B., Sánchez-Esguevillas, A.: A semantic autonomous video surveillance system for dense camera networks in smart cities. Sensors (2012)
39. Madkour, M., Maach, A.: Ontology-based context modeling for vehicle context-aware services. J. Theor. Appl. Inf. Technol. (2011)
40. Hamilton, A., González, E.J., Acosta, L., Arnay, R., Espelosín, J.: Semantic-based approach for route determination and ontology updating. Eng. Appl. Artif. Intell. (2013)

41. Feld, M., Müller, C.: The automotive ontology: managing knowledge inside the vehicle and sharing it between cars. In: International Conference on Automotive User Interfaces and Interactive Vehicular Applications (Conference Rank Not Found). ACM (2011)
42. Wang, J., Wang, X.: An ontology-based traffic accident risk mapping framework. In: International Symposium on Spatial and Temporal Databases. Springer (2011)
43. Hülsen, M., Zöllner, J.M., Weiss, C.: Traffic intersection situation description ontology for advanced driver assistance. In: Intelligent Vehicles Symposium (IV). IEEE (2011)
44. Berdier, C.: Road system ontology: organisation and feedback. In: Ontologies in Urban Development Projects. Springer (2011)
45. Kannan, S., Thangavelu, A., Kalivaradhan, R.: An intelligent driver assistance system (I-DAS) for vehicle safety modelling using ontology approach. In: International Journal of UbiComp. UbiComp, A-Rank Conference (2010)
46. Baumgartner, N., Gottesheim, W., Mitsch, S., Retschitzegger, W., Schwinger, W.: BeAware!—situation awareness, the ontology-driven way. Elsevier Data Knowl. Eng. J. (IF: 1.583 in 2018) (2010)
47. Liu, C.-H., Chang, K.-L., Chen, J.J.-Y., Hung, S.-C.: Ontology-based context representation and reasoning using OWL and SWRL. In: Conference on Communication Networks and Services Research (CNSR, B-Rank conference). IEEE (2010)
48. Niaraki, A.S, Kim, K.: Ontology based personalized route planning system using a multi-criteria decision making approach. Elsevier Expert Syst. Appl. J. (IF: 4.292 in 2018) (2009)
49. Yue, D., Wang, S., Zhao, A.: Traffic accidents knowledge management based on ontology. In: International Conference on Fuzzy Systems and Knowledge Discovery (FSKD, B-Rank conference). IEEE (2009)
50. Zhai, J., Chen, Y., Yu, Y., Liang, Y., Jiang, J.: Fuzzy semantic retrieval for traffic information based on fuzzy ontology and RDF on the semantic web. JSW (2009)
51. Sun, J., Wu, Z.-h., Pan, G.: Context-aware smart car: from model to prototype. Springer J. Zhejiang Univ.-Sci. A (2009)
52. Belhadef, H., Kholladi, M.: Urban ontology-based geographical information system. J. Theor. Appl. Inf. Technol. (2009)
53. Regele, R.: Using ontology-based traffic models for more efficient decision making of autonomous vehicles. In: International Conference on Autonomic and Autonomous Systems (ICAS, B-Rank conference). IEEE (2008)
54. Eigner, R., Lutz, G.: Collision avoidance in VANETs—an application for ontological context models. In: International Conference on Pervasive Computing and Communications (PerCom, A-Rank Conference). IEEE (2008)
55. Cheng, G., Du, Q., Ma, H.: The design and implementation of ontology and rules based knowledge base for transportation. In: International Conference on Computer Science and Software Engineering (CASCON, B-Rank conference). IEEE (2008)
56. Hu, Y., Wu, Z., Guo, M.: Ontology driven adaptive data processing in wireless sensor networks. In: International Conference on Scalable Information Systems. ICST (Institute for Computer Sciences, Social-Informatics) (2007)
57. Lorenz, B., Ohlbach, H.J., Yang, L.: Ontology of Transportation Networks (2005)
58. Budgen, D., Brereton, .: Performing systematic literature reviews in software engineering. In: International Conference on Software Engineering. ACM (2006)
59. Kitchenham, B., Pretorius, R., Budgen, D., Brereton, O.P., Turner, M., Niazi, M., Linkman, S.: Systematic literature reviews in software engineering—a tertiary study. Inf. Softw. Technol. (2010)
60. Rizzo, G., Tomassetti, F., Vetro, A., Ardito, L., Torchiano, M., Morisio, M., Troncy, R.: Semantic enrichment for recommendation of primary studies in a systematic literature review. Digit. Scholarship Humanit. (2017)
61. Noura, M., Gyrard, A., Heil, S., Gaedke, M.: Automatic Knowledge Extraction to Build Semantic Web of Things Applications (2019)
62. Compton, M., Barnaghi, P., Bermudez, L., Garcia-Castro, R., Corcho, O., Cox, S., Graybeal, J., Hauswirth, M., Henson, C., Herzog, A., et al.: The ssn ontology of the w3c semantic sensor network incubator group. Sci. Serv. Agents World Wide Web Web Semant. (2012)

63. Haller, A., Janowicz, K., Cox, S., Le Phuoc, D., Taylor, K., Lefrançois, M.: Semantic Sensor Network Ontology. W3C Recommendation (2017). [Online]. Available: https://www.w3.org/TR/2017/CR-vocab-ssn-20170711/
64. Daniele, L., Solanki, M., den Hartog, F., Roes, J.: Interoperability for smart appliances in the iot world. In: International Semantic Web Conference. Springer (2016)

IoT Semantic Interoperability for Active and Healthy Ageing

Regel Gonzalez-Usach, Matilde Julian, Manuel Esteve, and Carlos E. Palau

Abstract Semantic interoperability among different systems and platforms in the Internet of Things (IoT) represents a major challenge with an inherent high complexity, but its application can lead to massive benefits. In particular, the application of IoT technology in Ambient Assisted Living (AAL) and Active and Healthy Ageing (AHA) is considered especially beneficial. The IoT has an immense potential to critically enhance the health and quality of life of aged individuals. The accelerated rise of aged population in modern society and their associated caring necessities is a very concerning and critical social problem that finds a potential solution on the leverage of modern technologies. The European project ACTIVAGE aims to multiply the benefits obtained through the use of IoT in Elderly Smart Homes by enabling semantic interoperability and co-operation across these Smart Home clusters located in 12 different cities and regions across Europe. This interoperability will enable significant AHA service improvement. Semantic interoperability is achieved performing translations across systems via a real-time semantic translator (Inter Platform Semantic Mediator).

1 Introduction

Semantic Interoperability denotes the ability of Information and Technology (IT) systems to exchange data with unambiguous meaning [1]. This ability of different technology systems implies that they are capable to establish communication among them, exchange data, and interpret correctly the exchanged information, being capable of using it effectively [2]. Interoperability at the semantic level is a requirement for the enablement of machine computable logic, knowledge discovery, inferencing, data federation among information systems and, additionally, it is key to unlock the potential of the Internet of Things. In this sense, the existence of semantic

R. Gonzalez-Usach (✉) · M. Julian · M. Esteve · C. E. Palau
Universitat Politècnica de València, Valecia, Spain
e-mail: regonus@doctor.upv.es

R. Pandey et al. (eds.), *Semantic IoT: Theory and Applications*, Studies in Computational Intelligence 941, https://doi.org/10.1007/978-3-030-64619-6_14

interoperability enables cooperation actions, intelligent sharing of relevant information, service improvement and the creation of new valuable services. Thus, the interconnection of IoT Systems enabling a common understanding of information can lead to massive technological, economic and social benefits and open up numerous remarkable possibilities [3].

However, semantic interoperability among different IoT systems and platforms represents a major challenge with an inherent high complexity. Typically, information systems are created standalone using different standards, technologies and ways of storing and communicating information, thus being unable to inter-operate, communicate and share information with other systems. In the specific field of IoT, this situation is aggravated as far as the nature of IoT is intrinsically heterogeneous, with no global de facto standard to follow. Each platform uses its own standards, formats, ontologies and data models, following completely different manners of expressing and understanding the information. Consequently, IoT systems are typically unable to communicate correctly with each other or to interoperate [4]. Additionally, IoT information typically flows on real-time massive streams of big data. The management and processing of these data has an inherent high complexity [5, 6]. Due to these reasons, the achievement of semantic interoperability among IoT Systems is a very complex endeavor. Thus, interoperability is considered one of the major challenges and highest concerns in IoT. It has been estimated that the lack of interoperability in IoT hampers 40% of the potential benefits of the use of IoT [7] and, therefore, it is considered a blocking technological issue that causes major technological and business setbacks [8]. For these reasons, the existence of a global reference standard for IoT would be remarkably helpful in this situation, as it would enable interoperability by providing uniformity to the encoding and management of IoT information. However, nowadays there is a lack of such a standard, which poses a relevant problem in the design, integration or inter-connection of IoT systems [5].

Nowadays, a major concern of the World Health Organization (WHO) is to find solutions to the need of Active and Healthy Ageing [9]. This concept refers to the process for the development and maintenance of the functional capability for the improvement of wellbeing in elderly population. AHA aims to optimize opportunities regarding health, participation, and safety for the enhancement of the quality of life for the aged population [10, 11]. The steep increase of the elderly population in our society, in addition to the caring necessities, is a very concerting and critical social problem [12]. The use of IoT on the AHA domain, a special field of Ambient Assisted Living, can provide independent living for aged population, safety at home, continuous control of medical conditions in a remote way, while always keeping caregivers and family real-time reported of any anomaly or situation that may require their attention [13]. Hence, the IoT has an immense potential to radically enhance the health and quality of life of aged individuals [12]. AHA-IoT solutions will lead elderly individuals to have longer and healthier lives, mitigation of social isolation and pain soothing, in addition to the reduction of caring expenses. For these reasons, the application of IoT technology in Health, AAL and AHA areas is considered especially beneficial and promising [14].

This chapter describes an AHA use case in which semantic interoperability is provided in order to improve different systems and service availability, in addition to allowing the exchange of devices and services between different platforms, as well as the incorporation of new services and platforms. The European project ACTIVAGE [15] aims to multiply the benefits obtained through the use of IoT in Elderly Smart Homes by enabling semantic interoperability and co-operation across these Smart Home clusters located in 12 different cities and regions across Europe. In this way, services from other clusters can be employed, multiplying the number of available services, current services can be enhanced, and new services can be built fruit of this interoperability. The enablement of semantic interoperability across these systems leads to a significant enhancement of the quality of life of aged population living at their smart homes.

Current methods that enable IoT interoperability are explained in this chapter, as well as the interoperability solution employed in this AHA case study. Technically, the interoperability solution applied is mainly composed by a semantic translator that performs translations to/from a central AHA ontology, managing real-time streams of data. Additionally, other elements are necessary: a syntactic translator and a communication broker. The syntactic translator transforms the message syntactic format of the message into a serialization of RDF to feed the semantic translator. The broker handles the communication of real-time data flows of data between platforms. By these means, communication and a common semantic understanding of the information are established between different IoT systems. This technical solution is described in a following section.

Moreover, an overview of the AHA case study is provided, as well as a description of the application of the semantic framework in the ACTIVAGE ecosystem, explaining its implication in the AHA systems and the benefits that provides. Finally, conclusions of this chapter are presented.

2 Semantic Interoperability

There are different levels of interoperability (namely, technical, syntactic and semantic), being the semantic interoperability the highest level or layer, which requires the fulfilment of the other types (technical and syntactic) (Fig. 1).

Technical interoperability refers to the capability to establish communication between different systems or applications to the extent of enabling message exchange, but without implying correct understanding of the content or even being able to read the data received [16]. This type of interoperability generally requires the enablement of machine-to-machine (M2M) communication. Network connectivity is therefore a requirement [17].

Syntactic interoperability refers to the system capability to appropriately interpret the message structure of information received from other system or external element. This fact implies the ability to read the message content, but not necessarily understanding the meaning of the information contained [18]. For example, syntactic

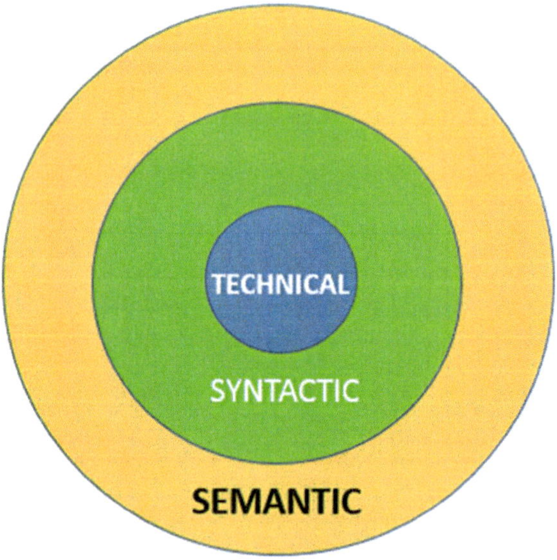

Fig. 1 Types of interoperability and their layered inter-relation

interoperability allows an e-Health system that receives data from an external source (e.g. a medical data center) to recognize the data format employed by the data source (e.g. XML) and correctly read the data (e.g. a set of values). However, it does not imply that this e-Health is aware of the meaning of those values (e.g. heart rate), and, as a consequence, it may not be able to utilize the received values with the proper context. The use of standardized data formats ensures unambiguous interpretation of the data structure and content, and facilitates the enablement of syntactic interoperability across entities.

Semantic interoperability [2, 12, 19, 20] allows systems to interpret correctly the meaning of the shared information, and requires previous technical and syntactic interoperability. As an example, the existence of semantic interoperability allows an AHA entity -that has correctly read the data received from smart home devices and extracted a set of values- to also interpret correctly the meaning and context associated to those values (presence in a certain room, use of the bed, etc.).

2.1 Methods for the Achievement of Semantic Interoperability

The use of semantic standards (i.e. a semantic ontology) and common semantics among different entities can enable semantic interoperability among them [21, 22]. Unfortunately, it is non-feasible across systems that already employ different semantics (which is the most typical situation [4, 23]). Systems that were not initially designed to inter-operate rarely present common semantics.

To enable semantic interoperability across non-interoperable systems, there are three different main techniques [24, 25]:

- **Ontology alignment**: refers to the set of semantic correspondences between several ontologies. Relations between different ontologies entities are defined in the alignment. Ontology alignments can be simple (among atomic entities) or have a higher complexity (across entity groups and entity sub-structures). An alignment includes predicates of similarity (i.e. matching) such as equivalence and subsumption axioms, or logical axioms (i.e. mapping). Frequently, tools for alignment definition report the level of accuracy and fiability of each inner correspondence.
- **Ontology merging**: this technique provides a semantic interoperability solution across systems by means of the combination of two or more ontologies into a common one, creating a new ontology that supports the knowledge from the original ontologies. The most simple application of ontology merging would be a resultant ontology consisting on the sets of axioms from all the original ontologies. The inclusion of new axioms derived from the relation between the source ontologies would create a more complete and complex global ontology. These new axioms are typically obtained via ontology alignments.
- **Ontology translation**: this technique for achieving semantic interoperability consists on the process of changing the semantics of a piece of information expressed through an ontology into the semantics of a different ontology. Thus, information described semantically in terms of a source ontology is transformed into information described in terms of a target ontology. The result contains information interpretable in the scope of the target ontology semantics. A requirement for a successful semantic translation is the preservation the original information, which must not suffer alterations on the meaning. Notionally, no information must be destroyed through translation. Also, in ideal terms, it should be possible to revert the translation process. Thus, the original content could be recovered by performing a reversed translation (from the final ontology into the source one). Those translations typically require previous ontology alignment.

Additionally, there are several recommendations and good practices for facilitating the achievement of semantic interoperability among systems [23] and reducing the heterogeneity of information models for a same domain:

- reuse of existing ontologies at the possible extent: reuse of existing knowledge, avoiding the construction of heterogeneous models that hinder interoperability. It also implies to use recommended core ontologies in order to create a new ontology covering the system needs, such as SSN in IoT domains.
- creation of ontologies following best practices [26].
- update and maintenance of ontologies [23].
- creation of catalogs of ontologies for easing the reuse of ontologies, and also their update and maintenance or the creation of new ontologies aligned with appropriate and core ones [23, 27].

The accomplishment of these recommendations makes more feasible that systems follow a common ontology and share common semantics. Also, it significantly reduces the effort required for applying methods for the enablement of semantic interoperability. It simplifies the process of ontology merging and ontology alignment (due to the fact that ontologies share more commonalties). Moreover, the fulfilment of these recommendations also helps to lower the complexity of semantic translations and to improve their quality.

2.2 Semantic Interoperability in IoT

The achievement of IoT semantic interoperability faces significant challenges [23] and remains one of the major challenges in IoT [2, 12].

First of all, IoT systems rarely follow common semantics. There is indeed a tendency that each project or system creates their own ontology [23, 27, 28] and does not seek consensus with other ontology developers [29], increasing the heterogeneity of the representations of the information.

Second, in general the good practices mentioned in previous section are not followed, adding complexity to potential enablers of semantic interoperability. Most IoT systems do not follow existing ontologies, and new ontologies generally do not reuse existing valid and well-established models. Regarding this point, W3C Semantic Sensor Network (SSN) ontology is considered the standard for generic modelling in IoT, and new ontologies should extend it adding the extra concepts needed in a particular IoT area [23, 30]. The use of a well-consolidated ontology such as SSN avoids to employ instead several ontologies for describing the same concepts [31]. However, despite of these recommendations, there is a lack of methodologies for modelling ontologies that fulfill the specific requirements for IoT data.

There is also an important need for ontology catalog creation barely covered, although the IoT community has manifested some concern for building catalogs of ontologies for some specific application domains [23, 27].

A third challenge is the heterogeneous formalism employed in IoT ontologies and vocabularies. Most vocabularies are described in markup languages or defined through UML artifacts; in most cases the formalism is not OWL. This fact hampers data integration from common vocabularies [23].

Another significant challenge encountered is the current high difficulty on unifying models, vocabularies and ontologies for semantically annotating the data. There is an important necessity of tools and approaches able of providing a unified semantic model aligned with vocabularies from IoT platforms [28].

In addition, due to the highly evolving nature of IoT, new IoT artifacts and features are constantly appearing, and ontologies require fast and timely updates [23, 27, 28], a fact that represents another challenge for IoT interoperability. By extension, also ontology catalogs should be timely updated and maintained [23, 27].

Moreover, it has to be considered that IoT data management requires the ability of handling large amounts of data (IoT big data) in real time, which requires special processing capabilities on the devices that handle it for performing semantic enrichment of data or for performing any other type of semantic management [2, 12, 32].

Another challenge is the lack of effective tools for ontology alignment or semantic translation. In literature at the time of publishing this book, Inter-IoT framework is considered the only existing tool for aligning ontologies and vocabularies in IoT environments [23, 28]. Furthermore, Inter-IoT provides the only known solution for performing real-time semantic translations [5, 23], which is the only possible option to provide semantic interoperability in real-time across IoT systems that do not share common semantics.

2.3 Semantic Interoperability in AHA

Semantic interoperability is critical in the e-Health and AHA sector, as it allows the inter-operation of the different elements and health systems that compose the AHA ecosystem, as well as the use in a seamless way of different data standards for the representation of AHA information. From an overall perspective, it enables the inter-operation and coordination of heterogeneous medical devices and systems, which typically follow different data standards and communication protocols, being this integration not trivial. Semantic interoperability makes compatible the use digital health information across diverse care settings and clinical software. Semantic inter-operability solves the problem of employing different electronic health record (EHR) or electronic medical record (EMR) systems, as well as the use of medical data associated to an specific patient provided by different care providers [16, 33]. Moreover, this common understanding of the information allows the existence of healthcare data analytics of data from multiple heterogeneous sources. In conclusion, semantic interoperability is crucial to allow inter-operation across caring systems and elements to seamlessly synthesize health information.

In order to achieve semantic interoperability in these specific domains, several ontologies have been developed for e-Health [5, 33, 34] (e.g. HL7[1]), while the first AHA-specific ontology has been recently created.

3 USE CASE: The ACTIVAGE European Smart Homes

3.1 Overview

The ACTIVAGE[2] project is a H2020 Large Scale Pilot that aims to improve the quality of life and autonomy of older adults through the use of IoT in the Active and

[1] https://www.hl7.org.

[2] http://www.activageproject.eu/.

Table 1 Initial DS of ACTIVAGE

DS	Platform
Galicia (Spain)	SOFIA2
Valencia (Spain)	FIWARE
Madrid (Spain)	universAAL
Region Emilia Romagna (Italy)	FIWARE
Greece	FIWARE, universAAL
Isére (France)	SensiNact
WoQuaz (Germany)	universAAL
Leeds (UK)	FIWARE, IoTivity + OpenIoT testbed
Finland	SENIORSOME, universAAL

Healthy Aging (AHA) domain [15, 35]. The use of IoT in this domain will support the independence of older adults while also responding to the needs of caregivers, service providers and public authorities, as well as contributing to the sustainability of the health systems.

ACTIVAGE has been designed as a Multi Centric Large Scale Pilot and involves a set of Deployment Sites (DS or Smart Home Clusters for aged population) [15]. A DS consists of a group of stakeholders working together in a particular geographical space, each of them covering different parts of the AHA value network. Hence, healthcare administration, AHA services, service providers, technology providers and end users (formal/informal caregivers and elderly people) are some of the possible participants in a DS. These DS provide AHA services to aged population. Initially, 9 DS, located in 7 countries, were defined in ACTIVAGE. With the second Open Call, 3 more DS were added (see Table 2). The 12 DS are located in 9 European countries (see Fig. 2) and use existing open and proprietary IoT platforms. Each DS is free to use the IoT platform (or platforms) that is more suitable for them as a base for their infrastructure. The original 9 DS use the following IoT platforms: FIWARE, SOFIA2, universAAL, SensiNact, OpenIoT, IoTivity and SENIORSOME (see Table 1).

The use of different IoT platforms requires an interoperability solution in order to enable co-operation and sharing of information at some degree among them, as well as to enable the creation of a European AHA ecosystem. Moreover, this interoperability framework should be extensible to new IoT platforms and AHA services.

Table 2 New DS of ACTIVAGE

DS	Platform
Lisbon (Portugal)	FIWARE
Sofia (Bulgaria)	universAAL, openHAB
Catalonia (Spain)	ekenku, ekauri

Fig. 2 AHA Interoperable DS of Smart Homes for elders

3.2 Interoperability Goals in ACTIVAGE

The following goals for interoperability have been defined in ACTIVAGE:

- **Intra-deployment site interoperability**: any service provided in a DS should be able to access all the necessary application data within the DS, even in the case of a DS that has more than one IoT platforms. This implies that the applications deployed in a DS should support all the platforms in the DS and they should be transferable between different platforms within a DS.
- **Inter-deployment site interoperability**: means that a DS should be able to incorporate automatically new services, which would have access to the necessary application data in that DS. These new services could be incorporated from other

DSs, which may use different platforms, or they could be designed to be used in any DS. Moreover, different DS should be able to exchange application data.
- **Interoperable external adopted solutions**: means that a DS will be able to incorporate new solutions, according to its needs, that will be interoperable with the other DSs in ACTIVAGE. These new solutions will become available in the ACTIVAGE Marketplace, which is an online repository of applications accessible to ACTIVAGE's partners and stakeholders.

To achieve the required interoperability, the applications must have access to the data from any of the platforms deployed in the DSs. This implies that the applications should use the same data format, which must be independent of the platforms being used by the DSs. Since multiple IoT platforms and AHA applications may coexist in a DS, interoperability among the platforms in the same DS is needed. This interoperability must be both syntactic and semantic to have a common understanding of the information shared among different platforms. Furthermore, applications developed for a DS can be adapted to AIoTES to make them usable by any other DS without the need of individual adaptations for each IoT platform.

3.3 AHA Scenarios

Several AHA scenarios that respond to specific user needs have been defined in ACTIVAGE. The aim of these scenarios is to improve the autonomy and quality of life of the elderly. The following AHA scenarios require interoperability:

Daily activity monitoring at home: this scenario is based on the use of presence, proximity and magnetic contact sensors in the home of the elderly person. The measurements are sent to the Cloud, where they are processed in order to estimate activities, trends and deviations. The formal and informal caregivers can receive information and alerts about deviations of the elderly person's habits, allowing early interventions in a way compatible with the independence of the elderly.

Integrated care for older adults under chronic conditions: this scenario combines the daily activity monitoring at home with the use of medical devices and e-Health solutions. This implies the integration of protocols from different care providers that usually work separately. Hence, the implementation of this scenario will promote the cooperation among care providers in order to offer joint response to emergencies, which will result in more effective interventions and better planning of resources. This will lead to an improvement of the quality of life of the people with chronic disease while at the same time reducing the associated costs.

Monitoring assisted persons outside home and controlling risky situations: the aim of this scenario is to promote socialization and activity outside home by combining the use of wearable and mobile devices and the Smart City infrastructure. For example, the wearable devices could be tracked in order to request help if a risk for the person is detected.

Emergency trigger: in this scenario, the system can detect critical situations and report them automatically. For the input, panic buttons connected to the emergency call-center system can be installed in the elderly person's home in order to provide them an easy way to call the emergency services. Emergencies can also be triggered based on the results of processing the data coming from different sensors, which allows alerting the emergency services when the person cannot do it (unconsciousness, fall, gas, etc.).

Exercise promotion for fall prevention and physical activeness, based on ambient sensors and wearable devices.

Cognitive stimulation for mental decline prevention: this scenario combines behavioral monitoring with interventions, such as the use of applications and peripheral devices in games that can promote mental or physical exercise. Its objective is to extend the independence of the elderly people.

Prevention of social isolation: this scenario is based on the use of applications connected to the Smart City infrastructure to provide data about events in order to promote mobility and social interaction and engagement. The information from the smartphone can be combined with the data coming from the home sensors in order to get a more complete view of the users' social activity. As a result, depression and decline can be avoided.

Comfort and safety at home, based on the use of smart home technologies to control temperature, light, energy consumption and perimeter safety.

Support for transportation and mobility: in this scenario, the system can use data from the Smart City infrastructure to determine routes that can be adapted based on the needs and goals of the elderly person, such as finding the easiest route or exercise promotion. Route planning can be applied both in a city and between different cities.

3.4 ACTIVAGE Technical Approach for Semantic Interoperability

The interoperability in ACTIVAGE has been achieved by using elements from the framework for interoperability provided by the H2020 project INTER-IoT.[3] These elements are the key components that perform the necessary syntactic and semantic translations to achieve the interoperability objectives described in the previous section.

The INTER-IoT project designed a solution to enable interoperability in IoT at different levels. The layered approach followed in INTER-IoT allows a more complete exploitation of the functionalities of each layer of an IoT system. More concretely, Inter-IoT offers interoperability solutions for the following

[3] https://inter-iot.eu/.

layers: Device-to-Device (D2D), Networking-to-Networking (N2N), Middleware-to-Middleware (MW2MW), Application and Services-to-Application and Services (AS2AS), Data and Semantics-to-Data and Semantics (DS2DS) [15].

3.4.1 Semantic Interoperability Layer

ACTIVAGE makes use of the following layers of INTER-IoT to achieve semantic interoperability among different IoT platforms:

- **Middleware-to-middleware layer (MW2MW)** [36, 37]: the interoperability solution for the middleware level is based on the use of an abstraction layer that connects to all the IoT platforms and provides access to all their data and resources through a common API. The core of the MW2MW layer is a broker, which handles the different communication flows in real-time. The MW2MW connects to the platforms using bridges that act as syntactic translators between the data formats of the platforms and the common Inter-IoT data format (JSON-LD) in addition to performing some communication functions. This layer also maintains a registry of all the connected platforms and devices and provides a common representation of these resources. The MW2MW solution, also known as Inter-MW, is responsible for providing syntactic interoperability and for enabling and managing the communication between platforms.
- **Semantics and Data layer (DS2DS)** [6]: the interoperability solution for the DS2DS layer provides semantic interoperability based on semantic translation between the different platforms' ontologies and a central modular ontology [18, 38]. This way, a common meaning of the data and information is achieved, thus enabling semantic interoperability among different platforms. The Inter Platform Semantic Mediator (IPSM) [6] is the component that performs the semantic translation of the data using ontology alignments.

The combination of both aforementioned solutions is called the Semantic Interoperability Layer (SIL) and is the key component that makes it possible for the ACTIVAGE deployment to enable and ensure semantic interoperability in DSs. **Inter**-MW (MW2MW Layer) provides a common API to access all the data and resources of the connected platforms and performs the syntactic translations between the common ACTIVAGE JSON-LD format and the platforms' formats. The data in JSON-LD (an RDF serialization) is sent to the IPSM (Inter-IoT DS2DS Layer) to perform the necessary semantic translations to/from a common central ontology to platform-specific semantics. This translation process is explained in more detail in the following section.

Since the SIL provides a common syntax and semantic representation of the data, the services and applications built on top of the SIL will be able to use the information coming from any of the connected platforms. Also, by using the SIL, the information can be shared among different platforms and its meaning can be understood by any of them. Once the data have been translated by the IPSM into the target platform semantic model, Inter-MW can perform the transformation from RDF to the platform-specific syntactic format, allowing platforms to receive understandable data and messages from applications or different platforms. The converted data would be used to update a virtual representation of the device in the target platform, which would be used by the native services.

Requirements for IPSM Semantic Interoperability Solution

The ACTIVAGE approach for achieving semantic interoperability among heterogeneous IoT platforms is based on:

- The definition of explicit OWL-demarcated semantics for each IoT platform in the ACTIVAGE ecosystem that would share data or co-operate with other elements of the ecosystem.
- The existence of a common central ontology specifically designed for AHA. This ontology is based on GOIoTP,[4] m3-lite[5] and other IoT ontologies, and covers all required AHA-related concepts in the ACTIVAGE ecosystem.
- A previous syntactic translation of the data messages, to transform the platform-specific syntactic format into RDF, which is capable of supporting semantic information. This transformation is performed by the Inter-MW component [36, 37], which performs the syntactic transformation programmed for a specific type of platform. This transformation is necessary in order to enable later a semantic translation, as the semantic translator component (IPSM) requires that incoming information is expressed in RDF (more specifically, in the INTER-IoT JSON-LD format).
- A real-time semantic translator, the IPSM component, capable of performing a real-time ontology-to-ontology translation between massive streams of IoT/AHA data from platforms [6]. The IPSM translates data messages from its native semantic format to the AHA ontology or vice versa (from the AHA ontology to the platform-specific semantics) [39]. By these means, it is possible to express the information from platforms into the common ACTIVAGE semantics. Also, platform-to-platform co-operation it is possible by using two sequential semantic translations of messages (from the sender platform to the central ontology and from the central ontology to the receiver platform). This IPSM double-translation approach makes feasible the communication of a high number of platforms, as far as it exceedingly simplifies the translation process and provides good scalability and extensibility. In comparison, direct translations require exponential effort for each additional platform, thus being an unfeasible approach in a system with 10 different platforms.

[4]https://inter-iot.github.io/ontology/.

[5]https://lov.linkeddata.es/dataset/lov/vocabs/m3lite.

- Semantic alignments between each platform's semantics and the central AHA ontology, which are needed by the IPSM to perform ontology-to-ontology translations [39]. These alignments specify the translation rules and matches between ontologies and are written using the IPSM-AF format [19], and have to be defined and stored in specific files before performing the translation process.

3.4.2 ACTIVAGE IoT Ecosystem Suite (AIoTES)

The ACTIVAGE IoT Ecosystem Suite (AIoTES) has been developed in order to provide the necessary interoperability for ACTIVAGE and to enable the security and privacy features needed for the use of AHA information. AIoTES is defined as a set of tools, techniques and methodology for interoperability between existing IoT platforms while ensuring security and privacy. Each DS will be able to deploy their devices using its own IoT platforms and the new AHA applications will be deployed over a local instance of AIoTES. The use of AIoTES will provide semantic interoperability by design and allow the integration of new services in the DS, such as remote or wearable-based health-care systems (Fig. 3).

Semantic Interoperability Layer (SIL): composed by the interoperability solution described previously. The SIL provides semantic interoperability across ACTIVAGE. Services and applications built on top of the SIL will be able to use the information coming from any of the connected platforms.

Security and privacy module: security and privacy mechanisms implemented across AIoTES to ensure the authenticity of the data and that only authorized users have access to the data, following the requirements of the GPRD.

Service Layer: provides a set of platform-independent functionalities and makes use of the semantic interoperability enabled by the SIL. The service layer includes the

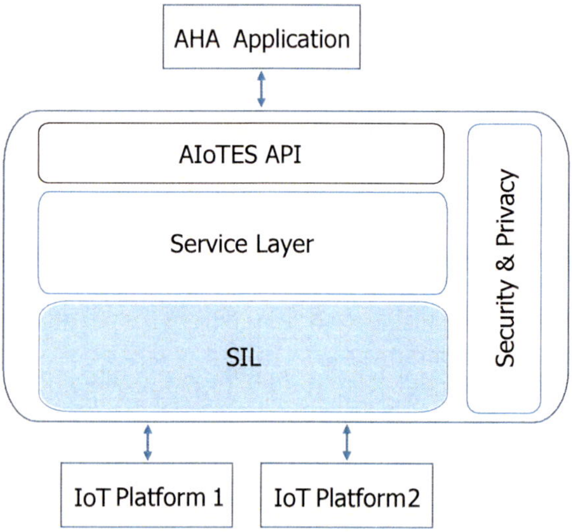

Fig. 3 AIoTES Architecture

Data Lake, a data analytics module and a set of deployment, development, management and visualization tools. The Data Lake allows access to the raw data collected by the devices, as well as the metadata used by the Data Analytics component. The Data Lake does not store the raw data from the sensors. Instead, it retrieves the data from the different platforms and makes use of the SIL to translate the data into the common ontology. Also, the Data Lake provides a metadata storage service for the parameters of the trained models produced and used by the data analytics module. The Data Analytics module is built on top of the Data Lake and provides a set of predictive data analytics methods that can be used on the data. These methods allow comparing parameters values against patterns or expected values in order to detect differences, changes or deviations. The results of these analyses can be retrieved from a REST API or represented using the visualization tools included in the Service Layer. Finally, the Service Layer includes several tools for the management of the devices and platforms of the DS and a set of development tools.

AIoTES API: all the components of AIoTES are accessible through a common API that allows a homogeneous and secure access to all its functionalities. This common API facilitates the development of an ecosystem of new applications and services based on the combination or exchange of data from different IoT platforms.

3.5 Application and Impact of the Semantic Framework

The enablement of semantic interoperability allows the improvement and exchange of services, as well as the creation of new services based on interoperability. AIoTES can translate the data in real time from one platform to another, which allows platform-specific applications and services to understand and use data coming from a different platform. Moreover, AIoTES offers a set platform-independent services that can be used by any DS or application. The following sections describe these new services.

3.5.1 Services Provided by AIoTES

The Service Layer of AIoTES provides a set of platform-independent services that rely on the semantic interoperability enabled by the SIL. Among these, the most important are the Data Lake services and the Data Analytics methods.

Data Lake

The Data Lake provides access to the raw data collected by the devices in a common format and semantics. The intended use of this component is not to store sensor data but to retrieve data from the platforms and make use of the SIL to translate it into the common ontology. Nevertheless, the Data Lake also provides storage for platforms unable to store their own data (for instance, platforms deployed in devices with storage restrictions). Hence, the Data Lake allows access to data from any DS in the common format to authorized users, while maintaining the security and privacy of the system.

The Data Lake provides a common interface to retrieve the data as if all the data were contained in the same database, to feed big data analytics methods to enable predictive analysis, or feed applications built on top. Thanks to semantic interoperability, the applications built on top of the Data Lake will receive the data in the common ontology and data model, regardless of their origin.

Data Analytics

The Data Analytics back end contains components for the analysis of the data retrieved from the Data Lake in order to extract meaningful information. These methods include feature extraction, feature selection, anomaly detection, prediction, clustering and hypothesis testing. The Data Analytics methods allow the analysis of several types of parameters and the identification of patterns and expected values, the comparison between users or the comparison of values from the same user at different contexts and the detection of deviations.

Since the Data Analytics component obtains the data in the common ontology and data model of ACTIVAGE, it offers services that are independent of the IoT platforms that collected the data. The Data Analytics methods are exposed through the AIoTES API and can be used by applications built on top of AIoTES. Moreover, these results can be displayed in a comprehensible way using the visualization tools included in the Service Layer of AIoTES, which also provide a graphical user interface for the data analytics methods.

3.5.2 Platform-Independent Applications

The existing AHA applications have been built on top of a specific IoT platform. This aspect limits the interoperability because the migration to a new platform would require to adapting the applications to the syntax and semantics of the new platform. Thus, without a common syntax and semantics, the effort required to integrate an application with a new DS or platform would increase exponentially. Similarly, without AIoTES, the creation of new multi-platform or multi-DS applications would be significantly more complex and prone to error, since they would need to communicate with different interfaces and deal with several syntactic and semantic representations of the data. For these reasons, the development of services that could be adopted by any DS would not be feasible without AIoTES.

Thanks to the interoperability framework used in ACTIVAGE, the inclusion of new platforms or DSs does not require any changes in the existing services and applications. Moreover, the interoperability approach followed in ACTIVAGE also enables the creation of platform-independent applications. As long as a common format is used, applications can be built on top of AIoTES and be independent from the platform and DS. These new applications can be uploaded to the ACTIVAGE

Marketplace[6] and made available for any DS interested in the services that they provide.

Service exchanges between Deployment Sites

The interoperability provided by AIoTES allows the exchange of devices and services between different DS that may use different platforms and data models without the need to modify the existing applications. Legacy native applications, typically platform-specific, can be used with a different platform thanks to the syntactic and semantic translations performed by the SIL. This way, services and applications can be exchanged between different DS without increasing considerably the complexity of the system. Thus, the DSs have incorporated new services and applications developed in a different one. Some examples of such exchanges are the following:

Bed sensor exchange
The DS of WoQuaz (Germany) has a native universAAL application that receives data from different sources, one of them being a bed sensor that provides presence information of a user, which is very valuable for detecting in a non-invasive way situations that may require the attention of caregivers (such as unusual long-time stays or lack of presence at night). They are interested in using a different sensor because their current option is expensive and not very reactive. The DS of Isére (France) has a different application that receives data from a bed sensor from a different vendor, but it is developed on sensiNact. Using the SIL to translate the data from sensiNact to universAAL, the first DS has been able to incorporate the bed sensor of the second DS without making changes in their application or having to develop a new driver. In the implemented solution, the new bed sensor is managed by sensiNact and a virtual representation of the device is created in universAAL to provide data to the native universAAL application. The SIL translates the bed sensor data in real time and updates the virtual sensor. This way, the application can make use of the bed sensor data, together with the data coming from other devices connected to universAAL.

Moreover, the DS of Isére can make use of the services developed in the DS of WoQuaz, such as presence detection in the bed and behavior patterns analysis and alerts based on the data obtained from the bed sensor (Fig. 4).

Health monitoring services exchange between Galicia and Greece
The DS of Galicia (Spain) offers health monitoring services developed on top of SOFIA2. The DS of Greece has a different set of health monitoring services, which are based on universAAL. Thanks to the semantic interoperability provided by AIoTES, these services can be used in both DS (see Fig. 5). For example, a SOFIA2 native service can use data from devices connected to universAAL without the need to

[6]https://marketplace.activage.iti.gr/.

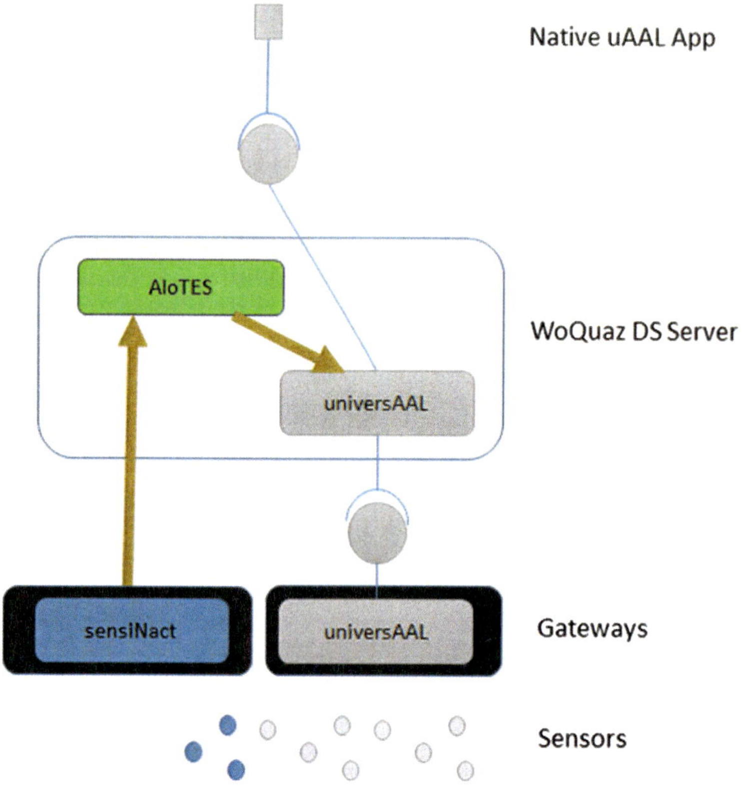

Fig. 4 Device (bed sensor) exchange between two different platforms using AIoTES

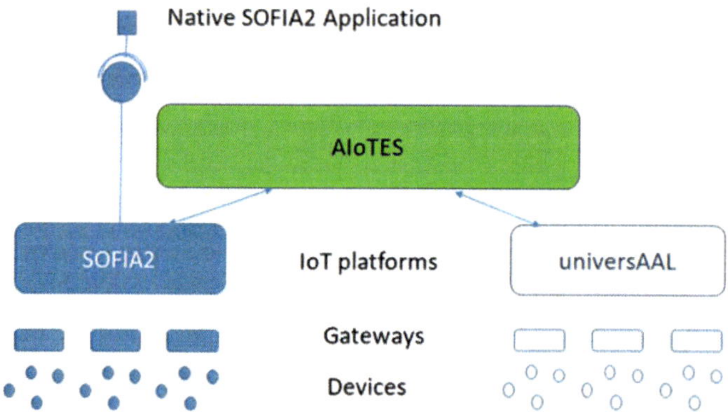

Fig. 5 Service exchange between two different platforms using AIoTES

adapt it to a different interface or data model. In that case, virtual representations of the universAAL devices would be created in SOFIA2. Then, the SIL would get the measurements from universAAL, perform the proper semantic and syntactic translations and send them to SOFIA2 to update the values of the virtual devices. Then, the application would be able to retrieve the data using the SOFIA2 API.

Exchange of activity monitoring services between Leeds and Valencia
The DS of Leeds (UK) and the DS of Valencia (Spain) have developed their own services based on FIWARE. Although each DS has a different data model, they can translate the data between both data models using AIoTES and make use of the services developed in the other DS, such as the indoor/outdoor position service of DS Valencia and the door status monitoring service of DS Leeds.

Gaming exchange between Finland and Madrid
The DS of Finland and the DS of Madrid (Spain) have implemented the use case of cognitive stimulation for mental decline prevention. Since semantic interoperability has been achieved, each of these DS can make use of the specific applications and services developed in the other DS.

New services incorporated in the Deployment Sites

In addition to the service exchanges, new services developed externally have been incorporated in the DSs. The following are some examples of such services:

Muvone
Muvone is a service for the prevention of osteoporosis based on the monitoring of physical activity and sunlight exposure using specific wearable devices and mobile applications based on FIWARE. Thanks to the semantic interoperability provided by AIoTES, two different DSs using universAAL and a third one using sensiNact have incorporated this new service.

Smartfloor
The DS of Greece has incorporated a fall detection/prevention service based on the data received from a smart floor. This smart floor also allows daily activity monitoring at home. These services have been developed on FIWARE. Using AIoTES, the smart floor services are being used in a DS with an infrastructure based on universAAL.

LOLA
LOLA is a solution for health monitoring and emergency detection based on the use of smartwatches. In an emergency situation, it can trigger and emergency call and provide information about the location and health of the person, which allows a more rapid and accurate response. This new service has been incorporated in the DS of Galicia.

Table 3 New DS

DS	Platform	Type of services
Lisbon (Portugal)	FIWARE	e-Home caring
Sofia (Bulgaria)	universAAL, openHAB	Activity monitoring
Catalonia (Spain)	ekenku, ekauri	e-Home caring, e-Health monitoring

MC CardioMonitor
The objectives of MC CardioMonitor are the monitoring of the cardiac function and physical parameters and the detection of arrhythmias. It uses a custom wearable acquisition device that provides a 3-lead ECG and motion monitoring through accelerometers and gyroscopes. This new service has been incorporated in a DS using FIWARE (Leeds) and in another one using SOFIA2 (Galicia).

3.5.3 New Deployment Sites

Three new DS from the second Open Call have been incorporated in ACTIVAGE. These new DS use different IoT platforms and, thanks to the semantic interoperability achieved by AIoTES, they will be able to incorporate any of the existing services and export their new services to other DSs. In this sense, the interoperability approach followed in Activage does not increase significantly the complexity of the system with each new platform added.

Information/details regarding these new DS can be found on Table 3.

4 Conclusions

In this chapter, a domain-specific case study for semantic technologies applicability has been described, focused on the AHA area. The ACTIVAGE project aims to enable semantic interoperability in European large-scale Smart Home Clusters for the ageing population across 12 cities or regions and 9 countries. The main objective beyond is the enhancement of the AHA services, on account of the fact that interoperability across systems can lead to very remarkable benefits.

The achievement of semantic interoperability across the different local IoT systems that are part of ACTIVAGE has to face major challenges. First of all, the enablement of semantic interoperability across IoT platforms, which follow different standards, formats, ontologies and data models, has an inherent high complexity. For this reason, interoperability is considered one of the most arduous challenges in IoT. Second, the semantic integration of a high number of IoT platforms is rarely feasible, as, in general, interoperability between IoT platforms is based on ad hoc solutions whose complexity increases exponentially with the number of IoT platforms. Third, the management of massive real-time streams of IoT data (big IoT data) has also high complexity and requires technology capable of effectively handling streaming data.

These important challenges have been solved in the ACTIVAGE project through the use of a real-time semantic translator (Inter Platform Semantic Mediator) capable of real-time management of large flows of data, conjointly with other complementary elements. The effectiveness of this solution has been proven across multiple AHA use cases, while part of the ecosystem development planned will be incorporated in the future.

This application of semantic technologies, enabling semantic interoperability, leads to an improvement of AHA services and functionality and, therefore, to an enhancement of the quality of life of ageing population. Important technical points to mention are the following. The interoperability solution adopted in ACTIVAGE provides a common semantic representation of the data in a growing AHA ecosystem. Moreover, the use of AIoTES makes the system easily scalable, which allows the exchange of services and applications among platforms through the use of virtual sensors handled by several platforms or systems simultaneously, and facilitates the incorporation of new services and platforms. In particular, the incorporation of new DSs, which make use of new IoT platforms that were not included in the initial design of ACTIVAGE, proves the scalability of the system. Thus, AIoTES could be easily adopted by other cities and regions independently of their existing Smart City and Smart Home technologies.

As has been explained in previous sections, the semantic interoperability solution used in ACTIVAGE allows the creation of platform-independent services and applications, which can be made available in the ACTIVAGE Marketplace. This would enable the creation of a new AHA IoT ecosystem that would facilitate the creation and adoption of new services and applications in AHA.

The use of this semantic interoperability framework would enable the creation of ecosystems based on data flows. Nevertheless, this goal presents challenges to face due the complexity associated with the management of real-time data streams.

Acknowledgements This research was partially supported by the European Union's "Horizon 2020" research and innovation programme as part of the "ACTivating InnoVative IoT smart living environments for AGEing well" (ACTIVAGE) project under Grant Agreement No. 732679.

References

1. ETSI: Interoperability Best Practices Solve the Challenge of Interoperability! www.etsi.org Supporting ICT Standardization www.plugtests.org EDITION 2 3 Interoperability Best Practices Content (2013)
2. Gonzalez-Usach, R., Yacchierma, D., Matilde, J., Palau, C.: Interoperability in IoT. In: IGI (ed.) Handbook of Research on Big Data and the IoT (2018)
3. World Economic Forum: Industrial Internet of Things: Unleashing the Potential of Connected Products and Services (2015)

4. Diallo, S.Y., Herencia-Zapana, H., Padilla, J.J., Tolk, A.: Understanding interoperability. In: Chinni, M.J., Weed D. (eds) 2011 Emerging M&S Applications in Industry and Academia Symposium (EAIA). 2011 Spring Simulation Multi-conference, SpringSim '11, Boston, MA, USA, April 03–07, 2011, Spring Simulation Conference, vol. 5, SCS/ACM, pp. 84–91 (2011)
5. Ganzha, M., Paprzycki, M., Pawlowski, W., Szmeja, P. and Wasielewska, K.: Semantic Technologies for the IoT - An Inter-IoT Perspective. In: 2016 IEEE First International Conference on Internet-of-Things Design and Implementation IoTDI 2016, Berlin, Germany, 4–8 April 2016. IEEE Computer Society, pp. 271–276 (2016). https://doi.org/10.1109/IoTDI.2015.22
6. Ganzha, M., Paprzycki, M., Pawłowski, W., Szmeja, P., Wasielewska, K.: Streaming semantic translations. In: 2017 21st International Conference on System Theory, Control and Computing, ICSTCC 2017, pp. 1–8. Institute of Electrical and Electronics Engineers Inc. (2017c). https://doi.org/10.1109/ICSTCC.2017.8107003
7. Mckinsey Global Institute: The Internet of Things: Mapping the Value Beyond the Hype (2015). https://www.mckinsey.com/business-functions/mckinsey-digital/our-insights/the-internet-of-things-the-value-of-digitizing-the-physical-world#
8. Aloi, G., Caliciuri, G., Fortino, G., Gravina, R., Pace, P., Russo, W., Savaglio, C.: Enabling IoT interoperability through opportunistic smartphone-based mobile gateways. J. Netw. Comput. Appl. **81**, 74–84 (2017). https://doi.org/10.1016/j.jnca.2016.10.013
9. World Health Organization: What is Healthy Ageing? (2018). https://www.who.int/ageing/healthy-ageing/en/ [Online]
10. Heitmann, B., Kinsella, S., Hayes, C., Decker, S.: Implementing semantic web applications: reference architecture and challenges. CEUR Workshop Proc. **524**, 16–30 (2009)
11. Paúl, C., Ribeiro, O., Teixeira, L.: Active ageing: An empirical approach to the WHO model. Curr. Gerontol. Geriatr. Res. (2012). https://doi.org/10.1155/2012/382972
12. Gonzalez-Usach, R., Yacchirema, D., Collado, V., Palau, C.: AmI open source system for the intelligent control of residences for the elderly. In: Lecture Notes of the Institute for Computer Sciences, Social-Informatics and Telecommunications Engineering, LNICST, vol. 242, pp. 46–52. Springer (2018). https://doi.org/10.1007/978-3-319-93797-7_6
13. Stavrotheodoros, S., Kaklanis, N., Votis, K., Tzovaras, D.: A Smart-Home IoT Infrastructure for the Support of Independent Living of Older Adults, pp. 238–249 (2018). https://doi.org/10.1007/978-3-319-92016-0_22
14. Gonzalez-Usach, R., Collado, V., Esteve, M., Palau, C.E.: AAL open source system for the monitoring and intelligent control of nursing homes. In: Proceedings of the 2017 IEEE 14th International Conference on Networking, Sensing and Control, ICNSC 2017, pp. 84–89. Institute of Electrical and Electronics Engineers Inc. (2017). https://doi.org/10.1109/ICNSC.2017.8000072
15. Gonzalez-Usach, R., Palau, C.E., Julian, M., Belsa, A., Llorente, M.A., Montesinos, M., Ganzha M., Wasielewska K., Sala, P.: Use cases, applications and implementation aspects for iot interoperability. In: Distributed Intelligence at the Edge and Human Machine-To-Machine Cooperation, River Publishers, pp. 139–173 (2018)
16. Molina, B., Palau, C.E., Fortino, G., Guerrieri, A., Savaglio, C.: Empowering smart cities through interoperable sensor network enablers. In: Conference Proceedings - IEEE International Conference on Systems, Man and Cybernetics, vol. 2014, pp. 7–12. Institute of Electrical and Electronics Engineers Inc. (2014, January). https://doi.org/10.1109/SMC.2014.6973876
17. van der Veer, H., Wiles, A.: Achieving Technical Interoperability—the ETSI Approach. In: Standards for Business. European Telecommunication Standards Institute (ETSI). Available via ETSI (2008). https://www.etsi.org/images/files/ETSIWhitePapers/IOP%20whitepaper%20Edition%203%20final.pdf. Accessed 27 Jan 2021
18. Gubbi, J., Buyya, R., Marusic, S., Palaniswami, M.: Internet of Things (IoT): a vision, architectural elements, and future directions. Future Gener. Comput. Syst. **29**(7), 1645–1660 (2013). https://doi.org/10.1016/j.future.2013.01.010
19. Ganzha, M., Paprzycki, M., Pawłowski, W., Szmeja, P., Wasielewska, K.: Semantic interoperability in the Internet of Things: an overview from the INTER-IoT perspective. J. Netw. Comput. Appl. **81**, 111–124 (2017). https://doi.org/10.1016/j.jnca.2016.08.007

20. Pileggi, S.F., Fernandez-Llatas, C.: Semantic Interoperability : Issues, Solutions, and Challenges. River Publishers (2012)
21. Elluri, L., Joshi, K.P.: A knowledge representation of cloud data controls for EU GDPR compliance. In 2018 IEEE World Congress on Services (SERVICES) (2018)
22. Hodges, J., García, K., Ray, S.: Semantic development and integration of standards for adoption and interoperability. Computer **50**(11), 26–36 (2017). https://doi.org/10.1109/MC.2017.4041353
23. Venceslau, A., Andrade, R., Vidal, V., Nogueira, T., Pequeno, V.: IoT semantic interoperability: a systematic mapping study. In: Proceedings of the 21st International Conference on Enterprise Information Systems, pp. 535–544. Heraklion, Crete, Greece: SCITEPRESS - Science and Technology Publications. (2019). https://doi.org/10.5220/0007732605350544
24. Ganzha, M., Paprzycki, M., Pawłowski, W., Szmeja, P., Wasielewska, K.: Towards Semantic Interoperability Between Internet of Things Platforms, pp. 103–127. Springer, Cham (2018). https://doi.org/10.1007/978-3-319-61300-0_6
25. Kalamaras, I., Kaklanis, N., Votis, K., Tzovaras, D.: Towards big data analytics in large-scale federations of semantically heterogeneous iot platforms. In IFIP Advances in Information and Communication Technology, vol. 520, pp. 13–23. Springer New York LLC (2018). https://doi.org/10.1007/978-3-319-92016-0_2
26. Gyrard, A., Serrano, M.: A unified semantic engine for Internet of Things and smart cities: from sensor data to end-users applications. In: 2015 IEEE International Conference on Data Science and Data Intensive Systems, pp. 718–725 (2015). https://doi.org/10.1109/DSDIS.2015.59
27. Gyrard, A., Zimmermann, A., Sheth, A.: Building IoT-based applications for smart cities: how can ontology catalogs help? IEEE Internet Things J. **5**(5), 3978–3990 (2018). https://doi.org/10.1109/JIOT.2018.2854278
28. Gyrard, A., Datta, S.K., Bonnet, C.: A survey and analysis of ontology-based software tools for semantic interoperability in IoT and WoT landscapes. In: 2018 IEEE 4th World Forum on Internet of Things (WF-IoT), pp. 86–91 (2018). https://doi.org/10.1109/WF-IoT.2018.8355091
29. Ganzha, M., Paprzycki, M., Pawłowski, W., Szmeja, P., Wasielewska, K.: Semantic interoperability in the Internet of Things: an overview from the INTER-IoT perspective. J. Netwo. Comput. Appl. **81**, 111–124 (2017). https://doi.org/10.1016/j.jnca.2016.08.007
30. Ganzha, M., Paprzycki, M., Pawlowski, W., Szmeja, P., Wasielewska, K.: Semantic technologies for the IoT—an Inter-IoT perspective. In: Proceedings—2016 IEEE 1st International Conference on Internet-of-Things Design and Implementation, IoTDI 2016, pp. 271–276. Institute of Electrical and Electronics Engineers Inc. (2016b). https://doi.org/10.1109/IoTDI.2015.22
31. Strassner, J., Diab, W. W.: A semantic interoperability architecture for Internet of Things data sharing and computing. In: 2016 IEEE 3rd World Forum on Internet of Things (WF-IoT), pp. 609–614 (2016). https://doi.org/10.1109/WF-IoT.2016.7845422
32. Davarzani, H., Fahimnia, B., Bell, M., Sarkis, J.: Greening ports and maritime logistics: a review. Trans. Res. Part D: Transp. and Environment **48**, 473–487 (2016). https://doi.org/10.1016/j.trd.2015.07.007
33. Iftikhar, S., Ali, W., Ahmad, F., Fatim, K.: Semantic interoperability in E-health for improved healthcare. In: Semantics in Action—Applications and Scenarios. InTechOpen (2012). https://doi.org/10.5772/36469
34. Adel, E., El-Sappagh, S., Barakat, S., Elmogy, M.: Ontology-based electronic health record semantic interoperability: a survey. In: U-Healthcare Monitoring Systems, pp. 315–352. Elsevier (2019). https://doi.org/10.1016/b978-0-12-815370-3.00013-x
35. Vermesan, O., Bacquet, J.: Cognitive Hyperconnected Digital Transformation: Internet of Things Intelligence Evolution, Riverpublishers, Denmark (2017)
36. Gonzalez-Usach, R., Esteve, M., Palau, C.: Interoperable dynamic lighting for port terminal containers. In TRA 2018 (2018)

37. Yacchirema, D., Gonzalez-Usach, R., Esteve, M.: Interoperability of IoT Platforms applied to the transport and logistics domain. In: Transport Arena Research 2018 (TRA2018), Vienna, 16–19 April 2018 (2018). https://doi.org/10.5281/ZENODO.1451428
38. Ganzha, M., Paprzycki, M., Pawłowski, W., Szmeja, P., Wasielewska, K.: Towards common vocabulary for IoT ecosystems—preliminary considerations. In: Asian Conference on Intelligent Information and Database Systems (ACIIDS 2017), Kanazawa, 3–5 April 2017. Lecture Notes in Computer Science, vol. 10191, pp. 35–45. Springer, Berlin (2017). https://doi.org/10.1007/978-3-319-54472-4_4
39. Moreira, J., Daniele, L., Ferreira Pires, L., Van Sinderen, M., Wasielewska, K., Szmeja, P., Pawlowski, W., Ganzha, M., Paprzycki, M.: Towards IoT platforms' integration: Semantic Translations between W3C SSN and ETSI SAREF (2017)

IoT in Provenance Management of Medical Data

Gennady Chuiko, Yaroslav Krainyk, Olga Dvornik, and Yevhen Darnapuk

Abstract In this chapter, we propose to investsigate the applicability of semantics in the context of Internet-of-Things (IoT) to trace the origins of medical data. As IoT-devices have become the first-order source of information in the field of healthcare in various systems, the challenge of correctness and reliability of retrieved data is becoming of tremendous importance. This challenge is directly connected with the quality of patient monitoring and treatment because the decision on the patient's state is made according to the set of measured parameters. Inaccuracy and low quality of measurements that may be caused by sensor malfunction, incorrect measurement procedure, etc. can lead to problems with comprehension of the current situation and affect further decisions. The photometric calibrating curves of Melatonin-sulfate in human urine were considered as a case-study. The Hill's equation was used for 'dose–response' relationship. The photometric calibrating graphs of Melatonin-sulfate in human urine were considered as a case study. Hill's equation imaged the 'dose–response' relation. The photometric transmittance of analyzed solutions was the response signal. The ordinary photometry of human urine can be in use as the simple ex-press-analysis of melatonin instead of expensive analyzes. If, sure, the accord-ant calibrators are reliable. The existing set of such calibrators yet unable warrants the trusty calibrating. Thus, the medical photometry of urinary Melatonin-sulfate is yet out of extensive use. The problem of reliable calibrators is mostly in the provenance of data.

Keywords Medical data · Semantic · IoT · Provenance · Ontology

G. Chuiko · Y. Krainyk (✉) · O. Dvornik · Y. Darnapuk
Petro Mohyla Black Sea National University, 68 Desantnykiv, 10, Mykolaiv 54003, Ukraine
e-mail: yaroslav.krainyk@chmnu.edu.ua

R. Pandey et al. (eds.), *Semantic IoT: Theory and Applications*, Studies in Computational Intelligence 941, https://doi.org/10.1007/978-3-030-64619-6_15

1 Introduction

The amount of medical data generated by Internet-of-Things (IoT) devices is constantly growing. Transition to electronic protocols and web-based systems for healthcare management allowed the appearance of complex tools for patient treatment, monitoring, on-line consultations, and other applications. Such systems store, process, and devise significant loads of information. While the storage of data is significant, the most important outcome is recommendations generated as a result of the analysis of input information. One of the instruments to improve recommendations is semantic technologies.

The primary goal of semantic technologies [1–4] is to provide new level of abstraction for machine data understanding. This empowers novel techniques for data interpretation and advanced means for recommendations. On the basic level, inclusion of semantic technologies is marked by annotation of incoming information and further processing using semantic engines [5, 6]. Annotated data have better level of meaningfulness and introduce additional interface for extended analysis.

Another problem that semantic technologies are trying to overcome is interoperability [6, 10–14, 20]. IoT itself is primarily characterized by high degree of heterogeneity. Devices employ different means for information transfer, different protocols, expose specific Application-Programming Interfaces (API) during their work, etc. [7]. All of these facts entail the issue of interoperability between IoT-devices and all other components. For instance, usage of two different communication protocols in temperature sensor node demands coordination on the level of server or gateway to observe measurements from both sensors simultaneously. Both hardware and software layer are concerned in this case [18].

It also has direct implications for medical IoT-devices as products from various vendors might be incompatible with a specific system, require separate application for monitoring, etc. As has been mentioned above, semantically annotated data can serve as an additional interface. From the point of view of interoperability, measurements from different devices can be aggregated to form a data frame about current patient state.

The final touching point of the presented paper is provenance [15–17] of medical data generated by IoT-devices. It is important to note that medical data should always be taken into account in combination with other factors e.g., previous measurements, observed patient state, patient's health card, parameters of device itself, etc. That creates the possibility to improve Quality-of-Service (QoS) for end users. For instance, that could be an early identification of the disease and prevention from its further development. Annotation of medical data with semantic information, in the following context, is an enabler of the mentioned functions.

At the same time, semantic annotation means are typically not available on the level of the device. They are deployed on the higher level of medical monitoring system in forms of software services.

This paper is organized as follows. The next section introduces review of contemporary scientific sources on the latest developments in the field of semantic technologies, IoT, and healthcare. Then, according to the analysis, we devise general recommendations for semantic-based healthcare system that utilizes IoT data. Finally, we demonstrate the use-case of data provenance for Melatonin-sulfate measurements.

2 Review of Related Work

In his paper [8], T. Burners-Lee envisioned Semantic Web as Web 3.0 and outlined the most significant advantages of inclusion of semantic data into existing web description. The leading outcome of this should be a transition in inter-machine communication that can be established on the semantic level rather than explicit communication using software commands. However, the pace of this innovation is not enough to declare semantic web as a ubiquitous technology and new developments are yet to come. The main culprit for this is, once again, a heterogeneity of approaches and semantic frameworks.

On the other side, valuable achievements provided in scientific researches and practical applications cannot be denied. First, standards for semantic data representation were developed. They include Resource Description Framework (RDF), Web Ontology Language (OWL), SPARQL Protocol, and RDF Query Language (SPARQL) among the most useful ones. They will be discussed later. Second, complex semantic frameworks and systems were designed to address problems of data management, analysis, system control in various fields. Healthcare system is not left overlooked and multiple research works are devoted to this field.

RDF constitutes basics for the storage and linking of semantics. All information described in RDF-format is stored as "triples". Triple consists of three elements:

- Subject
- Relationship
- Object.

Single triple defines how subject and object relate to each other. Element defined in one triple can be referenced from another one. Complex networks or graphs are formed this way. Being quite simple data representation format, triples provide a powerful instrument to define specific knowledge or activity domain when combined into a graph structure.

OWL is facilitates creation of ontologies [9]. While it is very similar to the RDF, OWL can be regarded as a higher-level entity that general rules for the specific domain.

SPARQL is a query language with SQL-like syntax designed specifically to query data stored in RDF and OWL-files. SPARQL is a key software component for processing of semantic data. SPARQL-queries can be stored as system rules that further evolves into knowledge extraction and analysis mechanism.

Despite the fact that huge number of ontologies have been developed and this process is still ongoing, they are facing a shortcoming of compatibility lack with each other. There is a tendency that when one ontology seems to be describing all necessary aspects of a domain needs to be combined with another one, multiple contradictions might be found. The mapping of instances from both domain to a third "unification" domain can solve it so they can exchange data without conflicts. However, that leads to development of the new ontology. Even one of the most well-known schemes, Semantic Sensor Networks (SSN), contains the most common types of information, does not overlap sensor domain completely and new ontologies are constantly appearing. The functionality available in the general-purpose framework is not enough to fulfill the solution for the narrow domain. At the same time, domain-specific ontologies will be appearing constantly to extend higher-level ones.

Project Ontobee from University of Michigan stands out among projects for medical ontologies. It concentrated an array of ontologies that relate to the healthcare. They contain terms about the disease, physical state, biological, biochemical, drug description, etc. The list is not limited to the mentioned domains and comprises 208 ontologies. They can be used in further developments that involve annotation of medical data in IoT system. However, while it does contains ontologies that overlap IoT system, there is not ready-to-use ontology for provenance of medical data.

We assume that at this point this development cannot be unified up to some level. Either way single ontology is not able to take into consideration all the peculiarities that may appear in the future. Hence, the ontology should provide extension points where new artifacts can be added. That implies the presence of abstract entities that serve as a base for the development of new types and relations.

One of the biggest semantic-related projects is M3 project [5, 14]. It consolidates contemporary researches and their software artifacts with approximately 550 ontologies collected. M3 also establishes its own solutions to common problems in the field of IoT. Speaking precisely, it is just one of the elements of semantic infrastructure. It covers all dataflow for sensor processing and even supports the generation of software for the specific sensor system. The software includes an application for acquiring sensor data, higher-level user interfaces, reasoning engine, and storage of the data.

The part of the M3 project that collects ontologies is Linked Open Vocabularies for IoT (LOV4IoT). The catalog is dynamic and constantly includes novel findings described in scientific researches with a link to the source code repository. The main goal of the subproject is to link existing vocabularies with each other and to establish fundamentals for the usage of the vocabularies. From the point of view of healthcare, the ontologies categorized as:

- General-purpose healthcare ontologies;
- Ambient assisted living;
- Wearable ontologies;
- Emotions ontologies;
- Activity recognition in smart home;
- Nutrition-related ontologies;

- Depression;
- Asthma;
- Obesity.

The review of general-purpose ontologies shows that LOV4IoT assembled 67 ontologies (the highest value among all catalogs) developed in the period from 2006 to 2018. It is also shown that proposed systems or models employ a diverse set of technologies. They vary by wireless connections, communication protocols, operating systems, target end-user devices, supported sensors, and deployment into computational infrastructure. Often projects from the catalog deal only with a single specific problem, e.g., Electro-Cardiograms (ECG) and their inclusion into other structure might be concerned with significant efforts. That fact, once again, proves that to cope with heterogeneity of the ontologies, it is necessary to provide extension interfaces so they can be reused for different purposes.

Researches on the topic of healthcare that involve IoT and semantic tools provide huge advances for the quality of life of patients. However, the observation regarding these researches is that they typically deal with single patient types having a particular type of disease (e.g. dementia, heart diseases) or peculiar data type for analysis (e.g. ECG). Therefore, a combination of the findings into a universal system is concerned with additional expenses of development time. Analysis of the resources also allows making an assumption about an architectural solution where devices can send data to the cloud environment and further activities take place in the cloud. The feedback is provided as a result of the analysis of the cloud tools or by an expert himself/herself.

Electronic healthcare systems exploit different data formats for storage and communications. When it comes to annotation of the raw data, it is necessary that data representation should support addition of semantic data at least as an ad-hoc function. If the protocol message cannot be altered with semantic data, it makes it unfeasible to use this protocol in the core of semantic system. Now, let us provide the review of the common data formats for healthcare systems and analysis of their suitability for mentioned purposes.

The most common problem associated with this process is that it increases overall size of the message that is important for lightweight applications. In more details attachment of the information to medical data is discussed hereinafter.

3 Medical Data Provenance

Despite the fact that semantic-enabled healthcare systems are one of the most common examples of technology application, the problem of medical data provenance has not gained necessary attention yet.

As data provenance is a general problem not only for the healthcare system, but also for other scientific fields that that face bulk amount of information, multiple endeavors were made in this direction. First, the W3C organization issued PROV Ontology (PROV-O) to deal with common issues of data provenance. The ontology establishes a basic framework to annotate data regardless of the source. However, the pitfall of the generic ontologies is that they lack tools for the specific domain. In this situation, it may cause not fully annotated data that notably limits possibilities for inference engine as multiple parameters are omitted and not provided to the system [19].

To provide further overview of the state-of-the-art technologies, first, we consider dataflow of medical data in general healthcare system, then, discuss types of medical data formats and, finally, provide our vision and possible solution to the problem of medical data provenance.

3.1 Dataflow of Medical Data

Let us first describe the general semantic-enabled information system with medical data. Perception layer is represented by devices that can capture physical parameters. They enlist devices that perform measurements in specific interval as well as devices that constantly observe state of the human. We also differentiate mass market wearable devices (e.g. smart watches with enriched functionality to monitor physical state) and specific medical sensors that is responsible for precise information capturing. This factor is directly concerned with provenance of the medical data. Medical data are submitted from multiple IoT-agents in concurrent manner. It makes possibly to aggregate data in a single storage and investigate them to devise complex dependencies between them.

At the edge level, devices support preliminary data processing and preparation for further transmissions to the top layers. Edge devices also facilitate storage capabilities and can even be used for the initial data annotation. However, this point of interest should be meticulously devised because it notably affects further flow of the data. To be precise, this situation is concerned with situation when decision about physical state is made on the early processing stage. While simple inference from the sensor measurements might demonstrate no problems, top level processing engine can generate the recommendation that contradicts to the previous one. It will result in ambiguous state of the system.

On the other side, edge layer is suitable for diagnosis of sensor devices. Edge-located computers are supposed to exchange diagnosis messages with sensors to control their state.

In opposite to streaming applications where decisions are generated exclusively regarding stream state with analysis of data on the interval of time, presence of the storage layer is obligatory for the semantic-enabled system. Availability of the storage layer provides an opportunity to improve semantic rule engine on the whole space of available data. As this paper deals with semantic data, we assume that storage is represented by sets of RDF-tuples with annotations added immediately before the write operation. Apparently, it entails that only basic information is available for the single sensor measurement. As soon as necessary samples from the sensor are collected, they can be regarded as an instance for learning, inference, and rules generation.

Annotations for incoming data are derived from the level of ontologies. According to that fact that system is supposed to be heterogeneity-agnostic, ontology layer is combined of multiple ontologies that may be connected with each other either via intermediate ontologies or independent ones responsible for separate field of knowledge.

Regarding the problem of data provenance, most notions will be established at the start of the system with minimal corrections during its work. Provenance notions can be placed into a single ontology and exploited when initially annotated RDF-tuples are stored to the database.

Abovementioned semantic rule engine is represented by SPARQL-queries. The queries incorporate application logic and form a core of the system from the point of view of application processing. Specific query is triggered whenever data related to its parameter arrive to the storage. For instance, let us consider an example where the patient reports that blood pressure test is performed. Storage engine writes this parameter to the RDF-store with time parameter. Then, corresponding rule is triggered that checks if it is performed in the recommended time interval. Violation of the recommended interval boundaries results that customer receives notification about this event.

As has been mentioned earlier, rules can be modified and added by an expert. In our case, a doctor plays a role of the expert. He/she accesses system's data via dedicated interface. While, in most cases, the system runs in autonomous mode, observation from the doctor's side is fundamental and control from the expert prevents situations when inference engine might cause malfunction or other severe consequences. Therefore, the doctor should approve each generated rule first.

The schematic presentation of semantic medical system with all mentioned components included is depicted in Fig. 1.

In this system, provenance of medical data is possible to establish on different system levels for different cases involving actors of the system. It has to provide a service that indicates wrong conclusions from correct data and prevents usage of incorrect data to make decisions. It is especially critical for the healthcare system responsible for the state of patients.

3.2 Medical Data Formats

Considering provenance of medical data, it is important to take into consideration presentation of medical data.

Obviously, that text-based formats are the preferred choice for usage in semantic system. As semantic origins imply that text information is prevalent in the system and semantic annotations themselves are performed in text format, it is easy to combine such representation with annotations in a meaningful. However, addition of the new

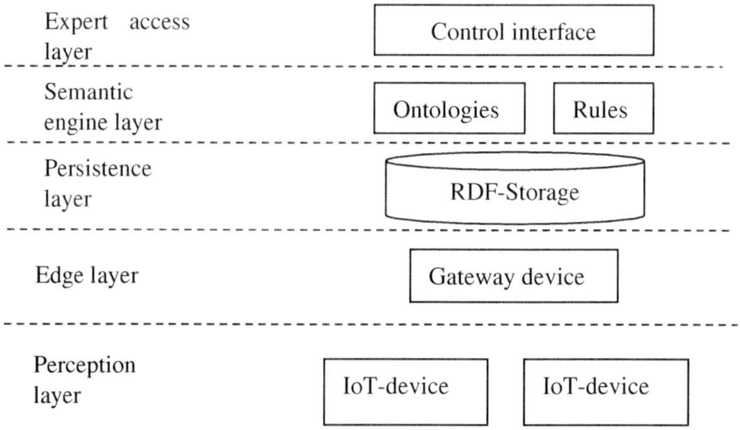

Fig. 1 General structure of semantic-enabled medical system

information into the message that is to be transmitted is always concerned with consistency in communication layer. Due to the fact that most IoT-systems are not designed for such modifications, the system should integrate annotation of data on higher levels.

The first data format to mention is Comma-Separated Values (CSV) format. It is a wide-spread format exploited in multiple areas of information systems. CSV presumes that stored values are separated by comma character and processing systems accesses necessary token by traversing string and cutting substring from the main string. On the other side, CSV is a sequential format and analysis of CSV file with further embedding of additional information is hindered by search over the file content. The process of embedding information itself is completely straightforward. In the final analysis, most devices employ other protocols that are more oriented on presentation of medical events.

One of the most well-known data formats for the field of healthcare is HL7 (Health Level Seven International). Speaking precisely, it is not only a data format, HL7 is a set of standards for development of information systems in the field of healthcare. It serves as a solution for multiple types of medical data transfer (e.g. HL7 aECG for exchange of ECG data). HL7 is based in on eXtensible Markup Language (XML) and, therefore, also belongs to the type of text formats. However, the main drawback of HL7 even though it is widely adapted and surely can be recommended for inclusion into architecture, is extensibility.

The shortcomings of HL7 protocol lead to the emergence of its improved version HL7 Fast Healthcare Interoperability Resources (FHIR). It allows mixing and adaptation to peculiar clinical context.

While it is not the main point of interest of this work, medical infrastructure also heavily relies on image information that is far more complex for automatic annotation than text data. The list of medical image data formats includes IntefFile, Analyze,

Nifti, Minc, DICOM. Addition of semantic data for such images is simply a linking between file itself and corresponding information from the ontologies and rule base.

3.3 *Analysis of the Proposed Solution for Medical Data Provenance*

First, let us state that on the lowest level the source of the erroneous data may be identified as

1. IoT-device.
2. The patient whose actions has direct effect on the received measurements. Execution that is not compliant with device instructions is an often reason for this event.

To extend the nomenclature of the data provenance, we consider IoT-device. Despite being a relatively simple computational device, it combines multiple technologies that should be taken into account. The following list includes general explanations of errors on the device level:

1. Sensor level (sensing element is not capturing data properly on the declared sensitivity range with necessary precision or other sources of erroneous data).
2. Hardware level (processing device cannot capture data and transfer them to the destination).
3. Power supply level (connected with two previous levels as can cause errors for both of them).
4. Software level (includes various aspects, e.g., protocol failure, software inconsistency, failures due to the specific software environment states, etc.).
5. Communication level (appears during transmission phase and partially dependent on the selected communication protocol).

This list is sufficient to provide control over most use-cases under consideration. We would like to devote particular attention to the sensor level malfunction. To denote sensors state we propose the following classification:

1. Sensor is in normal state under normal circumstances.
2. Sensor acquires data that is biased from the previous history and the patient is aware that conditions are normal.
3. Sensor is capturing data with output values located out of the limit on the sensitivity range (applied impact is not suitable for current sensor).
4. Sensor acquires a priori incorrect data under known conditions.
5. Sensor cannot capture a sample.

The second case attracts the most attention, as it is common for the most cases and relates to both correct and incorrect data. Slight bias from the measured value might indicate further trends in the patient's physical state. Hence, this information is helpful for prevention of possible negative impacts. At the same time, small discrepancy in amplitude value is not necessarily means cautious state. Therefore, it allows informing patient in advance about possible outputs according to sensor measurements and to recommend preventive activities to him/her.

In opposite to the device level, generalization of the patient's actions that affect medical data a priori cannot include all possible options. Moreover, automatic retrieval of this information is not possible unless IoT-device supports this feature. Therefore, this part should also facilitate interface when the patient adds evidence about improper device usage. The user enters feedback about the incorrect measurements and also marks time interval when data are probably recorded with these circumstances. The result of this activity is decreased trustworthiness of the data recorded during the marked period. That also entails that the data cannot be used during the inference process or, at least, their impact is not so strong.

The case under consideration is tightly connected with capabilities of semantic technologies, IoT-devices specification, and their limitations. The challenge that arises when patient is permitted to enter his own explanations is that semantic reasoning is based on the direct match between tokens in the patient's response and data from ontologies. If no match is found, no semantic data can be linked to this message automatically and it requires manual adjustment. Viable solution for this challenge is a list with options to select to explain the reason of failed measurement. However, it should be designed for each sensor/device specifically and even list of these options have to contain option that allows entering reasons that have not been predicted during list preparation. Information fetched from the individual sensor has its own peculiarities and its interpretation should strictly follow documentation.

Nevertheless, there is still probability that some data are left not annotated and excluded from queries and inference. To overcome this issue we propose a procedure when doctor annotates data manually and links are created as a result of initial manual augmentation.

On the other side, we can face situation when single sensor measured value is actually correct and it indicates cautious physical state of the patient. This scenario is critical for consideration and automatic denial of the sample could lead to critical consequences. Thus, we propose to exploit the notion "scenario" that describes possible consequences from the analysis of the retrieved data. In this case, multiple concurrent scenarios exist in the system. At least one of them marked as "critical" scenario depicts situation where action should be applied immediately. Typically, two parallel scenarios will be created. The second case assumes that data might be received from erroneous source and further analysis is required.

Critical scenario also means that doctor or medical surveillance service are notified about the possible patient's state. At the same time, patient is also notified about his/her critical state and has an option to approve or deny this fact. Hence, two-side communication mechanism is required by the system. Possibility of such scenario is the main reason why system design allows non-autonomous communication between patient and doctor. In our opinion, this indispensable option needs to be present in the medical healthcare system. Even though semantic technologies significantly enrich capabilities of the system, they cannot substitute experts experience completely or to avoid incorrect judgements according to limited information space.

Another variant for implementation of critical scenario processing is its deployment on the edge-device level without reactive actions required from the core system. However, as edge device observes only initial signal they only provide response according to the markers embedded in the software (e.g. hard-coded values of recommended sensor output boundaries where significant bias enforces pre-programmed actions).

The final level of the data provenance for the healthcare system is consolidated entirely in a semantic engine. As data is stored in the persistence layer, obtained information serves as a fundamental basis for semantic provenance management.

To demonstrate the importance of the semantic system design for medical data provenance, we investigate the case of measurements of Melationin-Sulfate level.

4 Provenance of Data and Reliability of Calibrators of Melatonin-Sulfate: Case Study

The health problems bound with the different breaks of circadian rhythms have 'a common root'—the melatonin [21]. Many of the widespread diseases such as cancer, metabolic function disorders, diabetes, cardiovascular system failures are tied to the abrupt phase shift level of melatonin as it is pointed out in review [21].

Melatonin called a "darkness hormone" is a versatile pineal hormone. Melatonin is controlling many physiological processes of humans and mammals generally [22]. The melatonin is affecting human physiology as a whole, regulates the sleep–wake cycle, via so-called the body's internal clock [21, 23]. The presence of melatonin in saliva or blood plasma, as well as the infiltration of the melatonin metabolite into the urine, are well-known facts [22, 23].

Enzyme-linked immunosorbent assays (ELISA) are the recognized, precise modern technique for the melatonin test in the urine. The reader can see it from early reports [24–27] as well as recent papers [28, 29]. That quite complicated test gives the presence and concentration of melatonin metabolite in urine exactly.

The use of calibrators is frequently the best solution in similar situations. The paper [30] comprises a detailed guide on how to use a few calibers, each one with the well-known dose of melatonin, to get a calibration graph. Authors of [30] not considered their results from the point of the data provenance. Meantime, its reliability is depending on that factor.

We are going to show and spread here the other point of view on the medical data. This approach takes into account the provenance of data based on modern technology of its obtaining, transmitting, and processing in conditions of real competencies of medical staff.

If one has a calibration graph (that is the "dose–response" curve), then the response of the sample sets the melatonin concentration (dose). The response, for instance, transmittance or the optical density of the sample, getting by photometer, allows the single finding of melatonin dosage [24–29, 31, 32]. Such inexpensive and fast probing can give a chance to less wealthy clinics and patients.

The well-known Hill's logistic curve is mostly in use for calibration graphs of a dose–response kind [33, 34]. There are options with four or five parameters, but the simpler one is more popular. We will revise here in the results [30] completed by more recent data [28, 29]. We intend to trace the effect of the provenance of data on the parameters of the calibration graph.

Calibers were samples of metabolite melatonin solutions in the urine. The definition of the concentration in samples has performed via two methods. The first of them was enzyme-linked immunosorbent assays (ELISA) [24–29]. Another was radio-immunoassay (RIA) that well-known as a highly accurate assay trial [31, 32].

The typical concentrations of melatonin metabolite in urine match the diapason (0–420) ng/mL. The usual numbers of calibers in a set are up 6–8 [24–29, 31, 32]. One can find the description of Hill's equation in [34]:

$$Y = \frac{a + (b - a)}{1 + \left(\frac{c}{x}\right)^d} \tag{1}$$

Here Y is the response matching the dose x; a, b, c, d are four parameters.

Parameter a is an asymptotical response at the condition $x \to \infty$. Another asymptote b is the stabilized response at the condition $x \to \infty$. Parameter c sets the inflection of the dosage-response curve, and it has various terms (e.g., EC50, ED50, LD50, IC50). The parameter d is the so-called Hill's slope [34]. Figure 2 shows Hill's graph with guidelines about parameters.

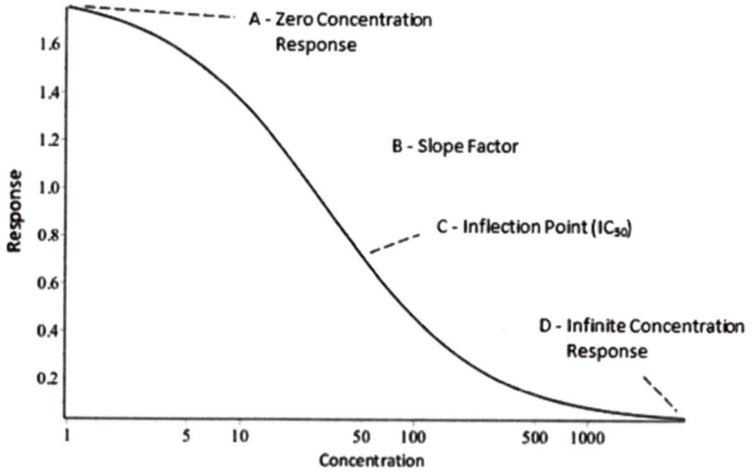

Fig. 2 Hill's logistic curve with 4 parameters [31]

Hill's equation can fit either the descending or ascending dependences. It depends on the sign of the Hill's slope ($d <> 0$). One has expected the descending trend, so and the negative slope [30, 33].

There are different ways of computing of Hill's parameters [33, 34]. Here is in use of the method [33] and the opportunities of the program package 'Statistics' of Maple 18.

Just a few different calibrators were used in laboratory investigations, as we said above. Two of them were described in [24–30], while others in [28, 31, 32]. RIA tested the calibrator of [31, 32], others—by ELISA [24–29].

Figure 3 shows a few dosage-response graphs. The optical transmittance of the analyte (B/B_0) serves here as the response. As one can see, all calibration curves are descending (that is d < 0 as it forecasted above).

Table 1 presents the complete set of computed Hill's parameters for each calibrating curve (that is each separate calibrator).

Figure 3 and Table 1 point out the notable divergence between data with a different origin. Compare data [24–27, 33] from first row of Table 1 and data [28, 31, 32] in other rows. The parameter c (also called EC50, ED50, LD50, IC50 etcetera) mostly makes this divergence. Its value is about six-time greater for the data of the first row than the others. That looks as "overly" for those accurate methods that were in working. The reasons for the mismatch are still hard to explain.

The b and d parameters not bad agreed among data of different origins. They have sensible and predictable magnitudes. The b value was near to 100, and the d was negative. However, with the recent data [29], this coefficient differs from the results [30], and that divergence also seems as unexpected enough.

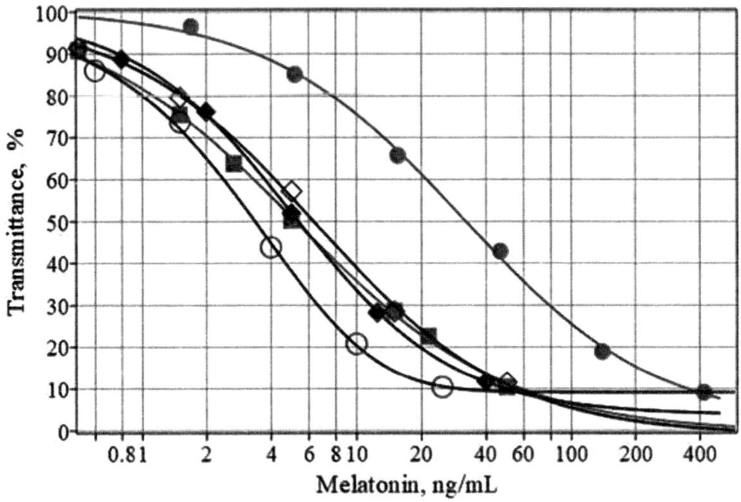

Fig. 3 The response-dosage curves: grey solid circles show the calibers of [24–27, 33]; the solid diamonds present data [28]; the grey sold boxes present data [31]; the circles present data [29] and diamonds—data [32]

Table 1 Hill's parameters

Method	A	B	c	d	Data source
ELISA	1.3	100.7	30.9	−0.96	[24–27, 33]
ELISA	3.7	99.7	5.1	−1.17	[28]
RIA	−0.5	100.7	5.2	−0.89	[31]
ELISA 2019	2.7	99.7	5.2	−1.03	[29]
RIA	−1.1	99.5	6.5	−0.98	[32]
Mean	2.3	100.1	10.6	−1.00	
Standard deviation	4.1	0.6	11.4	0.10	

Pay attention, the calibrator and the way of calibrating (ELISA) are the same in these as though independent sources [24–28, 33]. In contrast, the methods (ELISA and RIA), as well as the sets of calibers, are various for data [28, 31, 32]. The unchanged here is the provenance from one laboratory. The results of one origin have good agreement between themselves.

We believe clinic decision making strictly depends on data provenance. It is especially right in IoT because many medical devices may be connected to the clinic database. Medical staff and a patient should not doubt the reliability of the initial data and efficacy of the clinic decision. Under these conditions, the question of standardization of the device information such as its reliability, security, provenance, and acceptance ones need to accompany medical data.

Therefore, each medic has to account for the provenance of used data. This demand must be mandatory, despite their real competence in the computer and data science. Meantime, the estimation of reliable and less reliable data and origins of data is the field of expertise for medical data scientists.

We reckon the more or less reliable calibrators of urinary melatonin metabolite are possible now, basing on the data [28, 31, 32]. So, the question of our early study [30] tends to the gradual closure with the accounting of the data provenance.

5 Conclusions

The model of the semantic-based system for medical data provenance has been proposed in this paper. We revised the whole set of available technologies to employ in the semantic engine and analyzed its behavior under different circumstances. We identified that IoT-devices sending sensor information to the main processing system can be regarded as the main source of data to control. Therefore, the provenance of medical data using semantic instruments is one of the possible solutions. We considered the use-case of Melatonin-Sulfate measurements and identified that different measurement approaches lead to significant bias in overall results which can be identified by a semantic-enabled engine for healthcare data provenance.

Acknowledgements This investigation has been performed within the framework of the topic "Development of hardware and software complex of non-invasive monitoring of blood pressure and heart rate for dual-purpose usage" (registration number 0120U101266) supported by Ministry of Education and Science of Ukraine.

References

1. Nambi, S.N.A.U., Sarkar, C., Prasad, R.V., Rahim, A.: A unified semantic knowledge base for IoT. In: 2014 IEEE World Forum on Internet of Things, WF-IoT 2014. pp 575–580. https://doi.org/10.1109/WF-IoT.2014.6803232
2. Mishra, N., Chang, H.T., Lin, C.C.: An IoT Knowledge reengineering framework for semantic knowledge analytics for BI-services. Math. Probl. Eng. (2015).https://doi.org/10.1155/2015/759428
3. Bonte, P., Ongenae, F., De Turck, F.: Generic semantic platform for the user-friendly development of intelligent IoT services. In: CEUR Workshop Proceedings, pp. 79–90 (2016)
4. Seydoux, N., Drira, K., Hernandez, N., Monteil, T.: Capturing the contributions of the semantic web to the IoT: a unifying vision. In: CEUR Workshop Proceedings (2017)
5. Serrano, M., Gyrard, A.: A review of tools for IoT semantics and data streaming analytics. In: Building Blocks for IoT Analytics Internet-of-Things Analytics, pp. 139–166 (2017)
6. Mishra, S., Jain, S.: Ontologies as a semantic model in IoT. Int. J. Comput. Appl. **40**, 1–18 (2018). https://doi.org/10.1080/1206212X.2018.1504461
7. Krainyk, Y., Davydenko, Y., Tomas, V.: Configurable control node for wireless sensor network. In: 2019 3rd International Conference on Advanced Information and Communications Technologies, AICT 2019—Proceedings, pp 258–262 (2019)
8. Berners-Lee, T., Hendler, J., Lassila, O.: The semantic web. Sci. Am. **284**, 34–43 (2001)
9. Szilagyi, I., Wira, P.: Ontologies and semantic web for the internet of things—a survey. In: IECON Proceedings (Industrial Electronics Conference), pp 6949–6954 (2016)
10. Murdock, P.: Semantic Interoperability for the Web of Things. ResearchGate, pp. 1–19 (2016). https://doi.org/10.13140/RG.2.2.25758.13122
11. Jabbar, S., Ullah, F., Khalid, S., et al.: Semantic interoperability in heterogeneous IoT infrastructure for healthcare. Wirel. Commun. Mob. Comput. (2017).https://doi.org/10.1155/2017/9731806
12. Mazayev, A., Martins, J.A., Correia, N.: Interoperability in IoT through the semantic profiling of objects. IEEE Access **6**, 19379–19385 (2017). https://doi.org/10.1109/ACCESS.2017.2763425
13. Bajaj, G., Agarwal, R., Singh, P., et al.: 4W1H in IoT semantics. IEEE Access **6**, 65488–65506 (2018). https://doi.org/10.1109/ACCESS.2018.2878100
14. Gyrard, A., Datta, S.K., Bonnet, C.: A survey and analysis of ontology-based software tools for semantic interoperability in IoT and WoT landscapes. In: IEEE World Forum on Internet of Things, WF-IoT 2018—Proceedings. pp 86–91 (2018)
15. Hartig, O.: Provenance information in the Web of data. In: CEUR Workshop Proceedings (2009)
16. Alkhalil, A., Ramadan, R.A.: IoT data provenance implementation challenges. Proc. Comput. Sci. **109**, 1134–1139 (2017)
17. Olufowobi, H., Engel, R., Baracaldo, N., et al.: Data provenance model for internet of things (IoT) Systems. In: Lecture Notes in Computer Science (Including Subseries Lecture Notes in Artificial Intelligence and Lecture Notes in Bioinformatics), pp 85–91 (2017)
18. Krainyk, Y., Razzhyvin, A., Bondarenko, O., Simakova, I.: Internet-of-things device set configuration for connection to wireless local area network. In: CEUR Workshop Proceedings, pp 885–896 (2019)

19. Sahoo, S.S., Valdez, J., Rueschman, M.: Scientific reproducibility in biomedical research: provenance metadata ontology for semantic annotation of study description. AMIA. Annu. Symp. Proc. AMIA Symp. **2016**, 1070–1079 (2016)
20. Jacoby, M., Antonić, A., Kreiner, K., et al.: Semantic interoperability as key to IoT platform federation. In: Lecture Notes in Computer Science (Including Subseries Lecture Notes in Artificial Intelligence and Lecture Notes in Bioinformatics), pp 3–19 (2017)
21. Arendt, J.: Melatonin: countering chaotic time cues. Front Endocrinol. (Lausanne) **10** (2019). https://doi.org/10.3389/fendo.2019.00391
22. De Almeida, E.A., Di Mascio, P., Harumi, T., et al.: Measurement of melatonin in body fluids: standards, protocols and procedures. Child's Nerv. Syst. **27**, 879–891 (2011)
23. Kunz, D., Mahlberg, R., Müller, C., et al.: Melatonin in patients with reduced REM sleep duration: two randomized controlled trials. J. Clin. Endocrinol. Metab. **89**, 128–134 (2004). https://doi.org/10.1210/jc.2002-021057
24. Melatonin-Sulfate (EIA-1432), Report of DRG International Inc., USA, Revised 12 Sept 2011 (Vers. 8.1). https://weldonbiotech.com/wp-content/uploads/2018/05/eia-1432.pdf. Last accessed 2020/01/21
25. Melatonin Sulfate ELISA (RE54031), Report of IBL International GMBH, Revised 19 June 2017. https://novamedline.com/files/523001e4-ac11-409b-9cf2-017adfb65556.pdf. Last accessed 2020/01/21
26. Melatonin Sulfate ELISA (40-371-25006), Report of GenWay Biotech, Inc., Revised 18 May 2017. https://www.genwaybio.com/media/custom/upload/File-1313509984.pdf. Last accessed 2020/01/21
27. 6-Sulfatoxymelatonin ELISA (79-STMHU-E01), Report of ALPCO, Revised 7 Dec 2016. https://pdf.medicalexpo.com/pdf/alpco/6-sulfatoxymelatonin-elisa/69512-187076.html. Last accessed 2020/01/21
28. 6-Sulfatoxymelatonin ELISA (EK-M6S), Report of Bühlmann Laboratories AG, Revised 18 Jan 2016. https://buhlmannlabs.com/wp-content/uploads/BUHLMANN-6-Sulfatoxymelatonin-ELISA_EK-M6S_160118_RUO-1.pdf. Last accessed 2020/01/21
29. Direct Saliva Melatonin ELISA (EK-DSM), Report of Bühlmann Laboratories AG, Revised 14 Jan 2019. https://www.buhlmannlabs.ch/wp-content/uploads/2015/01/EK-DSM_IFU-CE_VA1-2019-01-14.pdf. Last accessed 2020/01/21
30. Chuiko, G.P., Dvornik, O.V., Shyian, I.A.: How reliable are calibrators for urinary melatonin sulfate? Med. Inform. Eng. (2016). https://doi.org/10.11603/mie.1996-1960.2016.3.6759
31. Melatonin RIA (RK-MEL), Report of Bühlmann Laboratories AG, Revised 20 Nov 2012. https://www.sceti.co.jp/images/psearch/pdf/BUL_RK-MEL2_p.pdf. Last accessed 2020/01/21
32. Direct Saliva Melatonin RIA (RK-DSM IFU), Report of Bühlmann Laboratories AG, Revised 20 Jan 2016. https://buhlmannlabs.com/wp-content/uploads/BUHLMANN-Direct-Saliva-Melatonin-RIA_RK-DSM.pdf. last accessed 2020/01/21
33. Khan, A.: Calibrating Response Curves for the Concentration of Melatonin Sulfate in Human Urine. https://www.maplesoft.com/applications/view.aspx?SID=154007. Last accessed 2020/01/21
34. Gadagkar, S.R., Call, G.B.: Computational tools for fitting the Hill equation to dose-response curves. J. Pharmacol. Toxicol. Methods **71**, 68–76 (2015). https://doi.org/10.1016/j.vascn.2014.08.006

Problem-Specific Applications

Semantic Localization for IoT

Matthew Weber and Edward A. Lee

Abstract Euclidean geometry and Newtonian time with floating point numbers are common computational models of the physical world. However, to achieve the kind of cyber-physical collaboration that arises in the IoT, such a literal representation of space and time may not be the best choice. In this chapter we survey location models from robotics, the internet, cyber-physical systems, and philosophy. The diversity in these models is justified by differing application demands and conceptualizations of space (spatial ontologies). To facilitate interoperability of spatial knowledge across representations, we propose a logical framework wherein a spatial ontology is defined as a model-theoretic structure. The logic language induced from a collection of such structures may be used to formally describe location in the IoT via semantic localization. Space-aware IoT services gain advantages for privacy and interoperability when they are designed for the most abstract spatial-ontologies as possible. We finish the chapter with definitions for open ontologies and logical inference.

1 Location as IoT Context

Today, we have mature theories of computation, developed over the last 80 years or so, and mature theories of physical structure and dynamics, developed over the last 300 years or so. But we have only the barest beginnings of theories that conjoin the two. One of the key points of friction is that the notion of location in space and time are central to a physical reality, but absent in a cyber reality. When the focus is mutual imitation, as in simulation, it is natural to construct cyber representations of space and time by approximating positions in a Euclidean geometry and Newtonian time with floating point numbers. But when the goal is the kind of cyber-physical

M. Weber (✉) · E. A. Lee
UC Berkeley, Berkeley, United States
e-mail: matt.weber@eecs.berkeley.edu

E. A. Lee
e-mail: eal@eecs.berkeley.edu

R. Pandey et al. (eds.), *Semantic IoT: Theory and Applications*, Studies in Computational Intelligence 941, https://doi.org/10.1007/978-3-030-64619-6_16

collaboration that arises in the Internet of Things (IoT), such a literal representation of space and time may not be the best choice.

When considering mobile devices and the IoT, applications often care more about logical spatial and temporal relationships than quantitative ones. To preserve security and privacy, for example, one device may be granted access to data held by another device only when the two devices are in the same room at the same time. The notion of "same room at the same time" is an example of what we call *semantic localization*. It is not so much about geometric location, but rather asserts a "semantic" spatial relationship.

In this chapter we survey IoT-relevant location models from robotics, the internet, cyber-physical systems (CPS)s, and philosophy. The diversity in these models is justified by differing application demands and conceptualizations of space (i.e. spatial ontologies). To facilitate interoperability of spatial knowledge across representations, we propose a logical framework wherein a spatial ontology is defined as a model theoretic structure. The logic language induced from a collection of such structures may be used to formally describe location in the IoT via semantic localization. Space aware IoT services gain advantages for privacy and interoperability when they are designed for the most abstract spatial-ontologies as possible. We finish the chapter with definitions for open ontologies and logical inference.

For all its importance to understanding IoT systems, localization, the challenge of determining the location of physical objects, remains an open problem. GPS, which has been a resounding success for outdoor localization, relies on direct line-of-sight signals from satellites, and is consequently ineffective for indoor environments or outdoor environments where obstructions, such as buildings, interfere with measurements. Researchers have been trying to address the indoor localization problem since the early 1990s with systems like Active Badges [1] and Cricket [2], and yet even to this day, a general purpose, accurate, cost effective, deployable system with the potential to reach the ubiquity of outdoor GPS remains elusive. A big part of what makes the problem difficult is the potential for interference in indoor environments where walls, furniture, and people, obstruct and reflect signals. Even something as simple as turning on a microwave oven causes interference to RF signals and might disrupt signal strength measurements for an indoor localization system operating in the 802.11 bands. Nevertheless, we are optimistic that in the near future, IoT applications will routinely have available a variety of types of location information with a range of quality. This chapter addresses how to organize and use that location information.

The most commonly articulated purpose for indoor positioning is indoor navigation. There is no doubt a market for apps that can help you find your way in whatever building you happen to be inside, but in our view this is probably a small market that dramatically understates the potential of contextual awareness in the IoT. The future of indoor and outdoor space-aware IoT systems involves scenarios where position in space is less important than spatial interrelationships. Consider a fleet of self-driving cars, where proximity in driving time, energy, and ride sharing opportunities are more useful criteria for control than geo-coordinates. Indoors, having awareness of which devices are in the same room may be more useful than measurements of their position in two or three-dimensional space. For such applications, different representations of space than a coordinate-system based physical map become appealing.

Relational ontologies of space aren't wildly foreign concepts; they can be found in some of today's apps. Take data from FourSquare, the app that lets users "check into" locations, as an example of a non-geometric representation of space. A user checked into a restaurant on FourSquare is known to be inside the establishment, but it would be a mistake to guess exact geocoordinates for him/her and plot them inside the restaurant's perimeter because they might be sitting at a table or standing by the door, and precise geocoordinates would suggest a false confidence as to the nature of unknown information. Unplotability doesn't make the FourSquare data somehow less accurate or reliable than a physical coordinate map, it just makes it different. We call this kind of *geometrically fuzzy yet logically precise spatial information* **semantic localization**.

1.1 Designing a Robo-Cafe

In collaboration with researchers at U Penn, Michigan, UW, CMU, and Berkeley, in 2015 we demonstrated a robotic delivery system at the DARPA "Wait, What?" conference where users could place an order on a smart phone localized by the ALPS Ultrasound Localization System [3] and have a desired snack delivered to their location by a roaming Scarab Robot [4]. The demo was designed to showcase integration and composability of IoT systems via accessors [5], but most relevant to this chapter is the spatial interaction needed between the Scarab and ALPS.

The Scarab comes equipped with a laser rangefinder which it uses with standard ROS packages to perform Simultaneous Localization and Mapping (SLAM) and to build an occupancy-grid map of its environment (see Fig. 1). An occupancy grid is a fairly simple data structure commonly used in robotics to represent an environment (modeled as a grid over 2D or 3D Euclidean space) that is essentially a big array with values from 0 to 100. A value of 0 indicates the robot is almost certain the cell does not contain an obstacle, and a value of 100 that the cell is almost certainly impassable. The robot also maintains an estimate of its pose (position and orientation) at the cell where it is currently located.

The second localization system, ALPS, uses ultrasonic beacons, and is also deployed in the DOP Center (Fig. 1). The system is deployed by placing beacons at known locations in a building and finding the correspondence between the beacons and coordinates on the building's floor plan. The beacons send time synchronized chirps of ultrasound in the 20–22 kHz bands that are beyond the range of human hearing but receivable at the standard sampling rate of a cell phone microphone. A smartphone with an ALPS app can locate itself on the floor plan's coordinate system.

When the robot is localized on its occupancy grid and the phone is localized on the floor plan, the Scarab uses ROS navigation packages to deliver a snack. However, there is a rather subtle challenge in the last step: the phone has known coordinates on the floor plan and the robot is at a known cell of the occupancy grid, but the two are, as

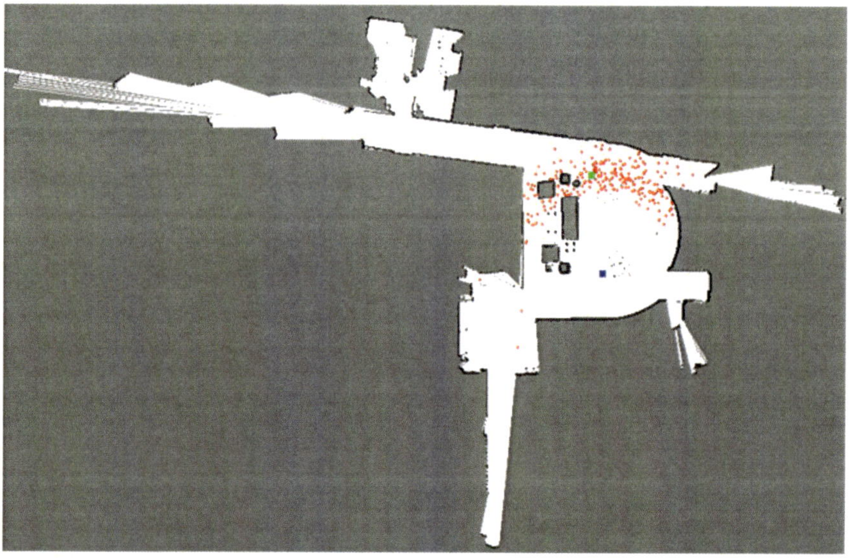

Fig. 1 Occupancy grid formed by a Scarab robot roving the DOP Center at Berkeley. This image shows use of a relatively poor distance sensor on the roving robot, measuring for example received signal strength from another object, and then applying a particle filtering algorithm constrained by the occupancy grid map to estimate the position of the other object. The red dots are the particles, the green square is the target, the blue square is the Scarab robot, and the black areas are occupied grid points as detected by the lidar rangefinder on the Scarab. The grey areas indicate where the occupancy grid has no information. Image courtesy of Ilge Akkaya

given, totally unrelated! Deployment of the Robo-cafe requires a coordinate system alignment phase in which ALPS's model of space is brought into concordance with the Scarab's model.

The coordinate system alignment problem in Robo-Cafe is in fact an instance of a general problem that must be addressed whenever two IoT systems seek to work across contextual ontologies. Usually when IoT systems are designed by different engineers working with different conceptualizations of space, spatial information cannot be shared between systems without additional translation. A central motivation for the modeling framework presented in this chapter is to formalize the structure of spatial ontologies for the development of mappings and relationships that enable heterogeneous mixtures of ontologies in IoT applications. We discuss a formalism for such cross-ontology reasoning in Sect. 2.1.

1.2 Spatial Ontologies

Location is one of the most important and challenging aspects of physical context. Location matters for the IoT in ways it does not for the Internet. There's a world of difference between illuminating a smart light bulb located in your home or one a thousand miles away. But while the physical location of a web server might affect the latency of communication or quality of service, it won't fundamentally change the content of the hosted page. For an IoT device, its physical relationship with the world has everything to do with what it can and cannot accomplish.

Any IoT system that seeks to interact with the physical world assumes a model of space, either explicitly or implicitly. Such a model is a spatial ontology. Broadly, the subject of ontology from philosophy is a study of the nature of existence, what it means for something to be and to be something. In computer science, ontology is usually about association of entities in a model with structured taxonomies, addressing questions like "is this object an instance or example of that class of objects?" (taxonomy) or "is this object *a part of* an instance or example of that class of objects?" (meronymy) relationships. In prior work, it has been shown that useful ontologies can be constrained to have a mathematical lattice structure, and that they thereby acquire enormous algorithmic and formal benefits that can be leveraged to compose ontologies, perform inference, and check correctness [6–8]. Such ontologies form a subset of commonly used ontology frameworks such as Web Ontology Language (OWL). Their mathematical structure resembles that of Hindley-Milner type systems, from which they inherit practical algorithms that scale to very large numbers of elements. For example, type inference maps into the problem of finding a fixed point of a monotonic function over a lattice.

Spatial ontologies have more diversity than just choice of coordinate system. A common dichotomy in ontologies is the distinction between "objects" and "fields" [9–11]. An "object" is an entity that is distinct, with a clear boundary, and in the language of [9] is "individual and fully deniable." Examples of objects include: an apple, a table, or a flashlight. A "field" describes phenomena without clearly defined boundaries that are "smooth, continuous and spatially varying" [9]. The magnetic field emanating from a hand-held bar magnet is a good example of this concept. From a certain pedantic perspective, the field is present everywhere in the universe, only its strength is almost everywhere so weak as to be negligible. Some geographical features like lakes have elements of both objects and fields because it can be hard to identify where they end.

Spatial ontologies can also vary with respect to their interpretation of entities with respect to time. SNAP and SPAN are two cooperative ontologies proposed by Grenon and Smith [10] to capture the distinction between "continuants," objects with an identity that persists across time, and "occurants," processes defined in part by their beginning and ending. Examples of continuants include the planet earth or a pair of shoes because it makes sense to consider their spatial properties at a particular snapshot of time. The same is not true for occurants like a volcanic eruption or the takeoff of a helicopter. Such occurants unquestionably have a spatial existence but their reality is best comprehended in four full dimensions; a sequence of 3D observations misses something essential about the nature of the process. There is clearly a strong interrelation between SNAP and SPAN ontologies. This point is not missed by Grenon and Smith, who devote a latter section of their paper to trans-ontology interrelations between SNAP and SPAN.

1.3 Semantic Technologies

The term "semantic technology" describes a collection of popular standards and technologies for representing and working with ontologies, be they spatial or otherwise.

The Semantic Web was proposed by Tim Berners-Lee in 2001 as an extension of the World Wide Web that would allow ordinary HTML web pages to be enhanced with special markup to label their semantic content. The hope is that when markup is combined with a collection of ontologies for web content and real-world objects, algorithms will be able to apply ontological reasoning to web elements and data. For example, an image of a bridge embedded in a web site could be labeled as such and found through a general search for "landmarks" by using the ontological information that a bridge *is* a landmark. According to the wikipedia article on the semantic web, by 2013 some 4 million web pages had been augmented with semantic web information. But this is done primarily through human intervention, which could account for the relatively modest penetration compared to the total number of web pages.

Semantic Web ontologies are expressed in Resource Description Framework (RDF), an abstract model for semantic data as sentence-like statements about the world in triples of subject, predicate, object. For example: the sentence "A cow" (subject) "isa" (predicate) "farm animal" (object), or "The mall parking lot" (subject) "has the number of free spaces" (predicate) "45" (object). As hinted at by these examples, triples can express both abstract information about classes (cows and farm animals) as well as facts about specific instances (the mall parking lot) and raw data values (45). A database designed and optimized for RDF data is known as a semantic repository or alternatively a triple store. The W3C SPARQL Protocol and RDF Query Language (SPARQL) recommendation [12] defines both a protocol and a query language for performing SQL-like operations on a semantic repository such as queries, inserts, updates, and deletes.

RDF is a natural way to express relational ontologies, as discussed in Sect. 2.3, for semantic localization. Additionally, some semantic repositories, like GraphDB, support geospatial plugins for efficient queries over geocoordinates (i.e. latitude and longitude pairs). If compatible with the GeoSPARQL standard [13], the semantic repository may also be able to automatically derive RCC (Region Connection Calculus) relationships, such as containment of one geospatial object within another, directly from the definitions of the objects themselves.

1.4 Standards for Spatial Representation

Many standards for spatial representation have been proposed in different domains, a sample of which is presented here.

According to Lieberman et al., as of 2007 the semantic web maintained at least seven varieties of spatial ontologies [14]. These include Geospatial Features, Feature

Types, Toponyms/Placenames, (Geo) Spatial Relationships, Coordinate Reference Systems, Geospatial Metadata, and (Geo) Web Services. The relationships between these, however, are highly unstructured and lacking in formal properties that can exploited algorithmically. More recently, geospatial ontologies like GeoDataOnt [15] have been developed to provide a unified ontology for this domain.

A popular (non-RDF) spatial ontology today is codified in a JSON schema called GeoJSON [16]. This is used by many location based services. In contrast to the semantic web, GeoJSON is good at representing geometries, but not higher level ontological concepts and relationships. It supports points, lines, polygons, and collections of polygons in 2D or 3D. Given the extensive support for GeoJSON in existing apps and software, it is a useful standard to leverage for geometric concepts. But restricting spatial ontologies to exclusively geometric concepts is a mistake. Spatial relationships are more complex.

On the opposite end of the complexity spectrum, the Open GIS Geography Markup Language (GML) Encoding Standard [17] is a 437 page specification document describing an XML schema for spatio-temporal ontologies. It follows the ISO 19101 definition of a feature as an "abstraction for real world phenomena" and represents the world as a collection of features defined as name, type, value triples. The increased complexity allows for the description of more sophisticated data such as spatial geometries, spatial topologies, time, coverages, and observations. The format can be extended to application schema such as IndoorGML [18] which is targeted for indoor navigation. IndoorGML focuses on layered graph representations of relationships such as adjacency and paths between semantic objects in indoor space. It models the world as a collection of cells representing geometry and topology via the Poincaré duality to achieve a "Multi-Layered Representation" of a given space in different contexts.

A variety of geometric data structures and algorithms are employed in the field of computational geometry when high performance is desired for computationally difficult spatial analysis [19]. For example, a doubly-connected edge list is used for the thematic map overlay problem, in which the overlay of spatial subdivisions is computed.[1] A trapezoidal map is another geometric data structure employed to solve point location queries: given the coordinates of a point and a map subdividing the plane into regions, determine which region contains the point.

Point clouds are another computationally useful format for spatial information in the domain of computer vision. Visually oriented sensors such as stereo cameras or time of flight cameras (e.g. Light Detection and Ranging, LiDAR) measure the location of individual 3-dimensional points in the world. These points represent sampled measurements of real-world objects. Once collected, software such as the open source point cloud library [20] can use a point cloud data set to reconstruct a sampled surface or perform segmentation to semantically identify objects.

The diversity of these standards for spatial representation is daunting. Yet it is easy to see how applications in the IoT with different purposes for spatial information and different sensors for collecting that data benefit from different data representations.

[1] Imagine overlaying two circles to form a venn-diagram, but with polygons instead of circles.

Engineers of spatial IoT systems are tasked with finding the best location model for their application. But IoT systems designed for interaction with other data sources must additionally be created with an understanding of how a selected location model is related to the location models of other sensors and IoT systems in the environment.

2 Location Modeling

The purpose of location modeling in this work is to support a logic for reasoning about spatial ontologies across independently designed IoT systems. By moving beyond just geometric position, this logic offers the possibility for a much richer set of applications than just navigation, including for example security (e.g. restricting access to some service to only devices in the same room); asset tracking (e.g. where is the remote control for this device, or the device for this remote); spatial search (e.g. find a temperature sensor in the same room as a mobile device); commissioning (e.g. deploying sensors and actuators without manually specifying their location); and context-aware services (e.g. lighting systems that automatically adjust to usage patterns of a room). We believe semantic repositories are a good start for this, but there is room for a larger suite of software components and services for creative application designers to use when reasoning about location information.

Such services could handle mobility (e.g. notification when a device is no longer in the same room) and superposition of disjoint maps constructed at different semantic and geometric layers (e.g., relating geometric information to "in the same room" semantic information).

2.1 Model Theory

Model theory is a domain of mathematical logic originally developed to analyze logical formulas regarding mathematical structures such as groups, graphs, and fields. The key observation behind model theory is that logical formulas can be written to express properties in a manner independent of the mathematical structures with respect to which they are evaluated. For example the formula $\exists n\ 0 < n < 1$ is true with respect to $\mathbb{R}$ or $\mathbb{Q}$, but not with respect to $\mathbb{Z}$ or $\mathbb{N}$. A model (or structure) specifies a domain, such as $\mathbb{R}$, and gives interpretations to the symbols 0, 1, and $<$ so that their particular relationship may be determined. We summarize the fundamentals of model theory relevant to CPS location modeling below. The main reference for the following definitions is [21], which may be referred to for a more comprehensive introduction to model theory.[2]

A **formula** is a logical statement constructed in the usual way from:

[2] A friendlier introduction can be found at https://plato.stanford.edu/entries/modeltheory-fo/

- logical symbols: ($\rightarrow$, $\leftrightarrow$, $\neg$, $\wedge$, $\vee$, $\forall$, $\exists$,), and
- variables (a countably infinite collection)
- function symbols (e.g. $+$ for a group operation)
- relation symbols (e.g. $\leq$ for the ordering relation on R)
- the relation symbol $=$, as the usual "equal sign"
- constant symbols which represent a particular element from the domain (e.g. 0 or π).

The arity of function and relation symbols is ≥ 1.

A **signature** is a particular set of function, relation, and constant symbols. The **language** of a signature is the set of well-formed formula expressible using functions, relations and constants from the signature. A variable v_0 is **bound** iff it appears in a subformula (i.e. a syntactically correct part of a formula) following $(\forall v_0)$ or $(\exists v_0)$. Otherwise the variable is **free**, and may be assigned a value separately. For example: formula ϕ with free variables v_0, v_1, ..., v_k may be written as $\phi(a_0, a_1, ..., a_k)$ to express the assignment of a_0 to v_0, a_1 to v_1, and so on.

The **sentences** of a language are formulas of the language with no unbound variables. A **structure** (or model) is a tuple $\mathbb{A} = \langle \mathbf{A}, I \rangle$ where **A** is a domain, i.e. a non-empty set, and I is an interpretation function. I maps function, relation, and constant symbols to functions defined over **A**, relations defined over **A**, and elements of **A** respectively. A structure $\mathbb{A}$ **models** a sentence S of a language when the interpretation of the sentence within the structure evaluates to true. This relationship is denoted by $\mathbb{A} \models S$ and its converse by $\mathbb{A} \not\models S$.

A language with a finite signature may be concisely written for example as, $\mathbb{L} = \{< , 0, 1\}$. Similarly, a model's domain and interpretation for that finite signature may be informally written as an analogous tuple, e.g. $\mathbb{A} = \langle \mathbb{R}, \mathbf{<}, \mathbf{0}, \mathbf{1} \rangle$. Here, $\mathbb{A}$ is the structure with domain $\mathbb{R}$ which interprets $\mathbb{L}$ with the strict ordering relation $<$, and constants 0 and 1.

Putting it all together, we may now formalize the motivating observation from the beginning of this section that the same formula may be true or false with respect to different domains. Regarding the example formula $\exists n\ 0 < n < 1$ we have $\mathbb{A} \models \exists n\ 0 < n < 1$, but for $\mathbb{B} = \langle \mathbb{N}, \mathbf{<}, \mathbf{0}, \mathbf{1} \rangle$, $\mathbb{B} \not\models \exists n\ 0 < n < 1$.

2.2 *Semantic Localization*

We propose using the concepts of model theory to formally describe location in IoT systems. A spatial ontology can be represented as a structure $\mathbb{A} = \langle \mathbf{A}, I \rangle$. For $\mathbb{A}$ to be useful as a model of the space, most likely the elements of **A** should be places or things located at places. Similarly, I should provide spatially meaningful interpretations of relations, functions, and constants. The language for such a structure will then consist of semantic localization statements.

Defining semantic localization as a model-theoretic language has the advantage of separating the specification of spatial reasoning from its implementation within a particular spatial ontology. Just as the formula $\exists n\ 0 < n < 1$ may be evaluated within different structures, so too might a semantic localization formula be evaluated within heterogeneous spatial ontologies. For example, let *contains* (a, b) be a binary relation which is true when room a contains person b, let $user1$ and $user2$ be constants for people, and let the variable *room* range over a set of rooms. The formula $\exists\ room\ contains(room, user1) \wedge contains(room, user2)$ expresses the spatial arrangement in which $user1$ and $user2$ are both within the same room, independently of a particular spatial ontology. We propose using semantic localization as a conceptual interface between location programming and location models in the IoT.

A semantic localization formula can be interpreted in one of two ways: either as an event condition or as a query into some spatial database. In the first case, the sentence acts as a predicate that triggers an event when it evaluates to true. In the second case, the formula can be evaluated against database entries to signify that the entries to return are those that cause the formula to evaluate to true when plugged into unbound variables. However, in either case a statement can only be evaluated in an ontology with a compatible signature.

Figure 2 represents a central idea governing location modeling, relating mathematical structures to the spatial connectives (relations) of physical objects in a CPS which they are capable of evaluating. Applying the concepts from model theory to CPS location modeling has the added advantage of enabling mathematical analysis to bring the theorems of model theory to bear on the relationships spatial ontologies have to one another.

As suggested in Fig. 2, spatial ontologies may be used to reason about spatial connectives, or spatial connectives may be discovered by sensors and used to construct mathematical structures. Relations represent the structural aspects of the

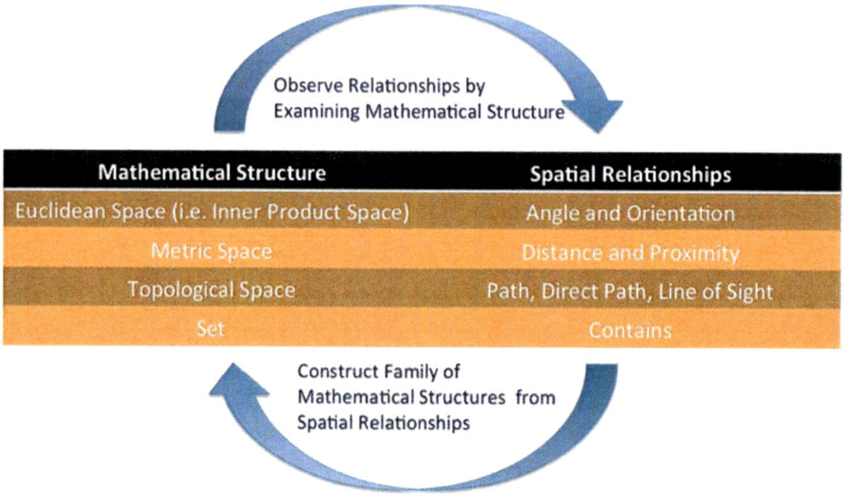

Fig. 2 A comparison between mathematical structures and corresponding evaluable spatial relationships as described in our previous work [32]

space (e.g. containment, path, proximity, angle, etc.), and functions define other structural aspects of the space (such as distance for a metric space) as appropriate. The values of constants, relations and functions are potentially time varying as the structure evolves. For instance, a topological ontology of an indoor space with doors opening and closing has a dynamic "path" relation. The quality and nature of the sensor data may constrain the level at which these ontologies may be constructed. For example, orientation information may simply not be available.

2.3 *Physical and Relational Ontologies*

In the previous section, a spatial ontology is a mathematical structure which can be used to evaluate a logical sentence that makes reference to spatial relationships. This notion of a spatial ontology is considerably more general than the usual notion of a printed paper map with a 2D representation of the road network of a city, for example. We will use the term "relational ontology" when we want to emphasize the abstracted nature of the spatial relationships that the map represents, but in this research, a spatial ontology is a mathematical object at any of these levels of abstraction, as long as it encodes some form of spatial relationships. For example, Fig. 3 shows a relational ontology that is a partial order induced by the containment relation between sets; the relational ontology does not say anything at all about geometric properties such as distance or orientation.

We have arrived at an important principle: *Space-aware services should be constructed for the signatures of the most abstract spatial ontologies as possible.* This will enable them to operate in more sensor-poor environments, to benefit from a greater variety of sources of spatial information, and to better preserve privacy by

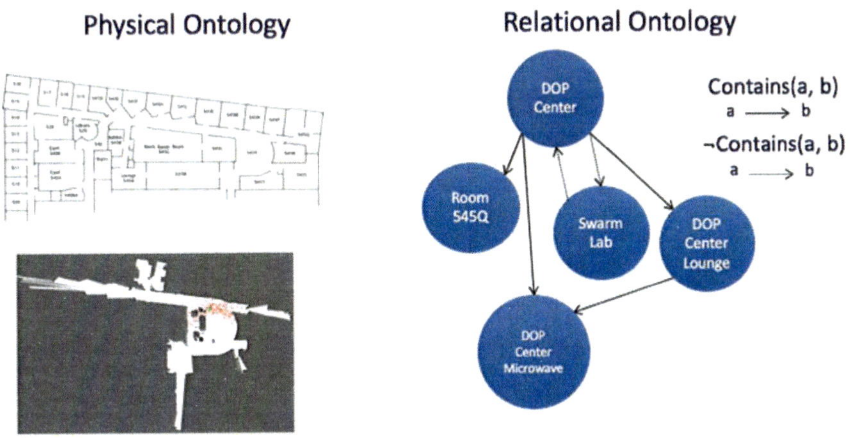

Fig. 3 Concrete examples of Euclidean-space ontologies vs. an abstracted relational ontology that represents only containment relations

not handling information that is not needed. For example, the set containment relation is all the map information necessary for the FourSquare localization example in the introduction, since the only information to be gained from checking in is containment.

Consider the advantages of applying this maximum abstraction principle to the spatial ontologies depicted in Fig. 3. All three ontologies, the occupancy grid, the floor plan, and the abstract graph, represent information about the same region of space. An IoT application could theoretically use any one of the ontologies to determine, for example, that room 545Q is inside the DOP Center. However, it takes a certain level of geometric understanding to extract that information from the less abstract physical ontologies. To use the occupancy grid, our IoT application must be equipped with an algorithm for parsing occupancy grids and determining when a collection of cells in a grid is contained by another collection of cells. In other words, effective use of the occupancy grid is restricted to IoT applications that are prepared in advance to interact with robotic maps. Similar limitations hold for IoT systems using the floor plan, or any other spatial ontology requiring geometric analysis.

But if the IoT system were designed to interact with map providers through an abstract notion of containment, the system wouldn't have to bother understanding the nuances of geometry in every spatial ontology it might come across. It may instead operate in terms of semantic localization. Perhaps the relational ontology was created by inspecting an occupancy grid, or maybe it was a floor plan. Either way the IoT application doesn't have to bother knowing the specifics. As long as it can pose the query regarding the DOP Center, room 545Q, and containment, the IoT application can treat the source of an abstracted spatial representation as a black box.

As they get more abstract, of course, relational ontologies lose the ability to evaluate some kinds of spatial relationships. This idea parallels the usual hierarchy of mathematical spaces. A Euclidean space has quite a lot of mathematical structure that may not match well with the information available sensors are able to deliver. A Euclidean-space ontology supports reasoning about angles and orientation, concepts that are not defined in the more abstract mathematical structures shown in Fig. 2.

The hierarchy of these mathematical spaces offers a starting point for reasoning about combinations of maps. For example, given a Euclidean-space map of an office space and a Set (containment) map of objects in the space, objects can be placed approximately, with known error bounds, onto the Euclidean-space map. But much more complicated mapping combinations will be required, since even two Euclidean-space maps may not use the same coordinate system. The concept of a spatial ontology becomes an essential feature of location modeling.

Topological spaces can be used to construct maps that represent paths through indoor settings. Navigation with graphs is a common concept in robotics [22], where nodes represent waypoints in a space and edges represent paths between waypoints. Such data structures are routinely used to construct sequences of actions to move a robot between nodes. Additionally, Ghrist et al. [23] show that algebraic topology can be used directly to relate the convex hull of a landmark set in a Euclidean space to a simplex of a simplicial complex. This provides a natural abstraction mechanism for topological maps.

Non-Euclidean metric maps are useful when the standard Euclidean metric does not really capture the interesting properties of a space. Consider a point x on the third floor of a building and the point y directly below it on the second floor. Points x and y are very close to each other in Euclidean space, but for the purposes of navigation, this misrepresents reality. We can instead define a metric space with metric D, where

$$D(x, y) = \begin{cases} \text{minimum length of a continuous path from x to y,} & \text{if there is such a path} \\ \infty, & \text{otherwise} \end{cases}$$

This is easily shown to be a metric (or even an ultrametric, for some graph-structured metric spaces). If stairways and elevators are not navigable open space for a particular robot, then this metric will yield $D(x, y) = \infty$, considerably more than the Euclidean distance.

An inner product space (of which a Euclidean space is a common variety) introduces the notion of angles. Angles can facilitate special kinds of analysis like trilateration, and the use of trigonometric angle measurements to localize objects in coordinate space.

As these examples illustrate, there are practical reasons to construct non-Euclidean ontologies. However each of these mathematical ontologies has the property that any map entity placed at a particular coordinate takes on all spatial relationships to other map coordinates implied by the structure of the space. This is undesirable when only a portion of those relations are positively known to be true and the rest are unknown. A key advantage of relational ontologies is the expression of *open ontologies*, where the absence of a relation does not imply its converse. This is analogous to ancient maps that provided useful navigation information despite significant distortions in the geometry and large gaps labeled "*terra incognita*". Open ontologies translate naturally into action plans that can deal with incomplete information.

This increased flexibility comes at the cost of a slightly more verbose vocabulary for relations. Consider the containment map on the right hand side of Fig. 3. Because this ontology is open, knowing that one place is not contained by another isn't enough to know they have no space in common. Another relation, "disjoint," is necessary to express that positive fact explicitly. We hope the reader can see a connection here to intuitionisitic logic, in that for open ontologies it is not enough to know a spatial relation is not not true to infer that it is true. Instead, relations must be constructively built up from known facts.

2.4 Formalizing Open Ontologies

An open ontology $\mathbb{A}$ is a way of expressing partial knowledge about a spatial structure. If we take the philosophical position that the unexpressed information in an open ontology is fundamentally unknowabable, there is nothing to be done to increase the amount of information represented in $\mathbb{A}$. However, if we assume the missing information is knowable and could be expressed in an idealized (but hypothetical[3]) ontology $\mathbb{A}^*$, we may consider logical inference as a means to obtain information available in $\mathbb{A}^*$ but not $\mathbb{A}$.

We formalize this notion below, but first some definitions. Let $\perp$ be the symbol for "unknown".[4]

Definition 1 (Partial Order on Functions)
We define a (pointwise) partial order on n-valued function f: $\mathbf{A}^n \rightarrow (\mathbf{A} \cup \perp)$ with $f \leq f'$ iff for $x \in \mathbf{A}$, $f(a_1, a_2, \ldots a_n) = x \rightarrow f'(a_1, a_2, \ldots a_n) = x$.

Observe this definition allows $f(a_1, a_2, \ldots a_n) = \perp$ with $f'(a_1, a_2, \ldots a_n) = x$. In other words, f agrees with f' everywhere where f is not unknown, but may disagree where f is unknown.

Let open ontology $\mathbb{A}$ and its idealized $\mathbb{A}^*$ both be structures with the same signature and the same domain. $\mathbb{A}$ may be missing some information available in $\mathbb{A}^*$.

Definition 2 (Partial Order on Open Ontologies)
We define a pointwise partial order on open ontologies $\mathbb{A}$ and $\mathbb{A}^*$ with the same signature and domain (**A**) by ordering relation $\sqsubseteq$. The $\sqsubseteq$ relation indicates $\mathbb{A}^*$ has more information than $\mathbb{A}$ when:

- $\mathbb{A}$'s functions may have unknown value ($\perp$) over some elements of the domain where $\mathbb{A}^*$'s functions are known. With f_{A} as the interpretation of function symbol f in $\mathbb{A}$ and $f_{\mathbb{A}^*}$ as the interpretation of f in $\mathbb{A}^*$, $f_{\mathrm{A}} \leq f_{\mathbb{A}^*}$.
- $\mathbb{A}$'s relations may be missing tuples which are available in the analogous relations of $\mathbb{A}^*$. For example, with r_{A} and $r_{\mathbb{A}^*}$ as interpretations of relation symbol r in models $\mathbb{A}$ and $\mathbb{A}^*$ respectively, $r_{\mathrm{A}} \subseteq r_{\mathbb{A}^*}$.
- $\mathbb{A}$'s interpretation of constant symbols may be less complete than the interpretation of $\mathbb{A}^*$. With k as the set of constant symbols in $\mathbb{A}$'s signature and c: $k \rightarrow (\mathbf{A} \cup \{\perp\})$, as the function mapping constant symbols to domain elements, $c_{\mathrm{A}} \leq c_{\mathbb{A}^*}$.

Not only does the ordering relation defined by $\sqsubseteq$ relate $\mathbb{A}$ to $\mathbb{A}^*$, it also relates $\mathbb{A}$ to a chain of non-idealized open ontologies $\mathbb{A} \sqsubseteq \mathbb{A}' \sqsubseteq \mathbb{A}'' \sqsubseteq \ldots \sqsubseteq \mathbb{A}^*$ with progressively more information than $\mathbb{A}$. Applying a logical inference procedure to $\mathbb{A}$, and filling in an unknown function, relation, or constant with a concrete value can be interpreted as finding an $\mathbb{A}'$ with $\mathbb{A} \sqsubseteq \mathbb{A}'$.

It may not be possible to definitively determine whether or not an open ontology models a formula which depends on unknown functions, relations, and constants. If the true/false value of a formula depends on evaluating a function where it is unknown, an unknown constant, or the negation of a relation which is not explicitly given in the model, the formula may not be evaluated with respect to the open model.

---

[3]Of course we don't actually know the contents of A* because it contains the information we currently don't know in A. But it is nevertheless useful to define A* as a model so we may make explicit our assumptions about the missing information.

[4]We do not always explicitly augment the domain of an open ontology to include $\perp$, but this may be assumed.

The advantage of an open ontology is the ability to evaluate formula regarding *known* information without being forced to make questionable assumptions on the unknown parts of the model.

The next section provides some examples of valid logical inference procedures for open relational ontologies.

3 Logical Inference on Ontologies

Consider the relational ontology on the left hand side of Fig. 4. Nodes in the map represent objects or places in the world, and dark edges signify a known upper bound on the distance between them in some metric given by the weight of the edge. Since this is an open map, the absence of a black edge does not signify the converse of proximity (which we might call "anti-proximity"); if we want to express anti-proximity in this graph we must explicitly designate it with a dashed line edge.

This being a metric space, we can apply the triangle inequality to the graph and note that if A and B are within 30 meters and if B and C are within 30 meters, then A and C must be within 60 meters. Before we add this edge to the graph as shown in the right hand side of Fig. 4, we may note that the triangle inequality applied to the edge from A to D and from D to C gives a tighter bound and express that A and C must in fact be within 40 meters of each other.

Next consider the example in Fig. 5 with an anti-proximity edge drawn from A to C. This indicates that A and C are known to be *at least* 10 m apart, whereas the proximity edge indicates that they are *at most* 40 m apart. Applying the contrapositive of the triangle inequality gives a relational ontology in which at least one of A or C must be more than 5 meters away from another node E. This matches the intuitive notion that for two objects known to be far away from each other; at least one of them must be somewhat distant from any third object. Note that this data structure is more than a simple graph now, since there is appended a disjunction between the two edges to E.

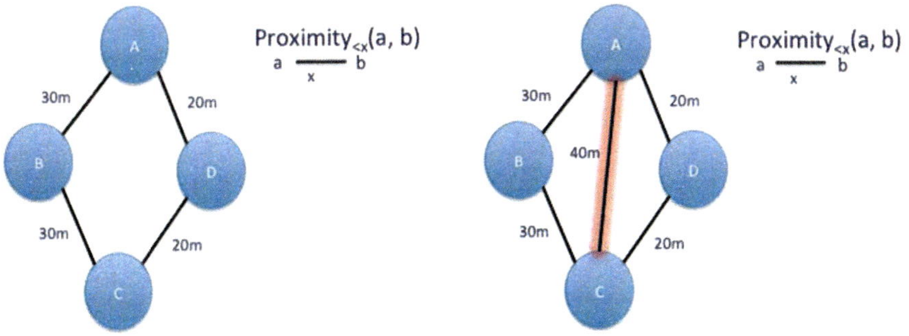

Fig. 4 An example of logical inference for a relational proximity map

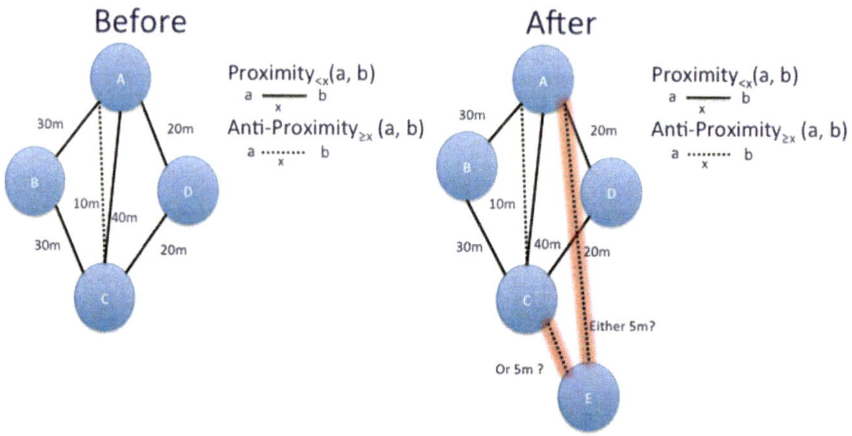

Fig. 5 Another example of logical inference for anti-proximity

A relational ontology of distance for a Euclidean space permits more sophisticated inference methods. A considerable amount of research has been undertaken in the sensor network community to find a Euclidean space embedding for a weighted undirected graph such that the Euclidean distances between nodes in the embedding match the edge weights in the graph. If such an embedding is successfully found, it is possible to infer internode distances not explicitly specified. One such algorithm [24] uses a process of iterative trilateration with robust quadrilaterals where three nodes with known Euclidean position (say A, B, and C) are used to establish the position of a connected node (D). Once the position of D is established, it can be used in the next iteration of the algorithm as a reference point to give the position of some other node E.

In addition to determining unknown inter-node distances, the properties of a Euclidean space also facilitate detection of inconsistent edges signifying outlier measurements. In prior work, [25] we expanded upon an algorithm given in [26] which uses graph rigidity theory to identify components of a graph that admit only a specific embedding. If a questionable edge is wildly inaccurate, it can be identified by considering other rigid subgraphs that are consistent with Euclidean geometry.

These sorts of Qualitative Spatial Reasoning (QSR) received significant research attention in the 1990s. The main focus of this work was the construction of formal algebras for inference on qualitative spatial relationships. For example, Frank's calculus for cardinal directions and informal distances such as "near" and "far" can infer such relationships for unknown cities given knowledge on how they are related to a known city network [27]. Arguably, the most notable outcome of QSR today is the Region Connection Calculus (RCC) for 2-dimensional mereology (the part whole relationship) and topology [28]. RCC laid the foundation for the GeoSPARQL standard [13], which is today widely (but incompletely) implemented by modern semantic repositories[5] to leverage RCC relationships for queries on geospatial data sets.

4 Related Formal Structures from AI and Robotics

Pereira's BigActor model [29] gives a formalism for mapping with many similarities to our approach. Specifically he defines two kinds of spatial structures: a logical-space model with a rough correspondence to what we would call a relational ontology, and a physical-space model corresponding to a coordinate-based physical map. Pereira requires the same relations hold true between the same objects in physical and logical space. The model-theoretic proposal for semantic localization in this chapter can be seen as a generalization of Pereira's approach to include more diverse kinds of spatial structures.

Similar ideas to relational ontologies have been around in the world of AI and robotics research for some time [22, 30, 31]. However, where semantic localization is designed to integrate modeling and programming for heterogenous IoT systems, the focus of research in this domain is commonly inference and autonomous decision making. As an additional point of contrast, spatial modeling in robotics is usually from the perspective of a robot as it moves from place to place, but spatial modeling for localization systems is usually from the perspective of a place as people (or robots) move within.[6]

The distinction between absolute and relative space is raised by Vieu [31]. The elements of a spatial ontology are Basic Entities (is the space composed of points or basic regions?), Primitive Notions (topology: relating to contact and part-whole relationships; orientation: absolute, intrinsic, and contextual; distance: metric functions and discrete distance notions), and bounded/unboundedness. Vieu goes on to overview actual approaches researchers have used to represent space. Vieu also examines the difference between 3D space composed with time and 4D views.

Kuipers [22], in a classic robotics paper, introduces an ontology for spatial information flow from sensor values to, ultimately, 2-D geometry. His ontology allows information to be incomplete at different levels. For example the graph-topological connections between different maps may be known even if each of the maps hasn't been entirely fleshed out.

An example of the advantages of combining physical maps with relational information for robotic localization was demonstrated by Atanasov et al. [30]. The authors use set-based identification of semantically interesting indoor objects such as chairs

[5]Essentially a semantic repository is a database for relational data.

[6]We refer to models of things moving through space as "Lagrangian Models" and models of space with things moving within as "Eulerian Models". The terminology comes from the analysis of fluid flows.

and doorways to localize their robot, instead of the more commonly used techniques that use edges and corners in the field of view without consideration for their semantics.

5 Conclusion

In this chapter we observe that spatial models used in IoT applications frequently have good reason to be domain specific. We propose semantic localization as a unifying interface between spatial modeling and spatial programming. This abstract approach is motivated by the need to reconcile diverse spatial representations for cross-domain interaction. By treating spatial models as mathematical structures from model theory, the language of mathematical logic becomes an effective tool for describing the qualitative spatial relationships important for developing contextually aware IoT services. When space aware services are designed for abstract spatial ontologies, they gain advantages in privacy and interoperability.

Semantic localization focuses our discussion of physical and relational ontologies in which information may be expressed through mathematical coordinates, spatial relationships, and non-Euclidian maps of an environment. We formalize the notion of an open ontology with partially unknown information, and give examples of logical inference on open ontologies. Open relational ontologies are promising for developing contextually aware IoT services, and have a conceptual match with semantic web technologies such as RDF, SPARQL, and semantic repositories. Semantic localization gives a principled foundation for location modeling and the design of spatially aware IoT systems.

Acknowledgements The work in this chapter was supported in part by the National Science Foundation (NSF), award #CNS-1836601 (Reconciling Safety with the Internet) and the iCyPhy Research Center (Industrial Cyber-Physical Systems), supported by Camozzi Industries, Denso, Siemens, and Toyota.

References

1. Want, R., et al.: The active badge location system. ACM Trans. Inf. Syst.(TOIS) **10**(1), 91–102 (1992)
2. Priyantha, N.B., Chakraborty, A., Balakrishnan, H.: The cricket location-support system. In: Proceedings of the 6th Annual International Conference on Mobile Computing and Networking, pp. 32–43 (2000)
3. Lazik, P., et al.: ALPS: a bluetooth and ultrasound platform for mapping and localization. In: ACM Press, pp. 73–84 (2015)
4. Michael, N., Fink, J., Kumar, V.: Experimental testbed for large multirobot teams. Robot. Autom. Mag. IEEE **15**(1), 53–61 (2008)
5. Latronico, E., et al.: A vision of swarmlets. Internet Comput IEEE **19**(2), 20–28 (2015)

6. Leung, J.M.-K., et al.: Scalable semantic annotation using latticebased ontologies. In: 12th International Conference on Model Driven Engineering Languages and Systems (MODELS 09), pp. 393–407. ACM/IEEE (2009)
7. Lickly, B.: Static Model Analysis with Lattice-based Ontologies. Tech. rep. UCB/EECS-2012-212. Ph.D. Thesis. EECS Department, University of California, Berkeley (2012)
8. Lickly, B., et al.: A practical ontology framework for static model analysis. In: International Conference on Embedded Software (EMSOFT), ACM, pp. 23–32 (2011)
9. Fonseca, F., Davis, C., Câmara, G.: Bridging ontologies and conceptual schemas in geographic information integration. GeoInformatica **7**(4), 355–378 (2003)
10. Grenon, P., Smith, B.: SNAP and SPAN: towards dynamic spatial ontology. Spat. Cogn. Comput. **4**(1), 69–104 (2004)
11. Spaccapietra, S., et al.: On Spatial Ontologies (2004). https://citeseerx.ist.psu.edu/viewdoc/summary?doi=10.1.1.88.7653
12. Harris, S., Seaborne, A., Prud'hommeaux, E.: SPARQL 1.1 Query Language. (2013). https://www.w3.org/TR/sparql11-query/
13. Battle, R. Kolas, D.: Enabling the geospatial semantic web with parliament and GeoSPARQL". In: Semantic Web 3.4, pp. 355–370 (2012)
14. Lieberman, J., Singh, R., Goad, C.: W3C Geospatial Ontologies. Tech. rep. (2007)
15. Sun, K. et al.: Geospatial data ontology: the semantic foundation of geospatial data integration and sharing. In: Big Earth Data 3.3, pp. 269–296 (2019)
16. Butler, H., et al.: The GeoJSON Format. Tech. Rep. (2015). https://tools.ietf.org/html/draft-butler-geojson-06
17. Portele, C.: OpenGIS® Geography Markup Language (GML) Implementation Specification, version. In: (2007). https://portal.opengeospatial.org/files/?artifact_id=20509
18. Lee, J., et al.: OGC IndoorGML version 1.0. In: (2014). https://docs.opengeospatial.org/is/14-005r5/14-005r5.html
19. de Berg, M., et al.: Computational Geometry: Algorithms and Applications, 3rd edn. Springer TELOS, Santa Clara, CA, USA (2008)
20. Rusu, R.B., Cousins, S.: 3D is here: point cloud library (PCL). In: IEEE International Conference on Robotics and Automation (ICRA). Shanghai, China (2011)
21. Weiss, W., D'Mello, C.: Fundamentals of Model Theory. Topology Atlas (2000)
22. Kuipers, B.: The spatial semantic hierarchy. Artif. Intell. **119**(1-2), 191–233 (2000)
23. Ghrist, R., et al.: Topological landmark-based navigation and mapping. In: University of Pennsylvania, Department of Mathematics, Tech. Rep 8 (2012)
24. Moore, D, et al.: Robust distributed network localization with noisy range measurements. In: Proceedings of the 2nd international Conference on Embedded Networked Sensor Systems. ACM, pp. 50–61 (2004)
25. Weber, M. et al.: Gordian: formal reasoning based outlier detection for secure localization. Tech. rep. UCB/EECS-2019-1. EECS Department, University of California, Berkeley (2019)
26. Yang, Z., et al.: Beyond triangle inequality: sifting noisy and outlier distance measurements for localization. In: ACM Transactions on Sensor Networks 9.2, pp. 1–20 (2013)
27. Frank, Andrew U.: Qualitative spatial reasoning about distances and directions in geographic space. J. Vis. Lang. Comput. **3**(4), 343–371 (1992)
28. Cohn, A.G., Renz, J., et al.: Qualitative spatial representation and reasoning. In: Handbook of Knowledge Representation, vol. 3, pp. 551–596 (2008)
29. Pereira, E.T.: Mobile reactive systems over bigraphical machines a programming model and its implementation. Ph.D. Thesis, University of California, Berkeley (2015)
30. Atanasov, N., et al.: Semantic localization via the matrix permanent. Robot. Sci. Syst. (2014)
31. Vieu, L.: Spatial and temporal reasoning. In: Stock, O. (ed.) Spatial Representation and Reasoning in Artificial Intelligence, pp. 5–41. Springer, Netherlands, Dordrecht (1997)
32. Weber, M., Lee, E.: A model for semantic localization. In: ACM Press, pp. 350–351 (2015)

IFTTT Rely Based a Semantic Web Approach to Simplifying Trigger-Action Programming for End-User Application with IoT Applications

Arun Kumar and Sharad Sharma

Abstract Introducing semantics into the Internet of Things (IoT) has been attracting increasing attention from researchers and industrial practitioners. Interoperability stays a critical weight to the engineers of the Internet of Things Systems. This is because of the way that the IoT gadgets are exceptionally heterogeneous as far as fundamental correspondence protocol, information organizations, and advances. Our spotlight in this section is on information semantics, liable for definition, the board, and preparing of information. The Internet of Things (IoT), true to form the foundation for the imagined idea of Smart structure, brings new conceivable outcomes for the structure of the executives. IoT vision presents promising and practical answers for gigantic information assortment and its investigation which can be applied in numerous areas thus cause them to work all the more productively. End-user development programming environments in the IoT enable end-users to customize their IoT objects' joint behavior, typically via IFTTT trigger-action rules. Aim of the chapter to propose an IoT based Semantic Interoperability Model with EUPont Semantic Web ontology to provide Semantic Interoperability among heterogeneous IoT devices for control of end-user applications.

Keywords IoT · IFTTT · Semantic IoT · Interoperability · EUPont Semantic

1 Introduction

Semantic interoperability encourages information trades with unambiguous, machine-reasonable significance. This level enhances the interoperability measures of the past level by giving unequivocal importance to the information in question. The main level takes into account unique frameworks to convey; the semantic level empowers them to comprehend what the information is 'stating'. The semantic level is fundamental for astute information differencing, information organization, and genuine machine insight [1–3]. By and large, gauges based interoperability binds

A. Kumar (✉) · S. Sharma
Maharishi Markandeshwar (Deemed to Be University), Mullana, India
e-mail: ranaarun1.ece@piet.co.in

R. Pandey et al. (eds.), *Semantic IoT: Theory and Applications*, Studies in Computational Intelligence 941, https://doi.org/10.1007/978-3-030-64619-6_17

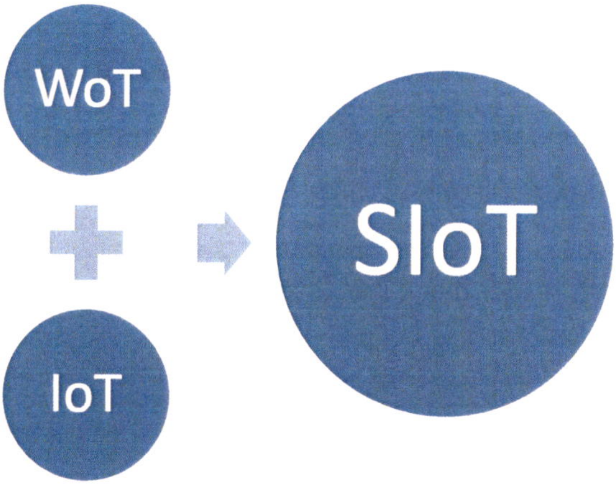

Fig. 1 Semantic technology for the Internet of Things [39]

together the approach of between framework correspondence, semantic interoperability actualizes machine knowledge of the information utilized, and continued interoperability keeps IoT systems going. Semantic interoperability, notwithstanding, is legitimately liable for the machine insight at the center of the robotization depicted in the above use cases and, without which, the IoT is simply one more utilization of huge information sans the convenient activity important to follow through on it Contingent upon the fast selection of wearable registering gadgets, for example, a smartwatch, savvy glasses, and wristband composed wellness tracker, we become ready to appreciate an IoT based shrewd life that can be associating the web anyplace, whenever and any gadget. IoT with the web of things constitutes semantic IoT as shown in Fig. 1. Most cell phones have worked in an assortment of sensors, for example, encompassing light sensor, nearness sensor, worldwide situating framework, accelerometer, compass, and gyroscopic sensor. Particularly, on-going delivered cell phones have contained increasingly ecological sensors, for example, temperature, stickiness, and gauge [4–7]. These sensors make things increasingly splendid and can make progressively clever applications, for instance, social protection applications, educational substance, and extended reality applications. The sensor is one of the most critical advancement segments to make "web of things", because of its screen the circumstance and can grow its helpfulness. There are four degrees of interoperability: (1) particular, (2) syntactic, (3) semantic, and (4) various leveled according to IERC AC4 (European Research Cluster on the Internet of Things) [8–12]. From the given scene, we hope to play out an examination of the present composition for semantics-based exercises for the IoT, WoT, and Smart City. From the examination, the key test appears to be on the most capable technique to ensure semantic interoperability among the current IoT endeavors, stages and conditions given that more than 380 way of thinking based (IoT) reaches out starting at presently exist [13, 14]. To meet these necessities, this chapter shows the open semantic IoT organization arranges advancement that guarantees the semantic interoperability of sensors or things' information and a segment of these benefits.

2 Background

The term semantic began by Berners-Lee in 1998, is a structure produced for automated arranging among people and machines through the translation of proposals to assets [15]. Genuinely, the objective of the Semantic Web is to engage the two people and machines to get a handle on through semantic interoperability dependent on very much depicted repercussions in the current web. At long last, levelheaded thinking is utilized to see new data-dependent on implications of information and relationship in ontologies [16]. Even though IoT advances [17] and the related usage/establishment systems [18], databases [19], prerequisites [20], and protection just as security angles [21], fabricating explicit engineering [22], and correspondence guidelines [23] have just been diagramed and the structures of tangible information get to, administration disclosure, design, and heterogeneity have likewise been introduced [24], the definite conversation of semantic models of IoT arrangements is yet to be had. Albeit a few parts of semantic sensor advancements have been diagramed effectively, a deliberate audit that follows the structure of IoT arrangements is yet to be directed. A chronicled investigation of the advancement of ontologies up until 2014 is introduced in [25]. The potential strategies for semantic explanation are diagramed in [26] which additionally thinks about significant level application-situated ontologies of setting the board. Heartbreakingly, contemporary EUD programming conditions get advancement subordinate depiction models that organize IoT contraptions and organizations by creator or brand. End customers must portray a couple of practically identical gauges to satisfy their needs, paying little heed to whether the standards play out the proportionate lucid movement.

3 IoT Gateways Architecture Based Semantic Web-Based Approach

The Semantic Web-based Information Annotation Framework on IoT Gateways IoT device components is divided primarily around three major elements: Sensor node, Gateway, and cloud platforms. Sensor nodes are usually the lowest stage and are comprised of a group of very small resource sensors and embedded systems whose main purpose is to gather data and transmit it to Gateways. Devices at the gateway level have more computing resources compared to the sensor at the node level. As such, this level works as a hub for aggregating sensory data and bridging connections between sink nodes and IoT cloud services. These platforms gather data from a variety of deployed gateway nodes and provide situational services to end-users customized as notification service, application, or graphical interface. These platforms gather data from a variety of deployed gateway nodes and provide situational services to end-users customized as notification service, application, or graphical interface. As illustrated in Fig. 2 our proposed approach is composed of three modules namely: Data preparation, Data annotation, and cloud interface modules.

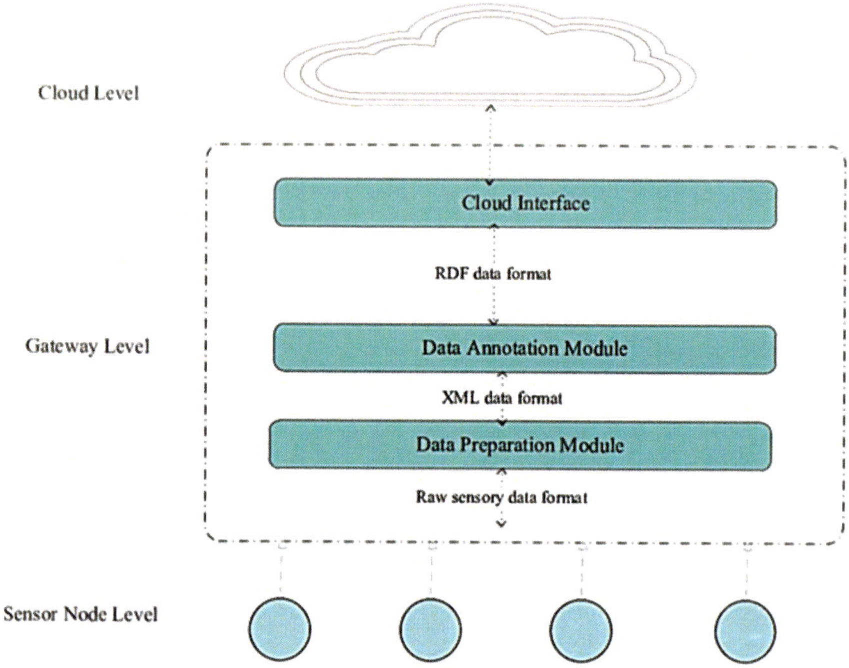

Fig. 2 Semantic IoT gateway architecture [25]

4 Communication Technologies and Protocol

Different technologies and protocol advancements bolster the IoT, and their description is explained below.

4.1 Exclusive Advances

Exclusive advances are numerous in the IoT area, the assortment of utilization cases, and application areas making a wide scope of necessities [23].

- EnOcean, which has just been presented, is a case of remote exclusive innovation.
- Phidget3 is a wired exclusive convention dependent on USB correspondence. In our model, a temperature and a radiance sensor are associated using this innovation.
- Z-Wave protocol is a short-run work that arranges remote correspondence innovation dependent on restrictive radio innovation. The nearness sensor of our utilization case conveys over Z-Wave.

4.2 Short-Extend Advancements

If important, different gadgets conveying locally at a short-range can make a work covering a wide territory [23].

- Bluetooth Low Energy (BLE) protocol is an augmentation of the Bluetooth correspondence innovation intended to have much lower control utilization. BLE is anyway founded on a similar worldview as Bluetooth, and just star topologies are permitted, with a focal ace and some fringe slaves.
- Zigbee protocol is a radio convention created by the Zigbee Alliance4. In opposition to BLE, Zigbee gadgets might be composed in a work. The fundamental attributes and use cases for Zigbee are very like Z-wave. Notwithstanding, since Zigbee is an open standard, more makers can deliver Zigbee gadgets. This makes a progressively assorted biological system; however, it produces interoperability issues among gadgets that should be founded on a similar innovation. The associated light introduced for the utilization case conveys over Zigbee.
- 6LowPan protocol is an abbreviation for "IPv6 over Low-Power Wireless Personal Area Networks", proposed in IETF5 RFC 49446. Conveying an IP arrange over low-control gadgets empowers the formation of work organize at the bundle level (in light of the OSI layered model7). BLE and Zigbee are Personal Area Networks advancements that may bolster 6LowPan systems.

4.3 Long-Extend Innovations

To have the option to actualize some utilization cases, for example, natural observing or horticulture, IoT gadgets must be conveyed over enormous regions, possibly not secured by conventional correspondence systems. A few advances have been created to give specially appointed systems that permit long-range and low-control correspondence [24].

- SigFox8 protocol is both a system administrator and a correspondence innovation sent by said administrator. SigFox gadgets speak with SigFox entryways that are associated with the Internet. Messages delivered by SigFox gadgets are along these lines put away on servers to be open through a Web interface from the customer side.
- Lora protocol is a correspondence innovation that is bolstered by the LoRa alliance9, and despite SigFox it isn't attached to an administrator: anybody may convey an impromptu LoRa arrange. The system topology empowered by LoRa is anyway very like SigFox: gadgets convey over LoRa with passages that are associated with "customary" systems and make the messages accessible to the client on devoted servers. At the point when a LoRa gadget awakens to communicate something specific, it is quickly conceivable to make an impression on it, empowering bi-directional correspondence.

5 Future of IFTTT (if This then that)

Linden Tibbets, IFTTT fellow benefactor, and boss structure official talked about the organization's future, his ongoing choice to step down as CEO, and an appalling remark made by an official at Google when it was declared the finish of its Works with Nest program. The center takeaway from our discussion was that IFTTT is changing its system, and we ought to hope to get familiar with that change before the finish of the summer.

IFTTT (which is another way to say "If this, at that point that") has been around since 2010; it was shaped as an approach to offer adaptability to the advanced articles and administrations we use every day around 700 services and some of the service as shown in Fig. 3. In one of my first discussions with Tibbets, he highlighted a mug around his work area and noticed how in the physical world we could utilize that mug to hold a drink; we could likewise utilize it as a paperweight [13]. Be that as it may, the virtual world is unique. In the virtual world, the product must be utilized for whatever reason its client has assembled it to perform. There's no simple method to take a bit of programming and repurpose it how one may repurpose a mug to turn into a

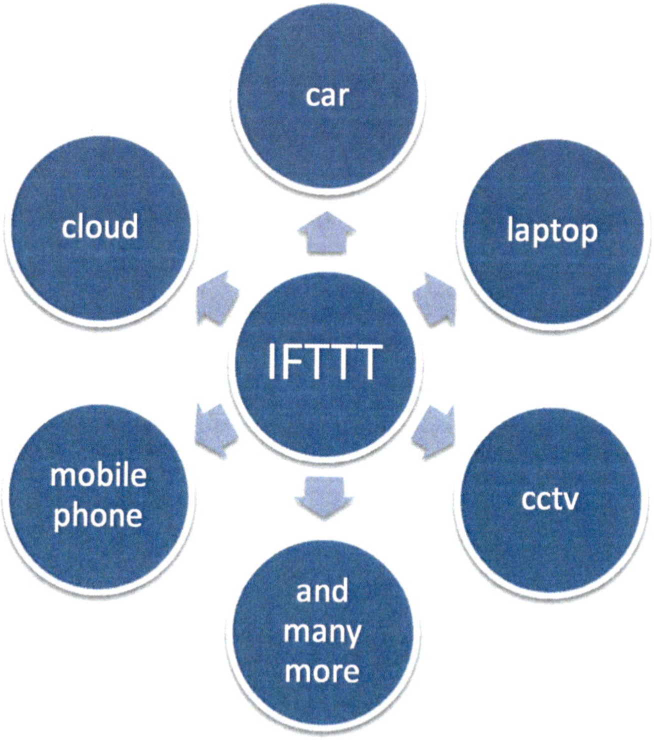

Fig. 3 IFTTT services for the users [39]

paperweight. However, that is the thing that IFTTT does. IFTTT takes the APIs related to electronic administrations and lets those go about as a trigger to make something different occur. For instance, in your email program, the appearance of an email may trigger a notice. Notwithstanding, if you connect that email program to IFTTT, the code that triggers a notice could be repurposed to likewise turn on a light. In its initial days, IFTTT attempted to take a portion of the data encoded in well-known advanced administrations and transform those into triggers that could bring about some sort of activity [14]. So IFTTT could turn your telephone's area to trigger an instant message, or it may utilize the presenting of the tweet on duplicate the sent tweet and consequently send it to Facebook. It did this by associating with an API gave by organizations, for example, Dropbox, Google, Honeywell, and so forth., and building interfacing code to cause the activity to occur on alternate assistance utilizing the other organization's APIs. As characterized by Lieberman et al. [15, 16, 27–31], EUD is "a lot of strategies, procedures, and devices that permit clients of programming frameworks, who are going about as non-proficient programming engineers, sooner or later to make, change or broaden a product curio". Among the business devices, IFTTT is broadly utilized and acknowledged. IFTTT is an accomplishment as far as the client understands ability and convenience, with more than 1 million standards made by its clients. It permits the organization of straightforward associations (named applets) between over 400 upheld IoT objects (named administrations) [32]. The bolstered articles run from business gadgets (e.g., the Nest indoor regulator), to the web or versatile administrations (e.g., Facebook). Applets, in any event in the free form, can incorporate a solitary trigger and a one of a kind activity and can be created by utilizing a wizard-based system (a screen capture has appeared in Fig. 4). For this reason, we created EUPont [2], a metaphysics that considers the elements of the IoT biological system dependent on their classes and abilities, and permits the meaning of theoretical trigger-activity decides that can be naturally adjusted to various logical circumstances [33–36]. EUPont is portrayed in the following Sections, alongside its assessment regarding helpfulness and expressiveness [37–39].

6 Dimensions for Interoperability

Interoperability is: "the limit of at any rate two structures or portions to exchange data and utilization information". This definition gives various challenges in the most ideal manner to:

- Get the data
- Exchange information, and
- Use the data in getting it and having the option to process it.

A basic portrayal of interoperability can be viewed as pursuing shown in Fig. 5:

Syntactical Interoperability is typically connected with equipment/programming segments, frameworks, and stages that empower machine-to-machine correspondence to occur [39]. This sort of interoperability is frequently

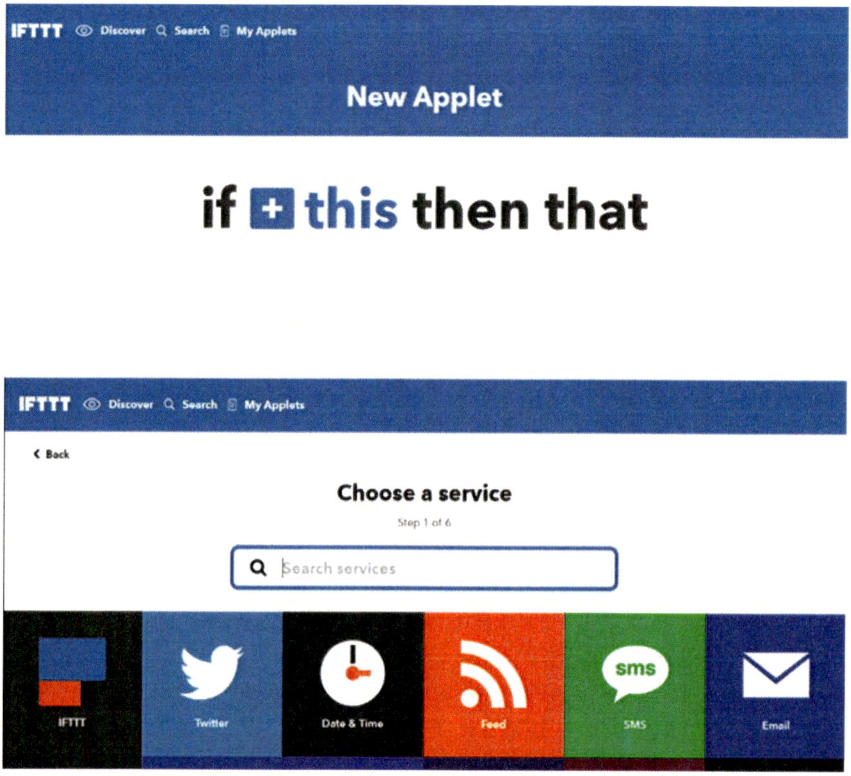

Fig. 4 Applet creation in IFTTT [32]

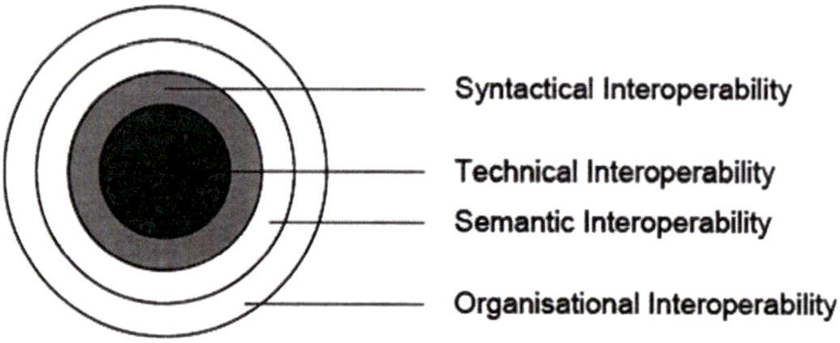

Fig. 5 Dimensions of interoperability [39]

focused on (correspondence) conventions and the foundation required for those conventions to work.

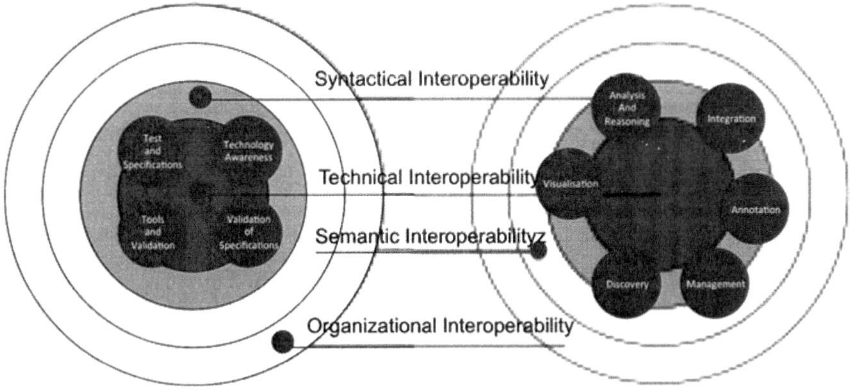

Fig. 6 Associated general challenges with scopes of interoperability [34]

Technical Interoperability is typically connected with **i**nformation positions. Surely, the messages moved by correspondence conventions need to have a well-characterized sentence structure and encoding, regardless of whether it is just as bit-tables. Be that as it may, numerous conventions convey information or substance, and this can be spoken to utilizing elevated level sentence structures, for example, HTML or XML. **Semantic Interoperability** is typically connected with the significance of substance and apprehensions of the social rather than device comprehension of the substance. Subsequently, interoperability in this situation infers for a run of the mill cognizance among people of the significance of the substance (information) being exchanged. **Organizational Interoperability** is the limit of relationship to satisfactorily give and move data regardless of the way that they may be using a wide scope of information structures over extensively different establishments, possibly transversely over different geographic regions and social orders [34]. Associated General Challenges with Scopes of Interoperability is shown in Fig. 6.

7 EUPont Based Semantic Model for END User Application

Specialized Interoperability is typically connected with equipment/programming segments, frameworks, and stages that empower M2M correspondence to occur. EUPont is organized as an ontology with four key blocks: trigger-action programming, Contextual Knowledge, IoT Ecosystem, and Semantic Reasoning, to allow the definition of high-level trigger-action rules that are technology/brand independent and can apply to different contextual circumstances as sown in Fig. 7. The semantic model makes it easy for the end-user development team to determine which IoT devices/services can perform a specific act or create a specific event. This semantic model can be used for the industry level to manage many applications.

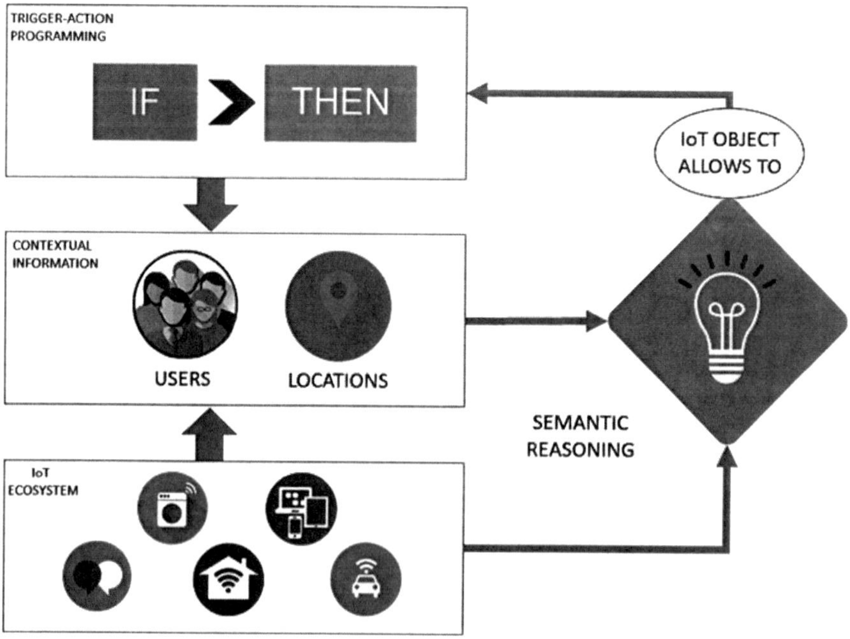

Fig. 7 EUPont ontology structure [35]

Surely, the messages moved by correspondence conventions need to have a well-characterized sentence structure and encoding, regardless of whether it is just as bit-tables. Be that as it may, numerous conventions convey information or substance, and this can be spoken to utilizing elevated level sentence structures, for example, HTML or XML Hierarchical Interoperability, as the name suggests, is the capacity of associations to adequately impart and move (important) information (data) although they might be utilizing a wide range of data frameworks over broadly various foundations, conceivably crosswise over various geographic districts and societies. Hierarchical interoperability relies upon effective specialized, grammatical, and semantic interoperability.

Figure 8 shows the chain of importance for some lighting-related activities. EUPont stages in the Internet of Things (IoT) enable clients to characterize and tailor joint practices between IoT gadgets and administrations in different zones, similar to the home, the vehicle, or for a sound way of life, regularly through trigger-activity rules. Such stages are effective as far as the client understandability and usability; however, they display different issues and difficulties as the quantity of accessible interconnected "things" develops. For example, a client can't make an IoT application that can be applied to all her associated lights, except if they are similarly marked, nor too different sorts of gadgets that may give inside lighting. EUPont is a Semantic Web philosophy that empowers clients to customize the joint practices

Fig. 8 Interface for composing trigger-action rules with EUPont [36]

between their IoT gadgets with less, more elevated level standards than contemporary stages. Such rules can be likewise adjusted to various logical circumstances and so far obscure IoT gadgets and administrations.

8 Conclusion and Future Scope

The essential goal of this chapter is to give interoperability among heterogeneous IoT devices by using semantically remarked on for various IoT sensors for end-user applications and also discussed EUPont to allow all end-user that can be adapted to different contextual situations. Use of, EUPont provides a very less number of rules to define their needs. The IoT Ecosystem square models IoT devices and organizations as substances that idea in any event one functionalities. Each client can have bearings to play out specific exercises or notification to enroll event crowd individuals. Impending IoT will display critical challenges similar to interoperability between different progressions and brands. Such challenges will similarly impact end customers' ability to change IoT devices and also impact Industry Application Control with IoT Scenario. We believe that EUPont could serve as the core information layer for future IoT end-user programming solutions.

Conflicts of Interest The authors declare that no conflicts of interest exist about the publication of this paper.

References

1. Rezaei, R., Chiew, T.K., Lee, S.P., Aliee, Z.S.: Interoperability evaluation models: a systematic review. Comput. Ind. **65**(1), 1–23 (2014)

2. Serrano, M., Barnaghi, P., Cousin, P.: Semantic interoperability: research challenges, best practices, solutions, and next steps, IERC AC4 manifesto. European Research Cluster on the Internet of Things, AC4, Tech. Rep. (2014)
3. Serrano, M., Barnaghi, P., Carrez, F., Cousin, P., Vermesan, O., Friess, P.: Internet of Things IoT semantic interoperability: research challenges, best practices, recommendations, and next steps. European Research Cluster on the Internet of Things, AC4, Tech. Rep. (2015)
4. Kumar, A., Salau, A.O., Gupta, S., Paliwal, K.: Recent trends in IoT and its requisition with IoT built engineering: a review. In: Advances in Signal Processing and Communication, pp. 15–25. Springer, Singapore (2019)
5. Rana, A.K., Sharma, S.: Enhanced energy-efficient heterogeneous routing protocols in WSNs for IoT application. IJEAT **9**(1), 4418–4415 (2019)
6. Kumar, K., Gupta, S., Rana, A.: Wireless sensor networks: a review on "Challenges and opportunities for the future world-LTE". Amity J. Comput. Sci. (AJCS) **1**(2) (2018). ISSN: 2456-6616
7. Rana, A.K., Krishna, R., Dhwan, S., Sharma, S., Gupta, R.: Review on artificial intelligence with Internet of Things-problems, challenges and opportunities. In: 2019 2nd International Conference on Power Energy, Environment and Intelligent Control (PEEIC), pp. 383–387. IEEE, October 2019. https://doi.org/10.1109/peeic47157.2019.8976588
8. Muralidharan, S., Yoo, B., Ko, H.: Designing a semantic digital twin model for IoT. In: 2020 IEEE International Conference on Consumer Electronics (ICCE), Las Vegas, NV, USA, pp. 1–2 (2020). https://doi.org/10.1109/icce46568.2020.9043088
9. Khan, M.N., Naseer, F.: IoT based university garbage monitoring system for healthy environment for students. In: 2020 IEEE 14th International Conference on Semantic Computing (ICSC), San Diego, CA, USA, pp. 354–358 (2020). https://doi.org/10.1109/icsc.2020.00071
10. Guo, C., Jia, J., Jie, Y., Liu, C.Z., Choo, K.R.: Enabling secure cross-modal retrieval over encrypted heterogeneous IoT databases with collective matrix factorization. IEEE Internet Things J. **7**(4), 3104–3113 (2020). https://doi.org/10.1109/JIOT.2020.2964412
11. Hu, L., Wu, G., Xing, Y., Wang, F.: Things2Vec: semantic modeling in the internet of things with graph representation learning. IEEE Internet Things J. **7**(3), 1939–1948 (2020). https://doi.org/10.1109/JIOT.2019.2962630
12. Li, Q., Cao, Z., Tanveer, M., Pandey, H.M., Wang, C.: A semantic collaboration method based on uniform knowledge graph. IEEE Internet Things J. **7**(5), 4473–4484 (2020). https://doi.org/10.1109/JIOT.2019.2960150
13. Gyrard, A., Bonnet, C., Boudaoud, K., Serrano, M.: LOV4IoT: a second life for ontology-based domain knowledge to build Semantic Web of Things applications. In: 4th International Conference on Future Internet of Things and Cloud (FiCloud). IEEE (2016)
14. Gyrard, A., Atemezing, G., Bonnet, C., Boudaoud, K., Serrano, M.: Reusing and unifying background knowledge for Internet of Things with LOV4IoT. In: 4th International Conference on Future Internet of Things and Cloud (FiCloud). IEEE (2016)
15. Branagh, P., Wang, W., Henson, C., Taylor, K.: Semantics for the Internet of Things: early progress and back to the future. Int. J. Sem. Web Inf. Syst. (IJSWIS) **8**, 1–21 (2016)
16. Skillen, K.L., Chen, L., Nugent, C.D., Donnelly, M.P., Burns, W., Solheim, I.: Ontological user modeling and semantic rule-based reasoning for personalization of Help-On-Demand services in pervasive environments (2017)
17. Whitmore, A., Agarwal, A., Da Xu, L.: The internet of things—a survey of topics and trends. Inf. Syst. Front. **17**(2), 261–274 (2015)
18. Rana, A.K., Sharma, S.: Industry 4.0 manufacturing based on IoT, cloud computing, and big data: manufacturing purpose scenario. In: Advances in Communication and Computational Technology, pp. 1109–1119. Springer, Singapore (2021)
19. Hellerstein, J.M., Hong, W., Madden, S.R.: The sensor spectrum: technology, trends, and requirements. SIGMOD Rec. **32**(4), 22–27 (2003)
20. Yaqoob, I., Ahmed, E., Hashem, I.A.T., et al.: Internet of things architecture: recent advances, taxonomy, requirements, and open challenges. IEEE Wirel. Commun. Mag. **24**(3), 10–16 (2017)

21. Lin, J., Yu, W., Zhang, N., Yang, X., Zhang, H., Zhao, W.: A survey on internet of things: architecture, enabling technologies, security and privacy, and applications. IEEE Internet Things J. **4**(5), 1125–1142 (2017)
22. Ehrlich, M., Wisniewski, L., Jasperneite, J.: State of the art and future applications of industrial wireless sensor networks. In: Kommunikation und Bildverarbeitung in der Automation, Technologien für die intelligente Automation, pp. 28–39. Springer, Berlin, Germany (2018)
23. Dizdarević, J., Carpio, F., Jukan, A., Masip-Bruin, X.: A survey of communication protocols for internet of things and related challenges of fog and cloud computing integration. ACM Comput. Surv. **1**, 1–27 (2018)
24. Mineraud, J., Mazhelis, O., Su, X., Tarkoma, S.: A gap analysis of internet-of-things platforms. Comput. Commun. **89**, 5–16 (2016)
25. Wang, X., Zhang, X., Li, M.: A survey on semantic sensor web: sensor ontology, mapping and query. Int. J. u- e- Serv. Sci. Technol. **8**(10), 325–342 (2015)
26. Rana, A.K., Sharma, S.: Contiki Cooja Security Solution (CCSS) with IPv6 routing protocol for low-power and lossy networks (RPL) in Internet of Things applications. In: Mobile Radio Communications and 5G Networks, pp. 251–259. Springer, Singapore (2020)
27. Corno, F., De Russia, L., MongeRoffarello, A.: A high-level approach towards end-user development in the IoT. In: Proceedings of the 2017 CHI conference extended abstracts on human factors in computing systems, Denver, CO, USA (2017)
28. Ur, B., Yong Ho, M.P., Brawner, S., Lee, J., Mennicken, S., Picard, N., Schulze, D., Littman, M.L., Lieberman, H., Paternò, F., Klann, M., Wulf, V.: End-user development: an emerging paradigm. In: End User Development, pp. 1–8 (2006)
29. Aggarwal, D., Bali, V., Mittal, S.: An insight into machine learning techniques for predictive analysis and feature selection. Int. J. Innov. Technol. Explor. Eng. (IJITEE) **8**(9S), 342–349 (2019). ISSN No. 2278-3075.
30. Lee, J., Garduño, L., Walker, E., Burleson, W.: A tangible programming tool for creation of context-aware applications. In: Proceedings of the 2013 ACM International Joint Conference on Pervasive and Ubiquitous Computing, Ubicomp '13 (2013)
31. Juneja, A., Juneja, S., Bali, V., Mahajan, S.: Multi-criterion decision making for wireless communication technologies adoption in IoT. Int. J. Syst. Dyn. Appl. (IJSDA) **10**(1), Article 1 (2020)
32. Ur, B., McManus, E., Pak Yong Ho, M., Littman, M.L.: Practical trigger-action programming in the smart home. In: Proceedings of the SIGCHI Conference of Human Factors in Computing Systems, CHI '14 (2014)
33. De Russis, L., Corno, F.: HomeRules: a tangible end-user programming interface for smart homes. In: Proceedings of the 33rd Annual ACM Conference Extended Abstracts on Human Factors in Computing Systems (2015)
34. Ghiani, G., Manca, M., Paternò, F.: Authoring context-dependent cross-device user interfaces based on trigger/action rules. In: Proceedings of the 14th International Conference on Mobile and Ubiquitous Multimedia (2015)
35. Kumar, A., Sharma, S.: Demur and routing protocols with application in underwater wireless sensor networks for smart city. In: Energy-Efficient Underwater Wireless Communications and Networking, pp. 262–278. IGI Global (2020)
36. Huang, J., Cakmak, M.: Supporting mental model accuracy in trigger-action programming. In: Proceedings of the 2015 ACM International Joint Conference on Pervasive and Ubiquitous Computing, Osaka, Japan (2015)
37. Barricelli, B.R., Valtolina, S.: Designing for end-user development in the Internet of Things. In: End-User Development, vol. 9083, pp. 9–24. Springer (2015)
38. Rana, A., Arora, O., Syal, N., Singh, P.: Holographic versatile disc: high speed information storage systems. In: International Congress on Ultra Modern Telecommunications and Control Systems, pp. 934–939. IEEE, October 2010
39. Berners-Lee, T., Hendler, J., Lassila, O.: The semantic web. Sci. Am. **284**(5), 28–37 (2001)

Semantic Internet of Things (IoT) Interoperability Using Software Defined Network (SDN) and Network Function Virtualization (NFV)

Jayashree R. Prasad, Shailesh P. Bendale, and Rajesh S. Prasad

Abstract In the coming few years, there is going to be rapid advancement in terms of technology like 5G, 6G, etc. When we think about 5G, the performance of the internet is going to increase multifold. The 5G i.e. fifth-generation network is going to be very heterogeneous. There is a need for a standardized solution to the issues in this technology. In this work, we are trying to understand the problems specifically in IoT (Internet of Things) area of 5G. In the literature survey, we found out that various solutions have been proposed in the area of the Internet of Things, but there is a lack of some generic solutions for all IoT projects. Every project works excellent in its closed and specified environment. When we try to connect multiple IoT projects, there is a big problem of interoperability. Some ICT standardization organizations have proposed some solutions to interoperability to overcome this scenario. Few solutions have been proposed by some authors to provide interoperability using semantic technologies. The solution for this problem of heterogeneous IoT can be provided using the semantic technologies in combination with that of SDN (Software Defined Network), NFV (Network Function Virtualization), and Cloud infrastructure.

1 Introduction

In this chapter, we are going to address the Semantic IoT interoperability issues of heterogeneous IoT devices in 5G, SDN, and NFV, cloud computing, solutions to overcome those interoperability issues.

J. R. Prasad (✉)
Sinhgad College of Engineering, Pune, India
e-mail: jrprasad.scoe@sinhgad.edu

S. P. Bendale
NBN Sinhgad School of Engineering, Pune, India
e-mail: bendale.shailesh@gmail.com

R. S. Prasad
Amity School of Engineering and Technology, Amity University Rajasthan, Jaipur, India
e-mail: rsprasad@jpr.amity.edu

R. Pandey et al. (eds.), *Semantic IoT: Theory and Applications*, Studies in Computational Intelligence 941, https://doi.org/10.1007/978-3-030-64619-6_18

2 Internet of Things (IoT)

IoT architecture is divided into 3 layers, IoT device layers, IoT oriented Cloud Networks and Platforms, and IoT devices. We are going to discuss all three layers in detail (Fig. 1).

IoT device layer

This layer consists of devices capable of identify, sense and actuate capabilities for interaction with that physical environment. These devices may consist of RFID tags, Wireless sensor motes, smart cars, industrial appliances, wearable IoT devices, etc.

IoT oriented Cloud Networks and Platforms

Due to the emergence of 5G or 6G networks in the industry and academia, a huge number of IoT devices have been launched into the market for sensing, actuation, accounting resource constraints, battery life, etc. To overcome this problem, two major solutions have been proposed to overcome this increasing demand and supply terminology. IPv6 based IoT protocol stack, Cloud, and edge computing.

Application Layer

This layer is used to provide the final service to the end-users. The major use cases in this layer will be related to smart cities, smart homes, military appliances, Health

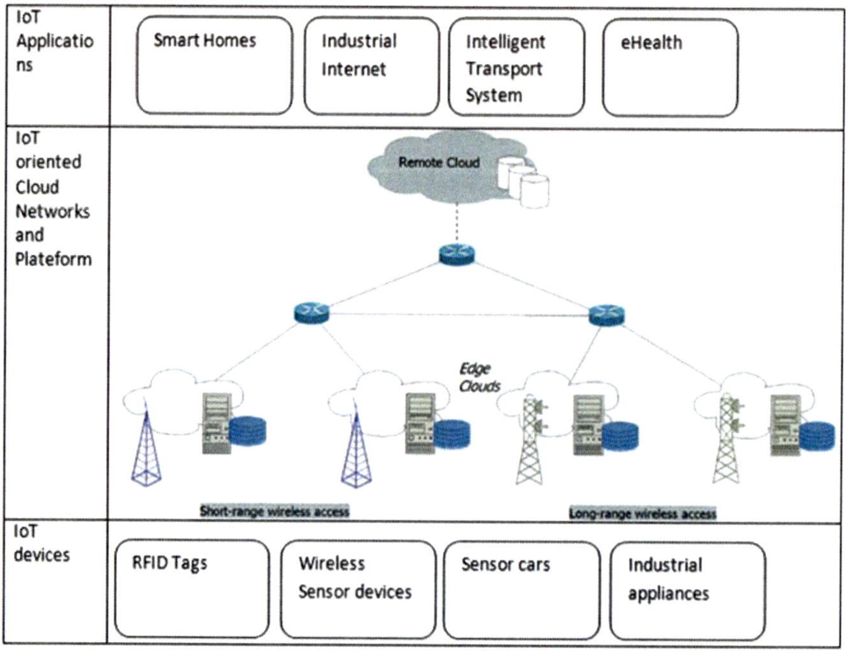

Fig. 1 IoT layers

and medical appliances, energy-conserving solutions, smart and intelligent transport mechanisms, etc.

3 Semantic IoT Interoperability Terminologies

The Semantic IoT consists of three different components Information model, data model, and ontology. The current IoT infrastructure is not interoperable with each other in different environments. Every infrastructure has its own semantics to understand information, data, and knowledge on and off the web. The Web of Things (WoT) uses the API to understand the system data and metadata. But these API does not speak the same languages between different environments which creates the problem of semantic IoT interoperability. Some standardized solutions like SDN, NFV, and Cloud can be used to provide the semantic IoT interoperability.

Some of the file formats used in semantic IoT interoperability are as follows XML, RDF, and JSON. XML: eXtensible Markup Language (XML) comes with an XML schema for the representation of IoT data in the proper format. It may consist of various types of nodes that are used for the storage of various types of attributes related to IoT devices which can be used to provide semantic interoperability between heterogeneous IoT devices. RDF: Resource Description Framework (RDF) comes with RDF schema which is used for the representation of information related to the connected devices in the IoT environment. RDF consists of triples i.e. subject, predicate, and the object. RDF graph or RDF datasets are the collection of these triples. JSON: JavaScript Object Notation is one of the data interchange format similar to the XML. XML is an older format as compared to JSON. It tries to get the features of the object. Some works have been proposed by some authors for proposing JSON schema. But it has not come as a standard yet.

Some languages which are used for the implementation of semantic interoperability Ontology: Ontology is used for the conversion of structured information crested by heterogeneous IoT devices into machine-understandable form. "Representation of a shared conceptualization of a particular domain" is the general definition for Ontologies. The four main components of Ontology are Classes, attributes, individuals, and relations. For defining more specific concepts of the domain, a Class and its one or more subclasses are used. With the help of their attributes, properties, and characteristics of Classes and subclasses are represented. Individuals are the instances of classes or their properties. Finally, relations are used for connection with the edges and all the presented components. Few of the ontology languages are OWL (Web Ontology Language), DAML (DARPA Agent Markup Language), and OIL (Ontology Interference Layer). Web Ontology Language (OWL): It is the mechanism used for the representation of knowledge gained from the concepts of the ontology. The knowledge may be generated by RDF/XML. NoSQL databases: NoSQL databases term is used commonly used in too many implementations. These include multi-paradigm, graph, document, object, triple store, key-value, and other types of databases. Need good semantic IoT interoperability solutions for it since it varies from vendor to vendor. eXtensible Access Control Markup Language: Attribute-Based Access Control (ABAC) policies are majorly used by eXtensible Access Control Markup Language (XACML). For its various internal formats usage of XML is done. SPARQL: It is one of the semantic Query search language which consist of SPARQL Protocol and RDF Query language.

Some terms related to semantic IoT required to be understood. Semantic Translation: Semantic translation is the mechanism that provides seamless communication (one direction or bi-direction) between two different platforms created for the IoT environment. Semantic interoperability: There is a lot of exchange of data, information, and knowledge in a meaningful manner between various applications, agents, and services. Meanwhile, this information is communicated within various types of heterogeneous IoT devices environment or ecosystem. This process is called semantic interoperability. Open interoperability: In open interoperability, various IoT devices from various heterogeneous backgrounds will be communicating with each other irrespective of the vendors. They will be successfully communicating with each other. Machine interpretability: In machine interpretability, various IoT devices will be communicating with each other in machine-understandable format, which can also be reused with the other environments to get faster results depending upon the knowledge derived from the communicating devices.

There are many solutions proposed to overcome the problem of semantic IoT interoperability till now. We will try to discuss some of the solutions in short.

The authors of the paper [1] focus on the Agent-Based Computing technique to support modeling, programming, and simulations for IoT. The Agents are useful for providing technical, syntactical, and semantic interoperability. But there are some limitations, in such kind of agent-based solutions like they cannot be applied for IoT everywhere. Agents are not the universal solution for semantic IoT interoperability. In [2], authors have proposed a solution using ontology alignment used for semantic translation for IoT with the help of Geospatial data. In [3], authors have proposed a solution for identifying common patterns under the Inter IoT project applicable for IoT artifact integration. In [4], authors have proposed a solution for semantic Interoperability between the IoT devices. In [5], authors have proposed a solution for simple and complex alignments under the Inter IoT project used for semantic translation. The usage of a semantic Software tool called Inter Platform Semantic Mediator (IPSM) was done. In [6], authors have proposed a solution for Semantic similarity using different dimensions.

In [7], authors have surveyed for semantic interoperability under the INTER IoT project for creating ontology from different data formats like RDF, XML, NoSQL databases, JSON, relational databases. In [8], authors have proposed a solution for semantic integration from a traditional approach to a cooperative approach for Dependable Embedded Wireless Infrastructure (DEWI) EU project. In [9], authors have proposed a solution with ontology-based management tools for semantic Attribute-Based Access Control policies. In [10], authors have proposed a solution for Identity management in IoT for overcoming the problem of semantic interoperability.

In [11], authors have proposed a solution for E/M health and transportation and logistics using semantic interoperability in the IoT environment under the INTER

IoT project. In [12], authors have proposed a solution for e-health using eXtensible Access Control Markup Language, privacy ontology, and HL7 security was used for the implementation of Policy Information Policy. In [13], authors have proposed a technique for semantic translations for Inter Platform Semantic Mediator (IPSM) under the INTER IoT project. In [14], authors have surveyed the practical implementation tools for ontology under the INTER IoT project. In [15], authors have discussed the 5 popular methods of ontology for the INTER IoT project as a solution to IoT interoperability. In [16], authors have proposed a semantic translation mechanism for the IoT ecosystem under the INTER IoT project. In [17], authors have discussed the standards for open interoperability for the IoT ecosystem in the INTER IoT project. In [18], authors have discussed the semantic interoperability for the IoT ecosystem in the INTER IoT project.

In [19], authors have proposed and discussed ontology and related it to Web Ontology Language (OWL), introduction to machine interpretability at context level. In [20], authors have proposed and discussed Web Ontology Language (OWL) related to machine interpretability using an ontology editor tool called Protégé. In [21], authors have proposed a solution related to the usage of SPARQL for OWL (Web Ontology Language)/RDF (Resource Description Format) for prepositions. In [22], authors have proposed a solution for semantic for SDN resource management.

In [23], authors have summarized some technologies like SDN, NFV, and cloud computing technologies to be used for the IoT environment for providing the security. So these concepts will be useful for 5G and 6G technology development in the future. We will discuss various solutions related to SDN, NFV and cloud technologies in this chapter. In [24], the authors have given the optical extension for BGP protocol which is used as a standard in the inter-domain routing. So such kinds of extension can be proposed for SDN, NFV, and cloud technologies together to provide the semantic interoperability between various IoT environments. Now the challenges related to SDN will be discussed. The authors of the paper [25] focus on the new use cases generated in SDN due to the COVID-19 pandemic situation. The same scenarios can be used for the IoT environment too for providing security. The main focus of authors in the papers [26, 27] is on providing the artificial intelligence in SDN using various machine learning techniques. It also opens the door for new future aspects in the field of SDN. The same kind of solutions can be extended in the IoT environment for interoperability. The authors of the paper [28] give insights into the SDN and NFV implementation aspects for providing security. The same kind of security aspects need to be considered while combining the solution for IoT interoperability. Apart from the aspects mentioned in the paper, the aspect of energy efficiency also needs to be considered while designing the new solutions.

We have discussed various solutions proposed by various authors to provide semantic IoT interoperability between various heterogeneous devices and environments. Now we discuss some terms for standardized solutions that can be used for providing semantic IoT interoperability.

4 SDN (Software Defined Network)

The Software-defined network is very helpful because of its centralized nature and programmable capability. It is having two separate planes, one data plane and another control plane. In SDN, we are separating both the control and data information from each other. So these two planes are decoupled from each other. The added advantage is we can implement a heterogeneous environment together with a centralized control. It also reduces the CAPEX (Capital Expenditure) and OPEX (Operational Expenditure) of the organization on a huge scale. But the disadvantage is of security of a centralized controller is very important. We can combine various heterogeneous IoT environments with this concept of SDN which is programmable. On top of this controller, we can apply various semantic technologies using the NFV (Fig. 2).

The architecture of SDN can be broken down into 3 planes-

Data Plane/Layer: The lowest layer of SDN architecture is also known as infrastructure plane. It contains a specific set of traffic forwarding or traffic processing resources including virtual switches and physical switches. Virtual switches refer to software-based switches whereas physical switches refer to hardware-based switches. The data plane lies on the south side of the architecture. This plane receives packets, takes actions on those packets, and updates counters. Types of actions performed by data planes include sending packets out a single port or multiple ports, modifying packet headers, and dropping the packet.

Control Plane/Layer: The brain of the SDN is the Control Plane. The control plane job is to provide the network-related core services and programmable interfaces for the networking nodes connected. All the protocols are placed in this plane. Control Plane functions includes the exchange of routing table information, management, and system configuration. The controller interacts with the other two planes using three communication interfaces - southbound, east/westbound, and northbound interfaces.

1. SBIs = The SBIs are southbound interfaces defined between the control plane and the data plane.
2. EBI/WBI = The east/westbound interfaces correspond to adjacent controllers in large SDN networks.

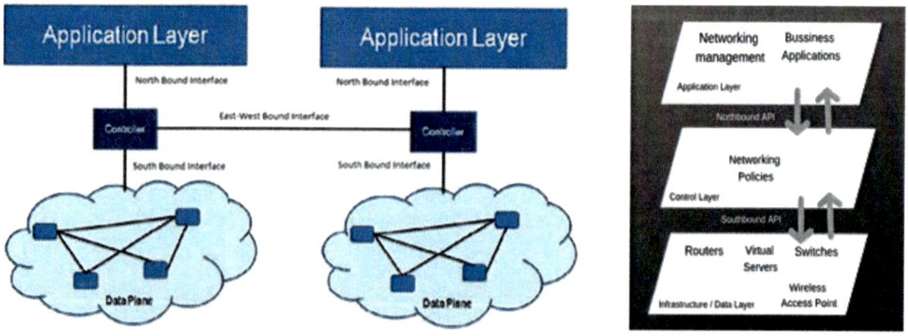

Fig. 2 SDN Architectures

3. NBIs = The NBIs are northbound interfaces defined between the control plane and the application plane.

Application Plane/Layer: This plane is placed at the top of SDN architecture. They consist of applications or programs per se that make use of an abstract view of the whole network for their various internal decision-making purposes.

5 NFV (Network Function Virtualization)

This NFV is used for performing various standalone operations like Firewall, Load balancers, Intrusion Detection Systems, etc. on different VM (Virtual Machines). Some functionality may also have various VNF (Virtual Network Functions') as Slices on a single VM.

VNF (Virtual Network Function): If one application is divided into several parts, then each part can be run as separate VNF for that dedicated task. It is also called a Network Slice.

NFV Architecture: The NFV is used to virtualize the network functions of the network devices such as switches, routers, etc. which would help to make the systems more scalable and traditional network devices instead (Fig. 3).

The NFV comprises of 3 main functionalities

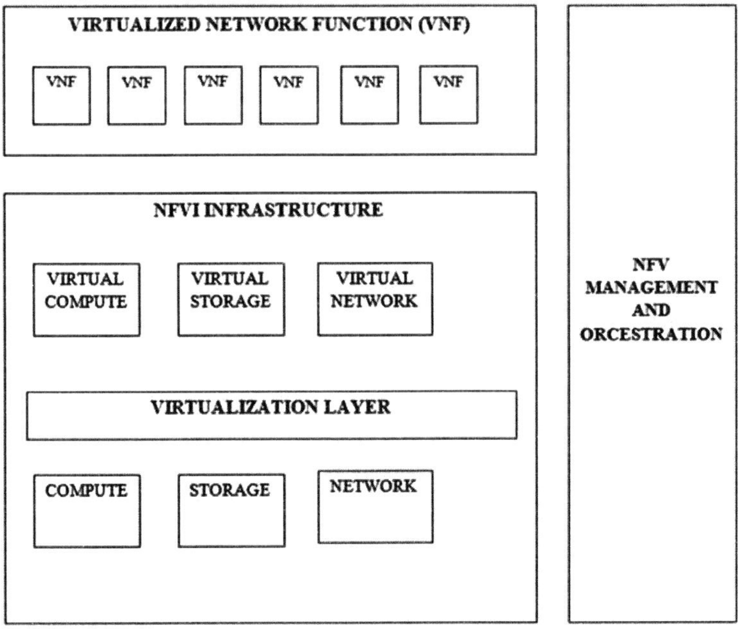

Fig. 3 NFV architecture

1. Virtualization
2. Softwarization
3. Orchestration and Automation.

The NFV architecture is briefly categorized into 3 parts they are the:

1. Network Function Virtualization Infrastructure (NFVI)
2. Virtualized Network Functions (VNF)
3. Management and Network Orchestration.

Network Function Virtualization Infrastructure (NFVI):

It is like the cloud-based data center for providing various hardware and software resources in NFV architecture. It mainly has 3 parts:

a. Hardware Resources.
b. Virtualization Layer
c. Virtualized Resources.

The hardware resource includes compute, storage, and network devices. The virtualization layer helps in decoupling the hardware resources and software resources. This can be done various proprietary resources like the KVM, VMWARE, etc. In the third part, virtualized resources include virtual compute, virtual networks.

Virtualized Network Functions (VNF):

The major building block of NFV architecture is VNF which performs actual virtualized functions like switching and routing. They are the software implementations of the virtual functions. E.g. Virtual firewall, load balancers, intrusion detection system, etc.

Management and Network Orchestration:

Management and orchestration is divided into 3 parts.

a. Virtualized Infrastructure Manager
b. VNF Manager
c. Orchestrator.

The virtualized Infrastructure Manager is used for the communication between the VNFI storage, compute and network for controlling and management purposes; various deployment and monitoring tools are made available by the virtualization layer. The VNF manager is used for the management of the VNFs life cycle, it includes the initialization, updating, querying, scaling, and terminating of the VNF instances. Various network services which include the instantiation, policy management, performance management, Monitoring are provided by orchestrator manager.

6 Cloud Technologies

Since all concepts are related to the programming environment and centralized storage. The best solution for us is to make the use of a cloud platform for it. Create a solution for the problem as a service.

7 Solutions for Semantic Interoperability for IoT

7.1 Cloud Based Solution

In [29], the author has tried to provide a cloud-based generic solution called Experimentation as a Service. He has created a testbed on the Internet of things for interoperability using semantic web technologies. He has also created a set of tools that will be helpful for workflow experimentation and IoT deployments irrespective of the independent domain. This solution is a cloud-based service which is providing increased scaling, cross-domain innovation, heterogeneity using various Application Programming Interfaces (APIs).

Due to many challenges in IoT domain like Industry 4.0, smart cities, smart home, etc. requires a unique solution which is having minimum changes to be done for applicability as a standard in the market. The semantic interoperability is the major challenge that is currently being faced by various new innovative solutions implemented in Industry and academia. Everyone is proposing solutions to the problem, but interoperability is not considered at the time of implementation. So to overcome this problem of interoperability one solution as Experimentation as a Service is proposed which can be utilized over multiple heterogeneous IoT testbeds (Table 1).

Table 1 Various solutions related to experimental testbeds

Experimental testbeds	Descriptions
Wisebed [30]	Wireless sensor network experimentation testbeds for simulation and emulation
FIT IoT lab [31]	Open experimental IoT testbed created for large scale
Fed4FIRE [32]	Future internet research and experiment (FIRE) facilities for experimentation on cloud and services
GENI [33]	Global environment for networking Innovation, virtual laboratory which provides experimentation for security and services
SmartSantander [34]	Smart city testbed for IoT experimentation
Livelab [35]	Experimentation testbeds for mobile sensing and behavioural analysis
Mobile sensing testbeds [36]	Mobile crowd sensing testbeds for IoT experimentation

Various experimental testbeds and their short descriptions are mentioned in the above table

Due to the above-mentioned techniques, experimentation becomes easy using the Application Programming Interface (API) and set of tools. The major advantage of using such techniques in the IoT environment is that we can have the cross-domain communication and various kinds of innovative solutions for providing the scalability and heterogeneity. Still, there is a need for solutions that can be used for the existing IoT environment. One solution proposed is Experimentation as a Service (EaaS). It consists of EaaS meta platform, Testbed APIs, different representation formats for RDF (e.g. JSON-LD, RDF/XML, turtle, etc.) for common ontology topologies, IoT registry, Experiment Result Storage (ERS), Experiment Management Console (EMC), Experiment Registry Module (ERM), Experiment Execution Engine (EEE), Semantic Annotator, Testbed Provider Service (TPS), Testbed and Resource Registration (TRR) module, Policy Enforcement Point (PEP), triplestore Database, Data Endpoints, Resource Broker(RB), Resource Manager (RM), SPARQL Endpoints, TDB query engine, web-based API, Results are captured in any data representation formats like JSON, CSV, XML, etc., Testbed Provider Interface(TPI), Testbed Provider Services (TPS). To provide the experimenter with Data Analysis using Knowledge Acquisition Toolkit (KAT) based Data Analytics web services (DAaaS). The detailed architecture and results can be found in the paper [29].

The solution which is provided in this chapter is not bounded to any ontology technique provided by semantic technology. The best part of these techniques is that the solutions are extendable and reusable. In the future, we can create some service for the cloud which will make the use of semantic technologies and will improve the performance using the SDN and NFV together.

7.2 SDN Based Solution

In [22], the author has tried to provide a solution specific to SDN. He performed some experiments and showed the applicability of SDN for Resource management. To implement this solution usage of OWL, RDF, SPARQL was been done. The Software of Protégé was used to replicate the results. In future various testbeds can be created for providing the semantic IoT interoperability using SDN (Fig. 4).

We have proposed the SDN based IoT interoperability solution as shown in the figure. The mentioned solution will be applicable for all kinds of IoT devices like WSN, WiFi, and cellular networks.

Protégé implementation of Semantic SDN has been demonstrated, so this kind of implementation can be replicated for Semantic IoT interoperability in the future using some emulation, simulation, or testbed (Fig. 5).

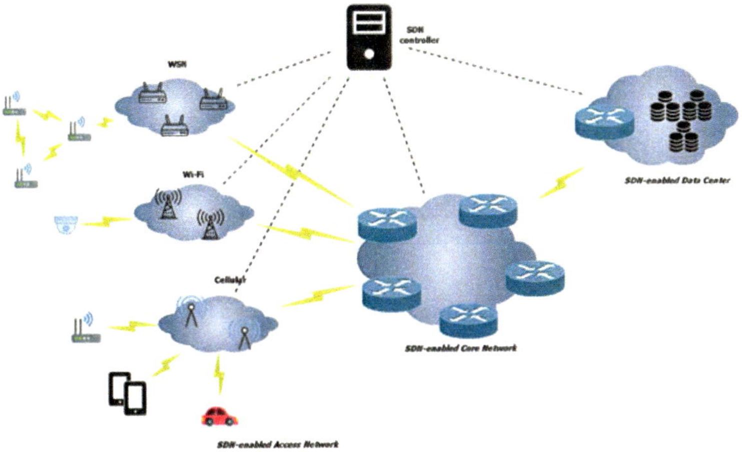

Fig. 4 SDN based IoT interoperability solution

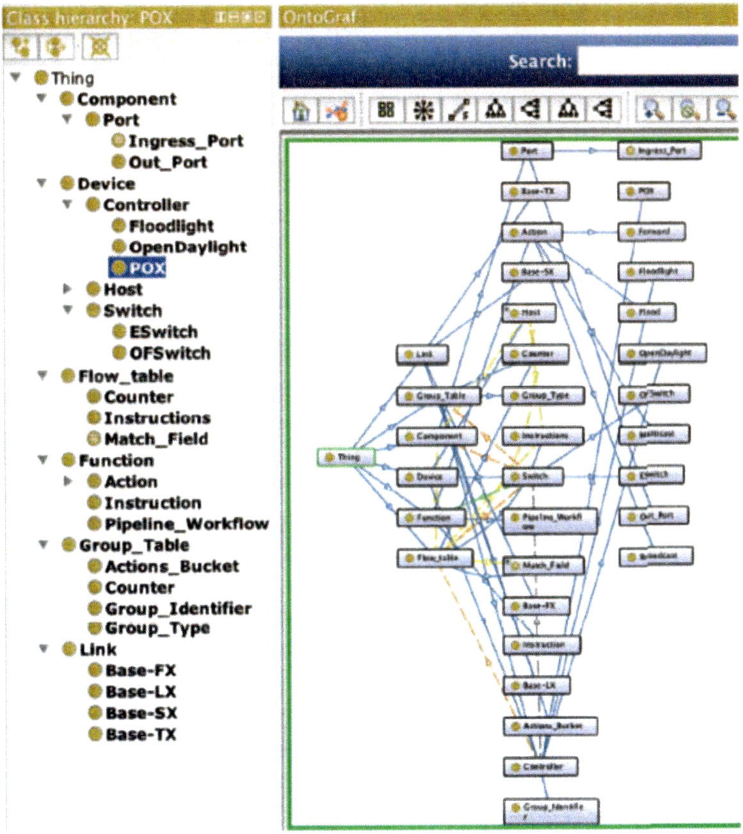

Fig. 5 SDN solution prepared in protégé [22]

7.3 NFV Based Solution

In [37], the author has tried to provide a solution specific to NFV. He performed some experiments to provide the semantic ontology for automatic network service generation and VNF management automation. To provide this semantic ontology the authors have used NSD (Network Service Description), VNFD (Virtual Network Function Description), VNFFG (VNF Forwarding graphs). To provide the NSD management automation TOSCA (Topology and Orchestration Specification for Cloud Applications) is used.

TOSCA consist of various kinds of Templates that are used to provide the semantic between the NFV nodes. The names of those templates and types are as follows: TOSCA service template, Topology template, Node template, Relationship template, Node types, Relationship types, and plans. Various kinds of annotation mechanisms are also discussed.

We have proposed NFV based IoT interoperability solution in the figure which is applicable for various kinds of Data Centers, Edge Data Center, Smart IoT devices, etc. the major role will be played by the ETSI MANO which will consist of NFV Orchestrator, VNF Manager, VIM and support for OSS (Operations Support system)-BSS (Business Support System) (Fig. 6).

In Protégé NSD solution was experimented and implemented for NFV, we can extend the implementation of NFV for semantic IoT interoperability in future (Fig. 7).

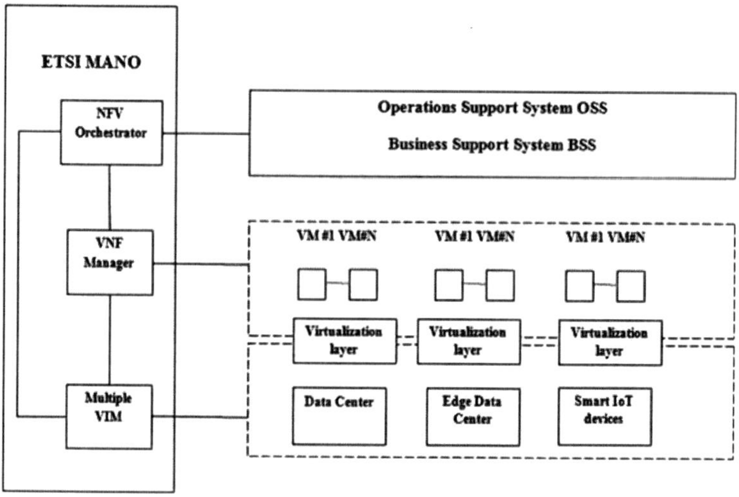

Fig. 6 NFV based IoT interoperability solution

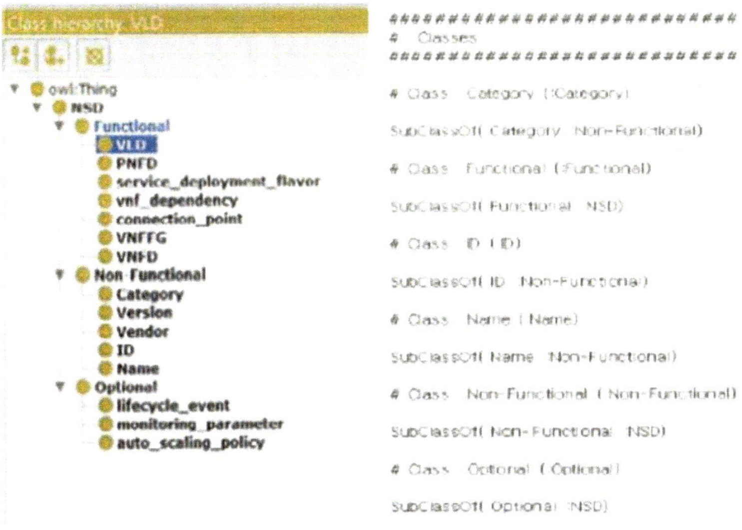

Fig. 7 NFV solution prepared in protégé [37]

7.4 Combined Solution of SDN and NFV

IoT environment can be created with the combination of both the solutions of SDN and NFV. Taking into consideration the advantages of both SDN and NFV we can provide the IoT environment using SDN and NFV in combination as shown in the (Fig. 8).

As discussed in the above sections about SDN and NFV solutions for semantic IoT interoperability, in future combination of both SDN and NFV as per there advantages can be utilized together for the implementation of emulation, simulation, or testbeds.

Some implementation observation and future directions related to SDN, NFV, and clouds are elaborated below:

Semantic Model-driven Approach to Deployment and Adaptivity of container-based applications in Fog Computing (SMADA-Fog) is proposed in [38], which is used to provide a solution to IoT and mobile devices. The solution consisted of containers and fog computing. A lot of work can be extended in the area of QoS, security, energy efficiency, and Unmanned Aerial Vehicle (UAV)-based Fog Computing. A Distributed Threat Analytics and Response System (DTARS) framework for IoT and 5G network were proposed in [39] for providing security using IoT, SDN, and NFV. There is still a lot of scope for new hybrid machine learning algorithms to be used for providing security in real-time scenarios at the wire level in the 5G network. In paper [40], authors have focused on the implementation aspects when combining the SDN, NFV, IoT, and Cloud. Various new research challenges and future directions are also discussed in detail. In paper [41], authors have proposed a solution for Semantic IoT interoperability by combining SDN, NFV, and IoT. The future

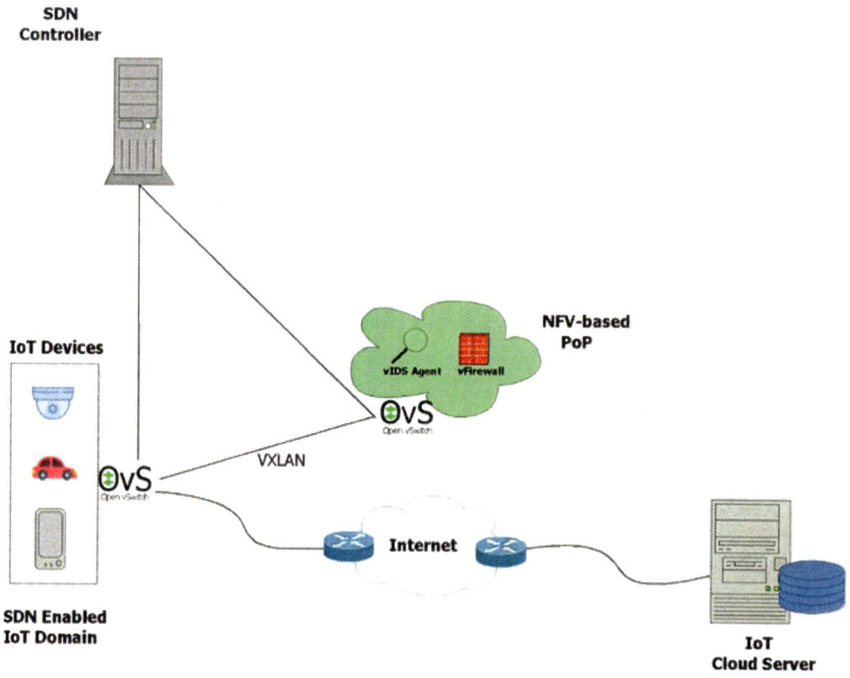

Fig. 8 SDN and NFV based IoT interoperability solution

directions of this paper are to provide a bridge between various IoT platforms and frameworks. In paper [42], authors have proposed a security solution for self healing and protection for the combined scenario of SDN, NFV, and IoT using semantic technologies.

7.5 *Graph Structure*

In [43], the author has tried to provide a solution for the IoT environment using the concept of graph structure. He proposed a framework which constructed the graph structure from semantic relations related to IoT data. It is used to measure the similarities between the meta path constructed in the graph structure. Afterward constructed cluster experiments for data similarities using matrix factorization.

Like this method of optimization using graph structure for semantic interoperability and interpretations, various other optimization graph structure techniques can be proposed for SDN, NFV, cloud technologies together and the optimization techniques for the graphs generated for the same. In the future, we can try to implement this graph structure solution to the combined solution which will be provided by SDN and NFV.

8 Conclusion

After reviewing in the area of semantic IoT, we came across a loophole of interoperability of heterogeneous IoT devices or platforms. Some standards and rules need to be identified which can be used to create a standard by making use of the three technologies SDN, NFV, and cloud in combination. We can combine the advantages of these three technologies and come with a new standard before the advent of the new era of 5G or 6G.

9 Future Work

As we have summarized various solutions for interoperability problems for IoT using SDN, NFV, Cloud technologies, or services. All these works are done separately without using their combinations. Some available solutions are shared in this chapter. We have seen various advantages of each technology. In future various testbeds can be created for the combination of SDN, NFV, and cloud services together and can be used for providing the solutions for IoT interoperability problems. These solutions which will be created can be very useful for the 5G and 6G upcoming eras.

References

1. Savaglio, C., Fortino, G., Ganzha, M., Paprzycki, M., Bădică, C., Ivanović, M.: Agent-Based computing in the internet of things: a survey. IDC Springer (2017)
2. Ganzha, M., Paprzycki, M., Pawłowskiy, W., Szmeja, P., Wasielewska, K.: Alignment-based semantic translation of geospatial data. In: IEEE ICACCA Conference (2017)
3. Tkaczyk, R., Wasielewska, K., Ganzhay, M., Paprzyckiz, M., Pawłowskix, W., Szmeja, P., Fortino, G.: Cataloging design patterns for internet of things artifact integration. IEEE ICC Workshop (2018)
4. Ganzha, M., Paprzycki, M., Pawłowski, W., Szmeja, P., Wasielewska, K.: Towards semantic interoperability between internet of things platforms. Springer ITTCC (2017)
5. Szmeja, P., Ganzha, M., Paprzycki, M., Pawłowski, M., Wasielewska, K.: Declarative ontology alignment format for semantic translation. IEEE IoT-SIU (2018)
6. Szmeja, P., Ganzha, M., Paprzycki, M., Pawlowski, W.: Dimensions of ontological similarity. IEEE ICSC (2016)
7. Ganzha, M., Paprzycki, M., Pawłowski, W., Szmeja, P., Wasielewska, K., Palau, C.E.: From implicit semantics towards ontologies—practical considerations from the INTER-IoT perspective. IEEE CCNC (2017)
8. Tkaczyk, R., Szmeja, P., Ganzha, M., Paprzycki, M., Solarz-Niesłuchowski, B.: From relational databases to an ontology—practical considerations. IEEE ICSTCC (2017)
9. Drozdowicz, M., Alwazir, M., Ganzha, M., Paprzycki, M.: Graphical Interface for ontology mapping with application to access control. In: ACIIDS 2017: Intelligent Information and Database Systems pp. 46–55 (2017)
10. Ganzha, M., Paprzycki, M., Pawlowski, W., Szmeja, P., Wasielewska, K.: Identifier management in semantic interoperability solutions for IoT. IEEE ICC Workshop (2018)

11. Ganzha, M., Paprzycki, M., Pawłowski, W., Szmeja, P., Wasielewska, K.: Semantic interoperability in the Internet of things; an overview from the INTER-IoT perspective. Sci. Dir. J. Netw. Comput. Appl. (2017)
12. Drozdowicz, M., Ganzha, M., Paprzycki, M.: Semantically enriched data access policies in eHealth. J. Med. Syst. (2016)
13. Ganzha, M., Paprzycki, M., Pawłowski, W., Szmeja, P., Wasielewska, K.: Streaming semantic translations. IEEE ICSTCC (2017)
14. Ganzha, M., Paprzycki, M., Pawłowski, W., Szmeja, P., Wasielewska, K., Fortino, G.: Tools for ontology matching—practical considerations from INTER-IoT perspective. In: IDCS (2016)
15. Ganzha, M., Paprzycki, M., Pawłowski, W., Szmeja, P., Wasielewska, K.: Towards common vocabulary for IoT ecosystems—preliminary considerations. In: ACIIDS (2017)
16. Ganzha, M., Paprzycki, M., Pawłowski, W., Szmeja, P., Wasielewska, K.: Towards high throughput semantic translation. In: Interoperability, Safety and Security in IoT (2017)
17. Fortino, G., Savaglio, C., Palau, C.E., de Puga, J.S., Ganzha, M., Paprzycki, M.: Towards multi-layer interoperability of heterogeneous IoT platforms: the INTER-IoT approach. In: Integration, Interconnection, and Interoperability of IoT Systems (2017)
18. Ganzha, M., Paprzycki, M., Pawłowski, W., Szmeja, P., Wasielewska, K.: Towards semantic interoperability between internet of things platforms. In: Integration, interconnection, and interoperability of IoT systems, pp. 103–127 (2017)
19. Pandey, R.K., Dwivedi, S.: Ontology description using OWL to support semantic web applications. Int. J. Comput. Appl. (2011)
20. Pandey, R., Dwivedi, S.K., Verma, P.: Univpeopleprogram ontology: a OWL based structural definition for semantic web. IEEE ICI&CT, 2013
21. Kumari, S., Pandey, R., Singh, A., Pathak, H.: SPARQL: semantic information retrieval by embedding prepositions. Int. J. Netw. Secur. Appl. (IJNSA) **6**(1) (2014)
22. Chen, X., Wu, T.: Towards the semantic web based northbound interface for SDN resource management. IEEE ICSC (2017)
23. Bendale, S.P., Prasad, J.R.: Security threats and challenges in future mobile wireless networks. In: 2018 IEEE Global Conference on Wireless Computing and Networking (GCWCN), Lonavala, India, pp. 146–150 (2018). https://doi.org/10.1109/gcwcn.2018.8668635
24. Bendale, S.P., Chowdhary, G.V.: Stable path selection and safe backup routing for optical border gateway protocol (OBGP) and extended optical border gateway protocol (OBGP+). In: 2012 International Conference on Communication, Information & Computing Technology (ICCICT), Mumbai, pp. 1–6 (2012). https://doi.org/10.1109/iccict.2012.6398201
25. Bendale, S., Prasad, J.R.: Preliminary study of software defined network on COVID-19 pandemic use cases (May 28 2020). Available at SSRN: https://ssrn.com/abstract=3612815 or http://dx.doi.org/10.2139/ssrn.3612815
26. Bendale, S.P., Prasad, J.R.: Security challenges to provide Intelligence in SDN with the help of machine learning or deep learning. IJAST **29**(05), 356–363 (2020)
27. Bendale, S.P.: Implications and application of artificial intelligence and machine learning concepts on software defined network and its future prospects. IJAST **29**(4), 1142–1152 (2020)
28. Shah, S., Bendale, S.P.: An intuitive study: intrusion detection systems and anomalies, How AI can be used as a tool to enable the majority, in 5G era. In: 2019 5th International Conference On Computing, Communication, Control And Automation (ICCUBEA), Pune, India, 2019, pp. 1–8 (2019). https://doi.org/10.1109/iccubea47591.2019.9128786
29. Lanza, J., Sanchez, L., Soldatos, J.: Experimentation as a service over semantically interoperable internet of things test beds. IEEE (2018)
30. Coulson, G. et al.: Flexible experimentation in wireless sensor networks. Commun. ACM **55**(1), 82–90 (2012)
31. Adjih, C., et al.: FIT IoT-LAB: a large scale open experimental IoT testbed. In: Proceedings of IEEE 2nd World Forum Internet Things (WF-IoT), pp. 459–464 (2015)
32. Vandenberghe, W., et al.: Architecture for the heterogeneous federation of future internet experimentation facilities. Proc. Future Netw. Mobile Summit 1–11 (2013)

33. Berman, M., et al.: GENI: a federated testbed for innovative network experiments. Comput. Netw. **61**, 5–23, (2014)
34. Sanchez, L., et al.: SmartSantander: IoT experimentation over a smart city testbed. Comput. Netw. **61**, 217–238 (2014)
35. Misra, A., Balan, R.K.: LiveLabs. ACM Sigmobile Mob. Comput. Commun. Rev. **17**(4), 47–59 (2013)
36. Cardone, G., Cirri, A., Corradi, A., Foschini, L.: The participact mobile crowd sensing living lab: the testbed for smart cities. IEEE Commun. Mag. **52**(10), 78–85 (2014)
37. Kim, S.I., Kim, H.S.: Semantic ontology-based NFV service modeling. IEEE (2018)
38. Petrovic, N., Tosic, M.: SMADA-Fog: semantic model driven approach to deployment and adaptivity in fog computing. Simul. Model. Pract. Theor. **101**, 102033 (2020). ISSN 1569-190X. https://doi.org/10.1016/j.simpat.2019.102033
39. Krishnan, P., Duttagupta, S., Achuthan, K.: SDN/NFV security framework for fog-to-things computing infrastructure. Softw. Pract Exper. **50**, 757–800 (2020). https://doi.org/10.1002/spe.2761
40. Alam, I., Sharif, K., Li, F., Latif, Z., Karim, M.M., Biswas, S., Nour, B., Wang, Y.: A survey of network virtualization techniques for internet of things using SDN and NFV. ACM Comput. Surv. **53**(2), 40 (2020). Article 35 (June 2020). https://doi.org/10.1145/3379444
41. Lakka, E., Petroulakis, N.E., Michalodimitrakis, E., Papoutsakis, E.: Validation of semantic interoperability between IoT platforms. In: 2020 Global Internet of Things Summit (GIoTS), Dublin, Ireland, pp. 1–6 (2020). https://doi.org/10.1109/giots49054.2020.9119517
42. Zarca, A.M., Bagaa, M., Bernabe, J.B., Taleb, T., Skarmeta, A.F.: Semantic-aware security orchestration in SDN/NFV-enabled IoT systems. Sensors **20**, 3622 (2020)
43. Hu, L., Gong, Y., Xing, Y., Wang, F.: Semantic representation with heterogeneous information network using matrix factorization for clustering in the internet of things. IEEE (2019)